Plates vs. Plumes

Plates vs. Plumes: A Geological Controversy

Gillian R. Foulger

A John Wiley & Sons, Ltd., Publication

Library of Congress Cataloguing-in-Publication Data

Foulger, Gillian R., 1952-
    Plates vs plumes : a geological controversy / Gillian R. Foulger.
        p. cm.
    Includes bibliographical references and index.
    ISBN 978-1-4443-3679-5 (hardcover : alk. paper) – ISBN 978-1-4051-6148-0 (pbk. : alk. paper)   1. Mantle plumes.   2. Plate tectonics.   I. Title.
    QE527.7.F68 2011
    551.21–dc22

                                                                                                                            2010012131

ISBN: 978-1-4443-3679-5 (hbk) 978-1-4051-6148-0 (pbk)

A catalogue record for this book is available from the British Library.

Set in 10.5 on 12.5pt Goudy by Toppan Best-set Premedia Limited

1   2010

Contents

Preface

The debate regarding whether or not melting anomalies are fuelled by hot diapirs from the deep mantle – the Plume hypothesis – or whether they arise from shallow processes ultimately related to plate tectonics – the Plate hypothesis – is an extraordinarily rich cross-disciplinary subject with seemingly endless ramifications. Relevant material is scattered far and wide, and because of this, over the last decade, scientists have come together and tried to address this problem using the website www.mantleplumes.org. It grew like Topsy following its founding on 5th March, 2003 and now includes thousands of items contributed by over 500 scientists from many parts of the world. Nevertheless, even when collected together in a single website, making sense of the vast ocean of data relevant to this fascinating subject still presented an extraordinary challenge to the aspiring student. A substantial, coherent work, organized by subject, was still lacking.

I was thus rather easily persuaded, in the summer of 2007, to write the present book. In it I have attempted to summarize where we stand today in some of the subjects most central to the debate. The Plume hypothesis makes five fundamental predictions that can be tested by studying the vertical motions of Earth's surface, magma volumes, the spatial and temporal pattern of volcanism, seismic imaging of the crust and mantle, and the temperature of the mantle source rocks. The Plate hypothesis makes contrasting predictions in these fields. I have devoted a chapter to each, focusing on how well the observations match the predictions. I include additionally a chapter on geochemistry, which has been extensively applied in an effort to reveal the physical and spatial origins of surface-erupted magmas. Deciding what to put in and what to leave out has not been easy, and important subjects such as the effect of pre-existing lithospheric structure on localizing volcanism, the physical origins of the melt source, bolide impacts, and planetary volcanism remain to be treated in more detail.

I am sometimes asked what my own background is, and by what path I arrived at the subject of plumes, "hot spots" and melting anomalies. I have worked my entire career as an earthquake seismologist, with a temporary residency in the parallel universe of GPS surveying applied to crustal deformations and plate motions. I was initiated into scientific research working on geothermal areas in Iceland, where I lived for several years, acquiring the language, an insider's knowledge of geological lore, and a conviction that cross-disciplinary work is essential for full understanding. I doubt it is really possible to ever truly appreciate the complexities of an area unless one has lived, breathed, and suffered there. What is understood usually finds its way into print, but what is not understood, wherein exciting new discoveries lurk, is only whispered in corridors, perhaps also in an unfamiliar language.

I received early a lesson in listening to the clamoring voices of data during my Ph.D. research. No shear-faulting mechanisms would fit my earthquake data, which seemed intransigent to interpretation. My supervisor told me I must have set the equipment up wrongly. I did not understand. I had calibrated my network in four independent ways and all the calibrations agreed. Six months later I realized that my earthquakes were not caused by shear faulting but by tensile cracks. Everything fell instantly into place. Thermal cracking was happening. An unknown heat source must exist, and a hitherto unknown geothermal reservoir. Geochemical and geological data were simultaneously announced that converged on the same conclusion. The euphoria will always remain with me.

That was the start of my career. Its latest phase has involved work on a much larger scale – the structure of the mantle. In this I was inspired by the pioneering work of H.M. Iyer, with whom I conceived an impossibly ambitious Icelandic project as early as 1989. In the mid-1990s I acquired a series of grants from the Natural Environmental Research Council, the National Science Foundation, and the European Community for a project to study "the Iceland plume". Had I applied for funding to test whether a plume existed, my applications would probably all have been rejected, just as I would have faced rejection in the 1980s had I applied to test whether earthquakes resulted from shear slip on faults.

In 1996 we installed the most ambitious seismic network of its kind yet deployed, covering all Iceland, and we operated it in the face of difficulties that sometimes seemed overwhelming. Nevertheless, a huge data set was recorded, which culminated in my sitting at my desk, late one dark, damp night in the last November of the 20th century, puzzling over a low-wave-speed anomaly that wasn't the shape we had expected. Everyone got the same answer, not everyone thought it mattered, but I don't like not understanding things.

It dawned on me that night what the data were shouting at us to hear. Although we could not see the bottom of the anomaly, its odd shape was telling us that it terminated in the transition zone – it did not extend deeper into the lower mantle. What did this mean? We rushed off a

paper to *Science*, only to find that this very week *Science* published a whole-mantle tomography paper spectacularly confirming our conclusion. I went to the December 1999 *American Geophysical Union* meeting in San Francisco with the message that "the Icelandic plume" was different from what we had hitherto thought. It was an "upper mantle plume".

Then Bruce Julian introduced me to Don Anderson.

"Maybe there isn't a plume there," he suggested. I was dumbstruck with astonishment.

Since that moment I have worked to see for myself whether such an hypothesis can stand – that Iceland does not owe its volcanism to a deep mantle plume but to processes rooted at shallow depth. And if such an hypothesis can stand for Iceland, can it also stand for other places traditionally assumed to be underlain by plumes? One does not lightly fly in the face of an almost-universally-assumed, cross-disciplinary paradigm of global importance that underpins concepts in almost every branch of Earth science. Nor is it easy to re-examine those concepts with access to little that is not selectively interpreted in terms of the traditional assumptions. I started by turning Iceland inside out. What did it imply for the celebrated "hot spot track"? Could we be certain that the high mantle temperatures assumed to exist there really do? How can we explain the high elevation, apparently vast magmatic volumes and the exotic geochemistry, claimed to come from the lower mantle? Most critically, if a deep mantle plume does not cause Iceland, then what does?

After a long journey, not easily traveled, I am persuaded by the data that melting anomalies arise from processes rooted in the shallow mantle, and not from deep mantle plumes. The data demand it too loudly and clearly to be denied. Extraordinarily few observations fit the Plume hypothesis – fewer, it seems, even than might be expected simply from random chance. The lengths to which the scientific community has had to go in order to cram the distorted plume foot into the glass data slipper, and the unfortunate departures from rigorous scientific practise that this has necessitated, are wholly consistent with this conclusion and have reinforced my conviction that the Plume hypothesis cannot be right.

I did not travel alone on my journey. Don and I were soon joined by others who also cannot bring themselves to embrace an hypothesis that does not fit the observations. A core group accreted – the PT group – that now numbers 16. Email and the internet, unavailable to most scientists before the 1990s, made possible a cross-disciplinary, international, collaboration unique in my experience, that grew to be characterized by the totally free, unrestricted, and unselfish exchange of data, ideas, and mutual support. My journey would have been impossible without the support of this group, and that of a wider community of scientists who want to know the truth. In this context my gratitude is due in particular to David Abt, Ercan Aldanmaz, Andrew Alden, Ken Bailey, Ajoy Baksi, Tiffany Barry, Erin Beutel, Axel Björnsson, Scott Bryan, Evgenii Burov, Maria Clara Castro, Françoise Chalot-Prat, Bob Christiansen, Peter Clift, Valerie Clouard, Piero Comin-Chiaramonti, Gerry Czamanske, Graham Dauncey, Jon Davidson, Don DePaulo, Arwen Deuss, Carlo Doglioni, Tony Doré, Adam Dziewonski, Linda Elkins-Tanton, Wolf Elston, Derek Fairhead, Trevor Falloon, Luca Ferrari, Carol Finn, Godfrey Fitton, Edward Garnero, Laurent Geoffroy, Laurent Gernigon, Phil Gibbard, William Glen, David Green, Jeff Gu, Gudmundur Gudfinnsson, Tristram Hales, Warren Hamilton, Karen Harpp, Robert Harris, Anne Hofmeister, Jack Holden, Bob Holdsworth, Peter Hooper, Stephanie Ingle, Ted Irving, Garrett Ito, Alexei Ivanov, Mark Jancin, Jeremy McCreary, Adrian Jones, Stephen Jones, Brennan Jordan, Fred Jourdan, Bruce Julian, Donna Jurdy, Mehmet Keskin, Scott King, Jun Korenaga, Lotte Melchior Larsen, Thorne Lay, Jean-Paul Liégeois, Erik Lundin, Michele Lustrino, Hidehisa Mashima, Rajat Mazumder, Greg McHone, Anders Meibom, Romain Meyer, Andrew Moore, Jason Morgan, Jim Natland, Ted Nield, Yaoling Niu, Ian Norton, John O'Connor, Mike O'Hara, Keith Orford, Giuliano Panza, Angelo Peccerillo, Carole Petit, Sébastien Pilet, Dean Presnall, Chris Reese, Jeroen Ritsema, Sergio Rocchi, Peter Rona, William Sager, Valenti Sallares, David Sandwell, Anders Schersten, James Sears, Hetu Sheth, Tom Sisson, Norman Sleep, Alan Smith, Alexander Sobolev, Carol Stein, Seth Stein, Martyn Stoker, William Stuart, Michael Summerfield, Benoit Tauzin, Marissa Tejada, Ingrid Ukstins Peate, Peter Vogt, Jerry Winterer, and Kerrie Yates.

Geological time scale

542 Ma to present	Phanerozoic Eon
66 Ma to 0 Ma	Cenozoic Era
2.6 to 0 Ma	Quaternary Period[1]
11,400 a to 0 a	Holocene Epoch
2.6 Ma to 11,400 a	Pleistocene Epoch
23 to 2.6 Ma	Neogene Period
5.3 to 2.6 Ma	Pliocene Epoch
23 to 5.3 Ma	Miocene Epoch
66 to 23 Ma	Paleogene Period
34 to 23 Ma	Oligocene Epoch
56 to 34 Ma	Eocene Epoch
66 to 56 Ma	Paleocene Epoch
245 to 66 Ma	Mesozoic Era
146 to 66 Ma	Cretaceous Period
201 to 146 Ma	Jurassic Period
251 to 200 Ma	Triassic Period
544 to 251 Ma	Paleozoic Era
299 to 251 Ma	Permian Period
318 to 299 Ma	Carboniferous Period (Pennsylvanian)
359 to 318 Ma	Carboniferous Period (Mississippian)
416 to 359 Ma	Devonian Period
444 to 416 Ma	Silurian Period
488 to 444 Ma	Ordovician Period
542 to 488 Ma	Cambrian Period
2500 to 542 Ma	Proterozoic Eon
1000 to 542 Ma	Neoproterozoic Era
1600 to 1000 Ma	Mesoproterozoic Era
2500 to 1600 Ma	Paleoproterozoic Era
3800 to 2500 Ma	Archaean Eon
4570 to 3800 Ma	Hadean Eon

[1] As a result of a review by the ICS of late Cenozoic subdivisions, and subsequent ratification by the IUGS, the Quaternary was formally made a Period co-terminus with the Pleistocene at 2.6 Ma, in 2009. This settles a long-standing confusion regarding terminology and subdivisions in the late Cenozoic (Head et al., 2008).

1

From plate tectonics to plumes, and back again

Je n'avais pas besoin de cette hypothèse-là.[1]

—Pierre-Simon Laplace (1749–1827)[1]

1.1 Volcanoes, and exceptional volcanoes

Volcanoes are among the most extraordinary natural phenomena on Earth. They are powerful shapers of the surface, they affect the make-up of the oceans, the atmosphere and the land on which we stand, and they ultimately incubate life itself. They have inspired fascination and speculation for centuries, and intense scientific study for decades, and it is thus astonishing that the ultimate origin of some of the greatest and most powerful of them is still not fully understood.

The reasons why spectacular volcanic provinces such as Hawaii, Iceland and Yellowstone exist are currently a major controversy. The fundamental question is the link between volcanism and dynamic processes in the mantle, the processes that make Earth unique in the solar system, and keep us alive. The hunt for the truth is extraordinarily cross-disciplinary and virtually every subject within Earth science bears on the problem. There is something for everyone in this remarkable subject and something that everyone can contribute.

[1] I have no need of that hypothesis.

The discovery of plate tectonics, hugely cross-disciplinary in itself, threw light on the causes and effects of many kinds of volcano, but it also threw into sharp focus that many of the largest and most remarkable ones seem to be exceptions to the general rule. It is the controversy over the origin of these volcanoes – the ones that seem to be exceptional – that is the focus of this book.

1.2 Early beginnings: Continental drift and its rejection

Speculations regarding the cause of volcanoes began early in the history of science. Prior to the emergence of the scientific method during the Renaissance, explanations for volcanic eruptions were based largely on religion. Mt Hekla, Iceland, was considered to be the gate of Hell. Eruptions occurred when the gate opened and the Devil dragged condemned souls out of Hell, cooling them on the snowfields of Iceland to prevent them from becoming used to the heat of Hell. Athanasius Kircher (1602–1680) provided an early pictorial representation of then-contemporary thought (Fig. 1.1) that has much in common with some theories still popular today (Fig. 1.2). The *agent provocateur* might be forgiven for wondering how much progress in fundamental understanding we have actually made over the last few centuries.

Plates vs. Plumes: A Geological Controversy, 1st edition. By Gillian R. Foulger.
Published 2010 by Blackwell Publishing Ltd.

Figure 1.1 Kircher's model of the fires of the interior of Earth, from his *Mundus Subterraneus*, published in 1664 (Kircher, 1664–1678).

The foundations of modern opinion about the origin of volcanoes were really laid by the work of Alfred Wegener (1880–1930). His pivotal book *Die Entstehung der Kontinente und Ozeane* (The Origin of Continents and Oceans; Wegener, 1924), first published in 1915, proposed that continents now widely separated had once been joined together in a single super-continent. According to Wegener, this super-continent broke up and the pieces separated and drifted apart by thousands of kilometers (Fig. 1.3). The idea was not new, but Wegener's treatment of it was, and his work ultimately led to one of the greatest paradigm shifts Earth science has ever seen. He assembled a powerful multi-disciplinary suite of scientific observations to support continental break-up, and developed ideas for the mechanism of drift and the forces

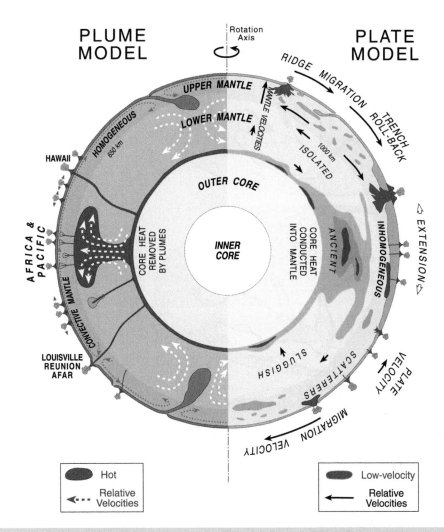

Figure 1.2 Schematic cross-section of the Earth showing the Plume model (left, modified from Courtillot et al., 2003) and the Plate model (right). The left side illustrates two proposed kinds of plumes – narrow tubes and giant upwellings. The deep mantle or core provides the material and the heat and large, isolated but accessible chemical reservoirs. Slabs penetrate deep. In the Plate model, depths of recycling are variable and volcanism is concentrated in extensional regions. The upper mantle is inhomogeneous and active and the lower mantle is isolated, sluggish, and inaccessible to surface volcanism. The locations of melting anomalies are governed by stress conditions and mantle fertility. The mantle down to ~1000 km contains recycled materials of various ages and on various scales (from Anderson, 2005). See Plate 1

that power it. He detailed correlations of fossils, mountain ranges, palaeoclimates and geological formations between continents and across wide oceans. He called the great mother super-continent Pangaea ("all land"). He was fired

with enthusiasm and energized by an inspired personal conviction of the rightness of his hypothesis.

Tragically, during his lifetime, Wegener's ideas received little support from mainstream geology

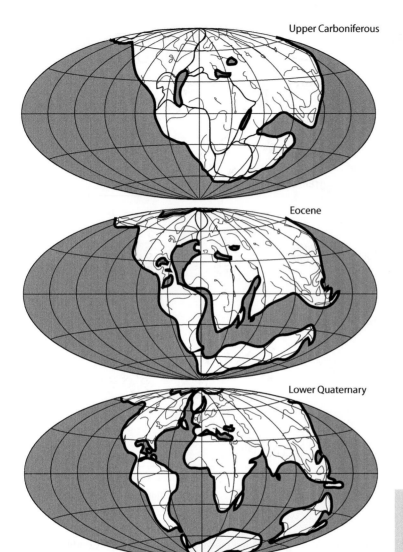

Upper Carboniferous

Eocene

Lower Quaternary

Figure 1.3 Wegener's original model for the break-up of the Pangaea supercontinent (from Wegener, 1924).

and physics. On the contrary, they attracted dismissal, ridicule, hostility and even contempt from influential contemporaries. Wegener's proposed driving mechanism for the continents was criticized. He suggested that the Earth's centrifugal and tidal forces drove them, an effect that geologists felt was implausibly small. Furthermore, although he emphasized that the sub-crustal region was viscous and could flow, a concept well established and already accepted as a result of knowledge of isostasy, the fate of the oceanic

crust was still a difficult problem. A critical missing piece of the jigsaw was that the continental and oceanic crusts were moving as one. Wegener envisaged the continents as somehow to be moving through the oceanic crust, but critics pointed out that evidence for the inevitable crustal deformations was lacking.

Wegener was not without influential supporters, however – scientists who were swayed by his evidence. The problem of mechanism was rapidly solved by Arthur Holmes (1890–1965).

Holmes was perhaps the greatest geologist of the 20th century, and one of the pioneers of the use of radioactivity to date rocks. Among his great achievements was establishing a geological time-scale and calculating the age of the Earth (Lewis, 2000). He had a remarkably broad knowledge of both physics and geology and was a genius at combining them to find new ways of advancing geology.

In 1929, Holmes proposed that the continents were transported by subcrustal convection currents. He further suggested that crust was recycled back into the interior of the Earth at the edges of continents by transformation to the dense mineral eclogite, and gravitational sinking (Fig. 1.4) (Holmes, 1929). His model bears an uncanny resemblance to the modern, plate-tectonic concept of subduction. Considering the vast suites of data that had to be assembled before most geologists accepted the subduction process, data that were unavailable to Holmes at the time, his intuition and empathy for the Earth were astounding.

It was not until half a century after publication of Wegener's first book that the hypothesis of continental drift finally became accepted by mainstream geology. The delay cannot be explained away as being due to incomplete details, or tenuously supported aspects of the drift mechanism (Oreskes, 1999). The case presented by Wegener was enormously strong and brilliantly cross-disciplinary. It included evidence from physics, geophysics, geology, geography, meteorology, climatology and biology. The power of drift theory to explain self-consistently a huge assemblage of otherwise baffling primary observational evidence is undeniable. After its final acceptance, its rejection must have caused many a conscientious Earth scientist pangs of guilt.

Wegener brilliantly and energetically defended his hypothesis throughout his lifetime, but this was prematurely cut short. He died in 1930 leading an heroic relief expedition across the Greenland icecap, and thus did not live to see his work finally accepted (McCoy, 2006). One can only speculate about how things might have turned out had he survived to press on indefatigably with his work.

Despite his premature demise, Wegener's ideas were not allowed to die. They were kept alive not least by Holmes who resolutely included a chapter on continental drift in every edition of his seminal textbook *Principles of Physical Geology* (Holmes, 1944). Notwithstanding this, the hypothesis continued to be regarded as eccentric, or even ludicrous, right up to the brink of its sudden and final acceptance in the

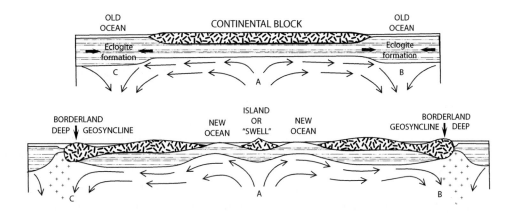

Figure 1.4 Holmes's original model for the convective system that enabled the continents to drift (from Holmes, 1944).

mid-1960s. Until then, innovative contributions and developments were often met with ridicule, rejection and hostility that suppressed progress and hurt careers. In 1954 Edward Irving submitted a Ph.D. thesis at Cambridge on palaeomagnetic measurements. His work showed that India had moved north by 6000 km and rotated by more than 30° counterclockwise, close to Wegener's prediction. His thesis was failed.[2] In 1963 a Canadian geologist, Lawrence W. Morley, submitted a paper proposing seafloor spreading, first to the journal *Nature* and subsequently to the *Journal of Geophysical Research*. It was rejected by both. As late as 1965 Warren Hamilton lectured at the California Institute of Technology on evidence for Permian continental drift. He was criticized on the grounds that continental drift was impossible, and later that year, at the annual student Christmas party, "Hamilton's moving continents" were ridiculed by students who masqueraded as continents and danced around the room.

Continental drift was finally accepted when new, independent corroborative observations emerged from fields entirely different to those from which Wegener had drawn. Palaeomagnetism showed a symmetrical pattern of normal and reversed magnetization in the rocks that make up the sea floor on either side of mid-ocean ridges. The widths of the bands were consistent with the known time-scale of magnetic reversals, if the rate of sea-floor production was constant at a particular ridge. Earthquake epicenters delineated narrow zones of activity along ocean ridges and trenches, with intervening areas being largely quiescent. Detailed bathymetry revealed transform faults and earthquake fault-plane solutions showed their sense of slip. A "paving stone hypothesis" was proposed, which suggested the Earth was covered with rigid plates that moved relative to one another, bearing the continents along with them (McKenzie and Parker, 1967). Senior geophysicists threw their

weight behind the hypothesis and the majority of Earth scientists fell quickly into line.

There has been much speculation over why it took the Earth science establishment a full half-century to accept that continental drift occurs (Oreskes, 1999). The reader is urged to read Wegener's work in its original form – to read what he actually wrote. Wegener had assembled diverse and overwhelming evidence, and a mechanism very similar to that envisaged today by plate tectonics that had been proposed by Holmes. The observations that finally swayed the majority of Earth scientists in the 1960s did not amount to explaining the cause of drift, the most harshly criticized and strongly emphasized weak spot in Wegener's hypothesis. They merely added an increment to the weight of observations that already supported drift.

It seems, therefore, that popular acceptance of a scientific hypothesis may sometimes be only weakly coupled to its merit. The popularity of an hypothesis may be more strongly influenced by faith in experts perceived as magisters than by direct personal assessment of the evidence by individuals (Glen, 2005). In highly cross-disciplinary fields where it is almost impossible for one person to be fully conversant with every related subject, this may seem to many to be the only practical way forward. However, the magisters, as the greatest will readily admit themselves, are not always right.

1.3 Emergence of the Plume hypothesis

Once continental drift and plate tectonics had become accepted, they proved spectacularly successful. A huge body of geological and geophysical data was reinterpreted, and numerous tests were made of the hypothesis. The basic predictions have been confirmed again and again, right up to the present day when satellite technology is used to measure annual plate movements by direct observation. Plate tectonics explains naturally much basic geology, the

[2] http://www.pnas.org/content/102/6/1819.full

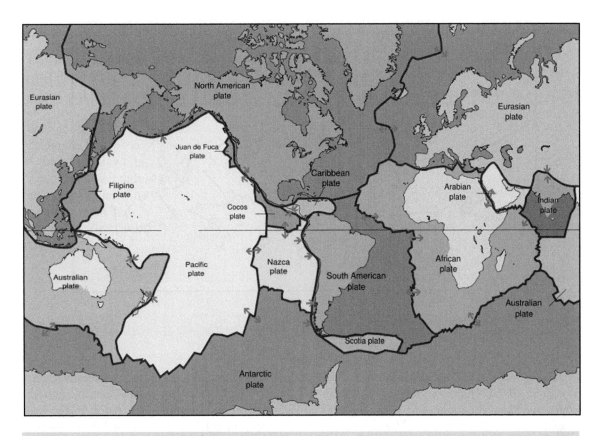

Figure 1.5 The main tectonic plates on the Earth's surface.[3]

origins of mountains and deep-sea trenches, topography, earthquake activity, and the vast majority of volcanism.

Plate tectonics views the Earth's surface as being divided up, like a jigsaw, into seven major, and many minor pieces ("plates"). Like the pieces of a jigsaw, the plates behave more or less like coherent units, that is, each moves as though it were a single entity (Fig. 1.5). Oceanic crust is created at mid-ocean ridges by a volcanic belt 60,000 km long – almost twice the circumference of the planet. It is destroyed in equal measure at subduction zones, where one plate dives beneath its neighbor and returns lithosphere back into the Earth's interior. As subduct-

ing plates sink, they heat up and dehydrate, fluxing the overlying material with volatiles and causing it to melt. This causes belts of volcanoes to form at the surface ahead of the subduction trench. In this way, processes near spreading ridges and subduction zones, where crust is formed or destroyed, account for 90% of the Earth's volcanism.

When a new hypothesis sweeps the board, and has been convincingly confirmed by testing, the things that immediately become the most interesting are the exceptions to the general rules because therein lie the next great discoveries. Following the widespread acceptance and development of plate tectonics, it rapidly became clear that several remarkable volcanic regions did not fit into the general picture. Most

[3] http://en.wikipedia.org/wiki/Plate_tectonics

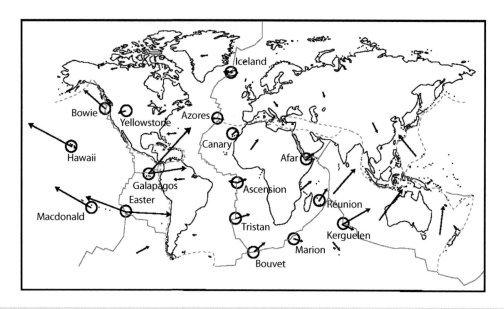

Figure 1.6 Morgan's original 16 plume localities (Morgan, 1971).

spectacular was the Big Island of Hawaii and its associated island chain, known to be time-progressive since the pioneering work of James Dwight Dana (1813–1895). The Hawaiian archipelago lies deep in the interior of the vast Pacific plate, thousands of kilometers from the nearest plate boundary of any kind (Fig. 1.5). J. Tuzo Wilson suggested that it arose from the motion of the Pacific sea floor over a hot region in the mantle (Wilson, 1963). Out of these beginnings the expression "hot spot" emerged, and the seed was sown for the modern hypothesis that any volcanism unusual in the context of plate tectonics results from local, exceptionally high temperatures in the mantle beneath.

The Plume hypothesis arrived eight years later in 1971 with the publication of a letter in *Nature* by W. Jason Morgan (Morgan, 1971).[4] Morgan suggested that there was not merely one but "about 20" hot spots on Earth. He further proposed that the "hot spots" were fixed relative to one another, and to explain this he suggested that they were sourced below the asthenosphere,

which was thought to be vigorously convecting. He postulated that they were in fact fueled from great depths within the Earth, through "pipes to the deep mantle", and he termed these systems "plumes". In this hypothesis, the motions of the surface plates overhead resulted in volcanic products being constantly carried away from their parent plumes, as though on a conveyor belt, resulting in time-progressive volcanic chains.

The sites that Morgan proposed to be underlain by plumes are Hawaii, the Macdonald seamount, Easter island, the Galapagos islands, Bowie, Yellowstone, Iceland, the Azores, the Canary islands, Ascension island, Tristan da Cuhna, the Bouvet triple junction, Marion-Prince Edward island, Réunion, Kerguelen and Afar (Fig. 1.6). He suggested additionally that plumes are the driving force of plate tectonics and that the material that they transport up from deep within the Earth's mantle is relatively primordial and compositionally different from that extracted from shallower depths, for example, at mid-ocean ridges.

The prediction that lavas at "hot spots" are compositionally different from mid-ocean ridge

[4] http://www.mantleplumes.org/Morgan1971.html

basalts was confirmed almost immediately. Jean-Guy Schilling found that rare-earth- and minor-element concentrations along the mid-ocean ridge in the north Atlantic vary in a way that is consistent with Iceland being fed by a compositionally distinct source (Schilling, 1973). He went on to suggest that melts from a plume beneath Iceland, and the shallow source of mid-ocean-ridge basalts (MORB), mix along the Reykjanes ridge south of Iceland.

The new Plume hypothesis met with both advocacy and skepticism early on. It offered an elegant explanation for volcanism away from plate boundaries, time-progressive volcanic chains, relative fixity between "hot spots", and their distinct geochemistry. However, scientists familiar with individual volcanic regions, or related aspects of Earth structure and dynamics, puzzled over how it accorded with the observations in detail. The suggestion that plumes are compositionally fertile was challenged on the grounds of density ("In other words, if fertile mantle plumes exist at all, they should be sinking, not rising"; O'Hara, 1975). The impossibility that two independent modes of convection can occur in the continuum of the Earth's mantle, that associated with plate movements, and an entirely separate plume mode, was pointed out ("It would be most helpful if someone would explain in terms that are meaningful to geophysicists in what respects the conventional geological pictures of rising magma differ from 'a thermal plume'."; Tozer, 1973). It was also pointed out that in fact "hot spots" are not fixed relative to one another, and would not be expected to be so in a mantle heated internally by radioactivity (McKenzie and Weiss, 1975), immediately removing one of the primary reasons for proposing the hypothesis in the first place.

During the first two decades following the initial proposal, alternative explanations for mid-plate volcanism were suggested and explored in detail (Anderson and Natland, 2005). Proposals included propagating cracks, internal plate deformation, membrane tectonics, self-perpetuating volcanic chains, recycled subducted slabs and continental breakup (Hieronymus and Bercovici, 1999; Jackson and Shaw, 1975; Jackson et al., 1975; Shaw, 1973; Turcotte, 1974). In the early 1990s, however, the Plume hypothesis received two major boosts that greatly increased its popularity, and interest in alternatives and debate temporarily waned.

The first boost came from laboratory tank experiments that continued earlier work by Ramberg (1967; 1981) and Belousov (1954; 1962) (Campbell and Griffiths, 1990; Griffiths and Campbell, 1990). Low-density fluid injected into the bottom of tanks full of higher-density fluid showed the development of rising compositional plumes (Fig. 1.7). Mushroom-like structures formed, comprising bulbous heads followed by narrow, stem-like conduits. Geologists were presented with powerful and compelling pictorial representations of a phenomenon that fitted well with the hypothesis that flood basalts such as the Deccan Traps in India represent "plume heads". Time-progressive volcanic trails were thus predicted to emanate from such flood basalts, representing a later "plume tail" stage. Although the experiments were conducted using fluids with compositional density differences, it was assumed that in nature the buoyancy of plumes would be thermal in origin, and that they would rise from a thermal boundary layer – a region where a large increase in temperature occurs across a relatively small depth interval. This is generally assumed to be at the core-mantle boundary, one of only two major thermal discontinuities known to exist in the Earth. Across the core-mantle boundary, the temperature increases abruptly by roughly 1000 °C. The other thermal discontinuity is, of course, the Earth's surface.

The second boost to the Plume hypothesis came from work on the noble gas helium (He) (Kellogg and Wasserburg, 1990). Earlier work by Craig and Lupton (1976; 1981) was developed to explain the unusually high $^3He/^4He$ ratios that had been observed in basalts from volcanic regions such as Hawaii and Iceland. In this

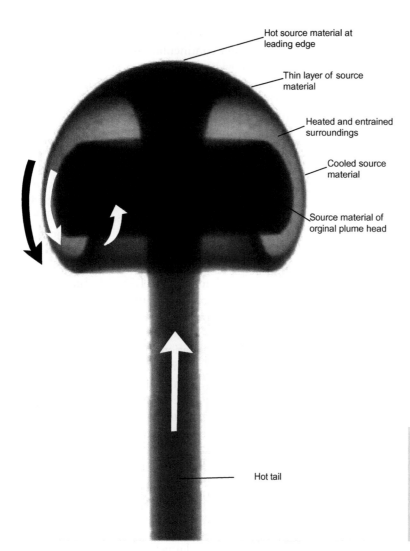

Hot source material at
leading edge

Thin layer of source
material

Heated and entrained
surroundings

Cooled source
material

Source material of
orginal plume head

Hot tail

Figure 1.7 Photograph of a thermal plume formed in a laboratory experiment by injecting warm syrup into the bottom of a tank full of cooler syrup (from Campbell et al., 1989).

model, the high ratios result from an excess of ^{3}He stored in the lower mantle, and thus high ^{3}He/^{4}He ratios observed in surface rocks were postulated to indicate a lower mantle provenance (Section 7.5.1).

This model contributed to the concept that plumes can be detected using geochemistry. ^{3}He is mostly a primordial isotope. That is, almost all the ^{3}He currently in the Earth was acquired when the planet formed, at ~4.5 Ga. In contrast, the Earth's ^{4}He inventory is continually increased by the decay of uranium (U) and thorium (Th). Rising magma transports both ^{3}He and ^{4}He to the surface, where it degasses to the atmosphere

and rapidly escapes to space. Volcanism thus constantly reduces the Earth's ^{3}He inventory and it is not significantly replenished. In contrast, ongoing radioactive decay maintains the Earth's stock of ^{4}He. The model that assigns a deep origin to high ^{3}He/^{4}He ratios views the lower mantle as having been much less depleted in ^{3}He than the upper mantle. This might be so if the upper mantle has repeatedly been tapped to feed the majority of the Earth's volcanism that occurs at mid-ocean ridges and subduction zones, while the lower mantle has been more isolated throughout Earth history. In this model, high ^{3}He/^{4}He ratios are a tracer for material from

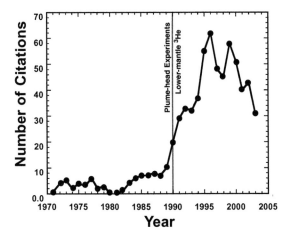

Figure 1.8 Number of published papers with "plume" in the title, in reference to mantle plumes vs. time, in the GeoRef data base[5] (from Anderson and Natland, 2005).

the deep lower mantle, and their detection supports the Plume hypothesis.

The popularity of the Plume hypothesis exploded in the early 1990s. Prior to about 1990, fewer than 10 published papers per year that refer to mantle plumes in their titles are listed by the data base GeoRef,[5] but this subsequently jumped by a factor of five or more (Fig. 1.8). During much of the last decade of the 20th century, the existence of mantle plumes was widely assumed, with little questioning by mainstream Earth science.

If the nineties was the decade of popularity of the Plume hypothesis, then the subsequent decade (the naughties) has been the decade of skepticism. The Plume hypothesis and its specific predictions have been re-examined on a fundamental basis. The most basic characteristic that makes an hypothesis scientific is its predictions, by which it can be tested. Without testable predictions, it is logically invulnerable to falsifica-

tion, it ceases to be scientific, and it degenerates to a faith-based belief (Popper, 1959).

1.4 Predictions of the Plume hypothesis

The specific predictions of the Plume hypothesis is a vexed question because of the wide variation in opinion that currently exists among scientists. The original hypothesis (Morgan, 1971) predicted that:

- Plumes are fixed relative to one-another,
- Time-progressive volcanic chains emanate from them,
- They are rooted in the deep mantle, whence they transport relatively primordial mantle upward.
- They break up continents.
- They drive plate tectonics, and
- They are hot.

In the three decades that followed the original proposal, the views of scientists evolved and diversified so that today many different visions exist of what the Plume hypothesis predicts.[6] So where, then, do we now stand? Recently, a clear, basic starting position has helpfully been stated (Campbell, 2006; 2007; Campbell and Davies, 2005; Campbell and Kerr, 2007; DePaolo and Manga, 2003).[7] According to this position, a plume is a thermal instability that rises from a layer at the bottom of the mantle, which is heated from below by the Earth's core. The instability comprises a large, bulbous head, followed by a relatively narrow tail, or feeder conduit. The tail is narrow because hot, low-viscosity plume material flows up the center of a pre-existing pathway of little resistance created by passage of the initial plume head. This modern version of the Plume hypothesis

[5] the online database of the American Geological Institute: http://csaweb113v.csa.com/factsheets/georef-set-c.php

[6] http://www.mantleplumes.org/DefinitionOfAPlume.html; http://www.mantleplumes.org/PlumeDLA.html
[7] http://www.mantleplumes.org/Plumes.html

Table 1.1 The predictions of the Plume and Plate hypotheses. Italics indicate the expectations of the respective model regarding the predictions of the other.

Plume	Plate
Current standard predictions (Campbell, 2006)	
precursory domal uplift	*not expected*
flood basalt eruption	*permitted but not required*
narrow conduit to the core-mantle boundary	*not expected*
time-progressive volcanic chain	*permitted but not required*
hot source	*no, or only modest mantle temperature variations expected*
Additional predictions made by the original model (Morgan,1971)	
melting anomalies fixed relative to one another	*permitted but not required*
compositionally distinct magmas	*expected*
break continents up	*magmatism may result from, but not cause, continental breakup*
drive plate tectonics	*not expected*
	The standard predictions (Foulger and Natland,2003)
radial extension expected to accompany precursory domal uplift	lithospheric extension
permitted but not required	fertile source

may be considered the current, standard model (Table 1.1). It makes the following five basic predictions:

1 *Precursory domal uplift:* Arrival of the plume head at the base of the lithosphere results in domal uplift of 500–1000 m, a few million years before flood basalt volcanism starts (Crough, 1983). The amplitude of the uplift depends on the temperature of the plume, and the area over which uplift is significant has a diameter of ~1000 km.

2 *Flood basalt eruption (the plume head):* The arriving plume head flattens to a disk at the base of the lithosphere, causes extension, and flood basalt eruptions occur rapidly over an area 2000–2500 km in diameter. The diameter of the volcanic region is dependent on the temperature difference between the plume and the surrounding mantle. If the plume head rises beneath continental lithos-

phere it may cause continental break-up and formation of volcanic margins.

3 *A narrow conduit to the core-mantle boundary (the plume tail):* Following flood basalt eruption, plume material continues to flow upward from the core-mantle boundary through a conduit 100–200 km in diameter.

4 *A time-progressive volcanic chain:* As the surface lithospheric plate above moves, continuous volcanism from the relatively fixed plume tail results in a trail of volcanism. The youngest volcanism occurs above the present-day location of the plume and older volcanism occurs progressively further along the trail.

5 *High temperatures:* The lavas associated with both the plume head and the plume tail formed at unusually high temperatures. Excess temperatures of $300 \pm 100\,°C$ occur at the center of the plume head, above the tail, reducing to ~100 °C further away, where cooler mantle material was entrained. Significant thermal anomalies persist below flood basalts for at least 100 Ma. Because of the high temperatures, picritic (high-MgO) basalts dominate early volcanism and the center of the plume head. Anomalously thick oceanic crust (volcanic margin) forms where continental break-up occurs.

1.5 Lists of plumes

The term "hot spot" carries with it the presumptions that the volcanism in question is fed by an unusually hot, highly localized source. Such features should be questioned, not assumed, and thus the term "hot spot" is not used in this book. Instead, the term "melting anomaly" is used, though it is itself not entirely satisfactory because what is an "anomaly" and what is merely a normal variation in a continuum is not easily decided.

Which melting anomalies are currently thought to be underlain by deep mantle plumes? In this basic question lies the first vexed problem. Over the years, many lists have been proposed for a global constellation of plumes.[8] The difficulty is that these lists vary radically both in length and content. This problem was ironically foreshadowed in the very first paper on the subject, where Morgan (1971) proposes that there are "about twenty deep mantle plumes" in the Earth, but plots only 16 in the accompanying figure (Fig. 1.6).

The numbers rose rapidly early on and within five years Burke and Wilson (1976) had proposed that there are 122 plumes. Lists were largely based on the observation of surface volcanism, but since the sizes of volcanic regions form a continuum ranging from the very large to the exceedingly small, where the cut-off line should be drawn is a subjective decision. Sleep (1990) listed 37 proposed plumes based on

[8] http://www.mantleplumes.org/CompleateHotspot.html

surface topographic anomalies ("swells") (Table 1.2), which he interpreted as the manifestations of hot plume material fluxing upward. Morgan's most recent list contains 69 proposed plumes, each assigned a degree of uncertainty (Morgan and Phipps Morgan, 2007) (Table 1.3). The world record for the plume population explosion is 5200, proposed on the basis of fractal arguments (Malamud and Turcotte, 1999).

Table 1.4 presents a recent summary of observations from 49 localities proposed to be underlain by plumes (Courtillot et al., 2003). For each locality the existence of the following observables is reviewed:

- A linear chain of dated volcanoes extending from the site of present volcanic activity;
- A flood basalt or oceanic plateau of the appropriate age at the older end of a volcanic chain;
- A high estimated buoyancy flux (in kg s^{-1}) and its reliability;
- High ^{3}He/^{4}He ratios in basalts; and
- Low seismic shear-wave speeds at 500 km depth beneath the present volcanically active site.

It can immediately be seen that these criteria correspond neither to the original criteria of Morgan (1971) nor to the modern standard criteria (Anderson, 2005a). This in itself illustrates the second problem – the diversity of diagnostic criteria which are, in practice, used.

Courtillot et al. (2003) categorized these 49 melting anomalies according to how many of the features listed above they display. A core-mantle-boundary origin was attributed to melting anomalies with high scores (9 localities), an origin at the base of the upper mantle at 650 km depth was assigned to anomalies with moderate scores (12 localities), and a lithospheric origin to those with low scores (28 localities) (Fig. 1.9). This approach is clearly subjective, as other criteria could have been used (Anderson, 2005a). Furthermore, it is curiously unscientific. An observation, for example of high ^{3}He/^{4}He, could

Table 1.2 Estimates of plume buoyancy fluxes.

Hotspot	Flux, Mg s^{-1} (from Sleep, 1990)	Reliability (from Sleep, 1990)	Flux, Mg s^{-1} (from Davies, 1988)
Afar	1.2	good	
Australia, East	0.9	fair	
Azores	1.1	fair	
Baja	0.3	poor	
Bermuda	1.1	good	1.5
Bouvet	0.4	fair	
Bowie	0.3	poor	0.8
Canary	1.0	fair	
Cape Verde	1.6	good	0.5
Caroline	1.6	poor	
Crozet	0.5	good	
Discovery	0.5	poor	0.4
Easter	3.3	fair	
Fernando	0.5	poor	0.9
Galapagos	1.0	poor	
Great Meteor	0.5	fair	0.4
Hawaii	8.7	poor	6.2
Hoggar	0.9	good	0.4
Iceland	1.4	good	
Juan de Fuca	0.3	fair	
Juan Fernandez	1.6	poor	1.7
Kerguelen	0.5	poor	0.7
Louisville	0.9	poor	3.0
Macdonald	3.3	fair	3.9
Marquesas	3.3	fair	4.6
Martin	0.5	poor	0.8
Meteor	0.5	poor	0.4
Pitcairn	3.3	fair	1.7
Réunion	1.9	good	0.9
St. Helena	0.5	poor	0.3
Samoa	1.6	poor	
San Felix	1.6	poor	2.3
Tahiti	3.3	fair	5.8
Tasman, Central	0.9	poor	
Tasman, East	0.9	poor	
Tristan	1.7	fair	0.5
Yellowstone	1.5	fair	
Total	54.9		

Table 1.3 Melting anomaly locations, empirical confidence estimates, and azimuths and rates of postulated plume tracks (from Morgan and Phipps Morgan, 2007). ND: not defined.

Plume	Tectonic plate	Lat (°N)	Long (°E)	Weight	Azimuth	±(°)	Rate (mm/yr)	±mm/ yr
Afar	African	7.0	39.5	0.2	30	±15	16	±8
Anyuy	N American	67.0	166.0	B-	ND	ND	ND	ND
Arago	Pacific	−23.4	−150.7	1.0	296	±4	120	±20
Ascension	S American	−7.9	−14.3	B	ND	ND	ND	ND
Azores	European	37.9	−26.0	0.5	110	±12	ND	ND
Azores	N American	37.9	−26.0	0.3	280	±15	ND	ND
Baikal	European	51.0	101.0	0.2	80	±15	ND	ND
Balleny	Antarctic	−67.6	164.8	0.2	325	±7	ND	ND
Bermuda	N American	32.6	−64.3	0.3	260	±15	ND	ND
Bowie	Pacific	53.0	−134.8	0.8	306	±4	40	±20
Cameroon	African	−2.0	5.1	0.3	32	±3	15	±5
Canary	African	28.2	−18.0	1.0	94	±8	20	±4
Cape Verde	African	16.0	−24.0	0.2	60	±30	ND	ND
Caroline	Pacific	4.8	164.4	1.0	289	±4	135	±20
Cobb	Pacific	46.0	−130.1	1.0	321	±5	43	±3
Cocos-Keeling	Australian	−17.0	94.5	0.2	28	±6	ND	ND
Comores	African	−11.5	43.3	0.5	118	±10	35	±10
Crough	Pacific	−26.9	−114.6	0.8	284	±2	ND	ND
Crozet	Antarctic	−46.1	50.2	0.8	109	±10	25	±13
Discovery	African	−43.0	−2.7	1.0	68	±3	ND	ND
Easter	Nazca	−26.4	−106.5	1.0	87	±3	95	±5
Eastern Aust.	Australian	−40.8	146.0	0.3	0	±15	65	±3
Eifel	European	50.2	6.7	1.0	82	±8	12	±2
Erebus	Antarctic	−77.5	167.2	A	ND	ND	ND	ND
Etna	European	37.8	15.0	A	ND	ND	ND	ND
Fernando Do Noron	S American	−3.8	−32.4	1.0	266	±7	ND	ND
Foundation	Pacific	−37.7	−111.1	1.0	292	±3	80	±6
Galapagos	Nazca	−0.4	−91.6	1.0	96	±5	55	±8
Galapagos	Cocos	−0.4	−91.6	0.5	45	±6	ND	ND
Gough	African	−40.3	−10.0	0.8	79	±5	18	±3
Great Meteor	African	29.4	−29.2	0.8	40	±10	ND	ND
Guadalupe	Pacific	27.7	−114.5	0.8	292	±5	80	±10
Guyana	S American	5.0	−61.0	B	ND	ND	ND	ND
Hainan	China	20.0	110.0	A	0	±15	ND	ND
Hawaii	Pacific	19.0	−155.2	1.0	304	±3	92	±3

Table 1.3 *Continued*

Plume	Tectonic plate	Lat (°N)	Long (°E)	Weight	Azimuth	±(°)	Rate (mm/yr)	±mm/yr
Heard	Antarctic	−53.1	73.5	0.2	30	±20	ND	ND
Hoggar	African	23.3	5.6	0.3	46	±12	ND	ND
Iceland	European	64.4	−17.3	0.8	75	±10	5	±3
Iceland	N American	64.4	−17.3	0.8	287	±10	15	±5
Jebel Marra	African	13.0	24.2	0.5	45	±8	ND	ND
Juan Fernandez	Nazca	−33.9	−81.8	1.0	84	±3	80	±20
Karisimbi	African	−1.5	29.4	B	ND	ND	ND	ND
Kerguelen	Antarctic	−49.6	69.0	0.2	50	±30	3	±1
Kilimanjaro	African	−3.0	37.5	B	ND	ND	ND	ND
Lord Howe	Australian	−34.7	159.8	0.8	351	±10	ND	ND
Louisville	Pacific	−53.6	−140.6	1.0	316	±5	67	±5
Macdonald	Pacific	−29.0	−140.3	1.0	289	±6	105	±10
Madeira	African	32.6	−17.3	0.3	55	±15	8	±3
Maria/S.Cook	Pacific	−22.2	−154.0	0.8	300	±4	ND	ND
Marion	Antarctic	−46.9	37.6	0.5	80	±12	ND	ND
Marquesas	Pacific	−10.5	−139.0	0.5	319	±8	93	±7
Martin Vaz	S American	−20.5	−28.8	1.0	264	±5	ND	ND
Massif Central	European	45.1	2.7	B	ND	ND	ND	ND
Mt Rungwe	African	−8.3	33.9	B+	ND	ND	ND	ND
N. Austral	Pacific	−25.6	−143.3	B	293	±3	75	±15
Ob-Lena	Antarctic	−52.2	40.0	1.0	108	±6	ND	ND
Peter Island	Antarctic	−68.8	−90.6	B	ND	ND	ND	ND
Pitcairn	Pacific	−25.4	−129.3	1.0	293	±3	90	±15
Raton	N American	36.8	−104.1	1.0	240	±4	30	±20
Reunion	African	−21.2	55.7	0.8	47	±10	40	±10
Samoa	Pacific	−14.5	−169.1	0.8	285	±5	95	±20
San Felix	Nazca	−26.4	−80.1	0.3	83	±8	ND	ND
Scott	Antarctic	−68.8	−178.8	0.2	346	±5	ND	ND
Shona	African	−51.4	1.0	0.3	74	±6	ND	ND
Society	Pacific	−18.2	−148.4	0.8	295	±5	109	±10
St Helena	African	−16.5	−9.5	1.0	78	±5	20	±3
Tasmantid	Australian	−40.4	155.5	0.8	7	±5	63	±5
Tibesti	African	20.8	17.5	0.2	30	±15	ND	ND
Tristan Da Cunha	African	−37.2	−12.3	A	ND	ND	ND	ND
Vema	African	−32.1	6.3	B	ND	ND	ND	ND
Yellowstone	N American	44.5	−110.4	0.8	235	±5	26	±5

Table 1.4 Melting anomalies and their features, from Courtillot et al. (2003). Columns give name, latitude and longitude, the existence or not of a chain of dated volcanoes, the existence and age of a flood basalt or oceanic plateau at the old end, buoyancy flux and its reliability, the existence or not of high $^3He/^4He$ ratios, the existence of low seismic shear-wave speeds at 500 km depth, and the total number of these five features observed at each locality.

Hotspot	Lat	Lon (°E)	Track	Flood/ plateau	Age (Ma)	Buoy. (10^3 kg s^{-1})	Reliab.	$^3He/^4He$	Tomo (500 km)	Count
Afar	10N	43	no	Ethiopia	30	1	good	high	slow	4
Ascension	8S	346	no	no	/	na	na	na	0	0+?
Australia E	38S	143	yes	no	/	0.9	fair	na	0	1+?
Azores	39N	332	no?	no	/	1.1	fair	high?	0	1+?
Baja/ Guadalupe	27N	247	yes?	no	/	0.3	poor	low	0	0+?
Balleny	67S	163	no	no	/	na	na	na	0	0+?
Bermuda	33N	293	no	no?	/	1.1	good	na	0	0+?
Bouvet	54S	2	no	no	/	0.4	fair	high	0	1+?
Bowie	53N	225	yes	no	/	0.3	poor	na	slow	2+?
Cameroon	4N	9	yes?	no	/	na	na	na	0	0+?
Canary	28N	340	no	no	/	1	fair	low	slow	2
Cape Verde	14N	340	no	no	/	1.6	poor	high	0	2
Caroline	5N	164	yes	no	/	2	poor	high	0	3
Comores	12S	43	no	no	/	na	na	na	0	0+?
Crozet/Pr. Edward	45S	50	yes?	Karoo?	183	0.5	good	na	0	0+?
Darfur	13N	24	yes?	no	/	na	poor	na	0	0+?
Discovery	42S	0	no?	no	/	0.5	poor	high	0	1+?
Easter	27S	250	yes	mid-Pac mnt?	100?	3	fair	high	slow	4+?
Eifel	50N	7	yes?	no	/	na	na	na	0	0+?
Fernando	4S	328	yes?	CAMP?	201?	0.5	poor	na	0	0+?
Galapagos	0	268	yes?	Caribbean?	90	1	fair	high	0	2+?
Great Meteor/ New England	28N	328	yes?	no?	/	0.5	poor	na	0	0+?
Hawaii	20N	204	yes	subducted?	>80?	8.7	good	high	slow	4+?
Hoggar	23N	6	no	no	/	0.9	poor	na	slow	1
Iceland	65N	340	yes?	Greenland	61	1.4	good	high	slow	4+?
Jan Mayen	71N	352	no?	yes?	/	na	poor	na	slow	1+?
Juan de Fuca/ Cobb	46N	230	yes	no	/	0.3	fair	na	slow	2+?

Table 1.4 *Continued*

Hotspot	Lat	Lon (°E)	Track	Flood/ plateau	Age (Ma)	Buoy. (10^3 kg s^{-1})	Reliab.	^{3}He/ ^{4}He	Tomo (500 km)	Count
Juan Fernandez	34S	277	yes?	no	/	1.6	poor	high	0	2+?
Kerguelen (Heard)	49S	69	yes	Rajmahal?	118	0.5	poor	high	0	2+?
Louisville	51S	219	yes	Ontong Java	122	0.9	poor	na	slow	3+?
Lord Howe (Tasman East)	33S	159	yes?	no	/	0.9	poor	na	slow	1+?
Macdonald (Cook-Austral)	30S	220	yes?	yes?	/	3.3	fair	high?	slow	2+?
Marion	47S	38	yes	Madagascar?	88	na	na	na	0	1+?
Marqueses	10S	222	yes	Shatski?	???	3.3	na	low	0	2+?
Martin/ Trindade	20S	331	yes?	no	/	0.5	poor	na	fast	0+?
Meteor	52S	1	yes?	no	/	0.5	poor	na	0	0+?
Pitcairn	26S	230	yes	no	/	3.3	fair	high?	0	2+?
Raton	37N	256	yes?	no	/	na	na	na	slow	1+?
Reunion	21S	56	yes	Deccan	65	1.9	poor	high	0	4
St. Helena	17S	340	yes	No	/	0.5	poor	low	0	1
Samoa	14S	190	yes	no?	14?	1.6	poor	high	slow	4
San Felix	26S	280	yes?	No	/	1.6	poor	na	0	1+?
Socorro	19N	249	no	No	/	na	poor	na	slow	1+?
Tahiti/Society	18S	210	yes	No	/	3.3	fair	high?	0	2+?
Tasmanid (Tasman central)	39S	156	yes	No	/	0.9	poor	na	slow	2
Tibesti	21N	17	yes?	No	/	na	poor	na	0	0+?
Tristan	37S	348	yes	Parana	133	1.7	poor	low	0	3
Vema	33S	4	yes?	yes? (Orange R.)	/	na	poor	na	0	0+?
Yellowstone	44N	249	yes?	Columbia?	16	1.5	fair	high	0	2+?

characterize one melting anomaly assigned a lithospheric origin (e.g., the Azores), but swell the score at another such that it is assigned a source at 650 km (e.g., Cape Verde). Since the rationale by which high ^{3}He/^{4}He is considered relevant to detecting plumes is based on the assumption that such ratios arise from the deep mantle, a scheme by which it could characterize a lithospheric anomaly, or decide on whether a melting anomaly is sourced in the lithosphere

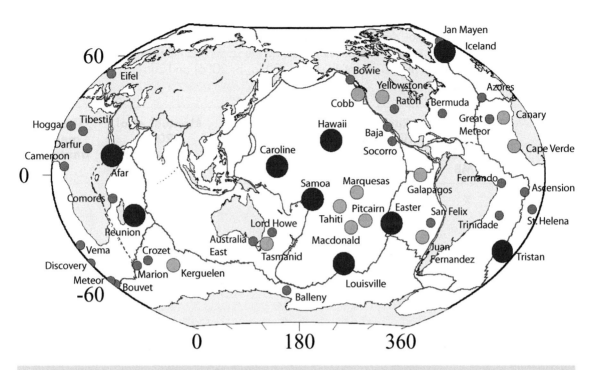

Figure 1.9 "Hotspots in the most cited catalogs" (from Courtillot et al., 2003).

or at 650 km depth, is clearly not rational. Nevertheless, Table 1.4 provides a handy summary of one group's current assessment of the basic global observations.

Several recent seismic experiments have been used to draw up lists of melting anomalies underlain by regions of low seismic wave speed, proposed to show images of plumes (Table 1.5). The list of Ritsema and Allen (2003) is based on the whole-mantle tomography model of Ritsema et al. (1999), which has a resolution of ~1000 km and is particularly reliable in the mantle transition zone, in the depth range 410–650 km. In this model, individual regions of low wave speed tend to be confined to either the upper or the lower mantle and do not traverse both. The list of possible plume localities drawn from this work comprises melting anomalies beneath which low wave speeds extend from the surface down to the base of the upper mantle at 650 km depth.

A relatively new method known as finite frequency ("banana-doughnut") tomography recently provided rather different seismic images of the mantle (Montelli et al., 2004a, b; 2006). This method involves dropping the simplifying assumption that seismic rays are infinitely narrow, and thus correcting for the finite wavelength of seismic waves. The resulting images contain smaller-scale structure and include features that traverse much of the mantle. These results have been challenged (van der Hilst and de Hoop, 2005),[9] but it is instructive, nonetheless, to compare the lists of proposed plumes from that work with other lists.

In the final column of Table 1.5, melting anomalies that erupt large volumes of tholeiitic basalt are indicated. Tholeiitic basalt is thought to result from large-degree partial melting –

[9] http://www.mantleplumes.org/BananaDoughnuts.html

Table 1.5 Melting anomalies defined as arising from the core-mantle boundary by Courtillot et al. (2003), underlain by seismic anomalies traversing the upper mantle (Ritsema and Allen, 2003), traversing the whole mantle (Montelli et al., 2004a, b; 2006) and currently erupting large volumes of tholeiitic lava.

	Courtillot et al. (2003) CMB	Ritsema & Allen (2003) TZ	Montelli et al. (2004a, b) (P waves) CMB	Montelli et al. (2006) (S waves) CMB	Tholeiitic melting anomalies
Afar	X	X			X
Ascension			X	X	
Azores			X	X	
Bowie		X			
Canaries			X	X	
Cape Verde				X	
Cook Island				X	
Crozet				X	
Easter	X	X	X	X	
Galapagos					X
Hawaii	X	X		X	X
Iceland	X	X			X
Kerguelen				X	
Louisville	X	X			
Macdonald		X			
Réunion	X				
Samoa		X	X	X	
Tahiti			X	X	
Tristan	X				
Yellowstone					X

perhaps 10–20% of peridotitic mantle or near-complete melting of eclogite. In this it contrasts with alkali basalt which is traditionally thought to arise ultimately from small degrees of partial melting – perhaps 1–2%. A current school of thought suggests that high degrees of partial melt require high temperatures and that plumes underlie only regions where large-volume tholeiitic lavas are currently being erupted.

The most striking feature of Table 1.5 is the lack of agreement between the different lists of melting anomalies proposed to be underlain by deep mantle plumes. Of the 20 melting anomalies listed, 15 appear on two or fewer of the proposed lists and none appears on every list. There is thus a fundamental difficulty in agreeing where plumes occur, even if just one criterion is used (seismology) and the list is restricted to the strongest candidates only. Afar, Iceland and Samoa, common favorites, appear on only three out of the five lists. Only Easter and Hawaii appear on four, but neither of these fulfill all the non-seismological predictions (Table 1.4).

1.6 Testing plume predictions

How may the predictions of the Plume hypothesis be tested? One of the great beauties of studying melting anomalies is the enormous variety of approaches within Earth science that can potentially contribute. On land, evidence for precursory domal uplift can be sought using stratigraphic mapping and fission track analysis. In the oceans, sedimentary layers sampled in marine drill cores testify to the water depth when they were deposited. Where a large volcanic province formed in the ocean, and was later transported away from its presumed mantle source by plate motion, initial uplift is expected to be matched by subsequent subsidence as the province drifted over cooler mantle.

The existence of plume-head-related flood volcanism can be investigated by geological mapping on land, though the total volume may be hard to assess if the thickness of the lavas cannot be estimated accurately or if much has been eroded away. Flood basalts are often referred to as "large igneous provinces" or "LIPs", a term that has been defined on the basis of the surface area covered by eruptives. Minimum areas of $100,000\,km^2$ and $50,000\,km^2$ have both been proposed (Bryan and Ernst, 2008; Coffin and Eldholm, 1992; 1993; Sheth, 2007b),[10] In the present book, the more general term "flood basalt" will be used, to avoid using leading, assumption-driven terminology. In the ocean, broad areas of unusually shallow sea floor ("plateaus") can be seen in bathymetric maps. They are usually assumed to indicate expanses of thickened crust, but this is not always true

(Vogt and Jung, 2007).[11] Plateaus also show up in gravity maps as regions of anomalously high gravity because of their excess mass. The thickness of the igneous layers can be probed using seismic methods both on land and at sea, using techniques designed to study crustal structure.

The regularity of time-progression in volcanic chains can be investigated using radiometric dating. Unfortunately, many of the most remarkable chains are on the ocean floor and formidable problems have to be overcome before they can be studied (Clouard and Bonneville, 2005). First, fresh samples can only be retrieved by drilling from ocean-going research ships, an expensive enterprise. Second, although many whole-rock potassium-argon (K-Ar) dates have been derived, this method is inaccurate (Section 4.2.1). It has been superceded by the potentially much more accurate Ar-Ar method, but this has not yet been widely applied. As a result, how time-progressive volcanic chains really are, and how fixed melting anomalies are relative to one another, is still surprisingly poorly known. Many studies deal with this by simply assuming that chains trending in the expected direction are regularly time-progressive. Obviously this is unsafe, and there have been some notable surprises when detailed investigations have been made (McNutt et al., 1997; Tarduno and Cottrell, 1997).

The question of how fixed melting anomalies are relative to one another is generally studied using dates, sample locations and knowledge of the relative motions between the plates on which the melting anomalies lie. Ideally, some fixed frame of reference would be used. The position of the Earth's magnetic pole is a candidate, since it is thought to have been roughly aligned with the rotation axis throughout geological time. Palaeomagnetism can then be used to measure the latitudes of lavas when they were erupted. A problem with this approach, however, is that it cannot yield longitude.

[10] http://www.mantleplumes.org/TopPages/LIPClassTop.html

[11] http://www.mantleplumes.org/Superswell.html; http://www.mantleplumes.org/Bermuda.html

Seismology is essentially the only method currently available that can test for conduits extending from the surface to the core-mantle boundary beneath melting anomalies. For this, techniques are needed that are powerful enough to image Earth structure much deeper than the crust, and extending throughout the mantle. Seismic tomography, using either local or global instrument arrays, is a commonly used technique. Unfortunately, it typically suffers from inadequate spatial resolution or inability to image sufficiently large depths. Tomography has thus been supplemented by methods designed to target localized regions of interest deep in the mantle. For example, the waveforms of carefully selected seismic waves that passed through the target region have been modeled in detail. The deep mantle and the core-mantle boundary beneath currently active volcanic regions are of particular interest because this is where the roots of plumes are hypothesized to lie.

High temperature is perhaps the most basic and fundamental of all the characteristics attributed to plumes. It can potentially be investigated using many different approaches, including seismology, petrology, vertical motions and heat flow (Foulger et al., 2005a). A significant initial challenge is, however, deciding on the norm against which the mantle temperatures beneath melting anomalies should be compared. Typically, the mantle temperature beneath mid-ocean ridges is chosen because they are assumed to have sampled "normal-temperature" mantle. Ridges have been widely studied using seismological and other geophysical methods, and the lavas erupted there have been sampled by dredging, so many data are available to study them. However, it is not clear that these can be compared with melt extracted from very different depths because they probably derive from within the surface conductive layer, and not from below it (Section 6.1; Presnall et al., 2010).

The plethora of methods available to estimate temperature vary hugely in precision, ambiguity, and how closely they approach a true measure of the temperature of the melt source. Seismic wave speeds are notoriously ambiguous. They vary not just with temperature but also with composition and the presence of partial melt. For example, even a trace of partial melt can radically lower seismic wave speed. The problem is that it is usually impossible, in practice, to separate out the contributions from individual effects. Seismology can tell us little about temperature, even though it is widely assumed to be the most powerful indicator of temperature for the deep mantle. Petrological methods suffer from the difficulty of acquiring a fresh sample confidently known to represent the original melt. Specimens are almost always modified by material lost during crystallization or gained during transport to the surface. Only glass samples devoid of crystals can confidently be assumed to correspond directly to an original liquid. Petrological thermometry has to make simplifying assumptions that are often unlikely to be realistic (Section 6.2.2).

Petrological and geochemical methods have also been widely applied in efforts to obtain the compositions of melt sources, their locations in the mantle, and their histories of formation. Deducing source composition from lava composition is inherently ambiguous, however, and geochemistry has almost no power to constrain depth of origin below ~100 km. Only if melting anomalies are shallow-sourced is geochemistry likely to be important in constraining their origins.

These are practical problems that present great challenges, but the picture heretofore painted is nevertheless straightforward. A precisely defined Plume hypothesis has been developed, clear, specific predictions have been made, and techniques exist, albeit inaccurate and in need of improvement, to test those predictions. If things were thus simple, and required only rigorous application of the scientific method, our problems would be only scientific ones. Unfortunately, there is another dimension to how plume science is, in reality, practiced. That is, every one of the five basic features predicted, and also the predictions of many variants of the hypothesis, is commonly either considered optional or its absence, even in the face of extensive searching, is considered to be inconclusive. This problem besets both theoretical and observational work, and renders the hypothesis essentially unfalsifiable.

From the theoretical point of view, the expectation of simple, domal precursory uplift has been brought into question by numerical modeling. For rheologicaly realistic lithospheres that include the viscoelastic layers that are known to exist, the pattern of vertical motion is predicted to be a complex mix of uplift and subsidence (Burov and Guillou-Frottier, 2005a).[12] A bulbous plume head and an initial flood basalt is not predicted if a plume has a higher viscosity than the mantle through which it rises. In that case, a distinct head does not develop (Davies, 1999, p. 303). The necessity of observing a narrow conduit extending to the core-mantle boundary is relaxed if the plume is postulated to pulse, be discontinuous, or arise from the base of the upper mantle at a depth of 650 km. The lack of time progression of volcanism can be explained away by postulating irregular lateral flow of plume material. How high a temperature anomaly is required at the surface is, in practice, vague. Hot, rising material is expected to entrain cooler mantle and to reduce the temperature anomaly in the shallow part of the plume where melting is thought to occur. Some estimates of plume temperature anomalies proposed are $<100\,°C$, and these fall within the normal variation in mantle temperature expected from place to place as a result of plate-tectonic-related processes such as subduction and continental insulation.

Technical, scientific problems can be addressed scientifically. However, immunity of an hypothesis to testing is something that science cannot deal with. Is the Plume hypothesis immune, and no longer scientific? This is not merely detached philosophical speculation but touches on the very issue of whether a fundamental scientific problem actually exists or not.

1.7 A quick tour of Hawaii and Iceland

Hawaii and Iceland exemplify the difficulties that a flexible hypothesis presents. In agreement with the predictions, the Hawaiian-Emperor volcanic chain is unidirectionally time-progressive, and picrite glass samples have been found and interpreted as indicating a high source temperature (Section 6.5.3). Nevertheless, the system is lacking a flood basalt at its old end, there is no evidence for precursory uplift there, and a conduit to the core-mantle boundary beneath the "Big Island" of Hawaii has not been observed (Fig. 1.10). On the other hand, a flood basalt is currently forming at the young end of the chain. Suggested solutions to these problems with the Plume hypothesis include the subduction and disappearance of an original plume-head-related volcanic plateau that preceded formation of the Emperor chain, and insufficient resolution in seismic tomography of the mantle beneath Hawaii. The recent surge in volcanic output has been explained by a pulsing plume.

In the Iceland region, the story is close to the reverse. Uplift accompanied flood basalt eruption early in the volcanic sequence. However, there is no time-progressive volcanic chain – volcanism has always been centered on the mid-ocean ridge, currently at Iceland (Foulger and Anderson, 2005; Foulger et al., 2005a; Lundin and Doré, 2005, 2004).[13] Also, a large suite of independent methods indicate strongly that the temperature of the mantle beneath the north Atlantic is possibly $50–100\,°C$ higher than is typical beneath mid-ocean ridges, but certainly not approaching the $200–300\,°C$ predicted for plumes. Iceland is more easily studied seismically than Hawaii, and multiple, independent, high-quality seismic studies leave little doubt that there is no low-wave-speed conduit extending down to the core-mantle boundary.

[12] http://www.mantleplumes.org/Burov2005.html

[13] http://www.mantleplumes.org/Iceland1.html
http://www.mantleplumes.org/Iceland2.html;

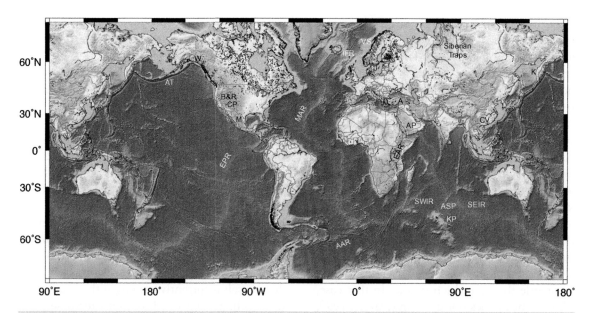

Figure 1.10 Global topography and bathymetry, from gravity measurements:[14] IFR – Iceland-Faeroe Ridge; VS – Vøring Spur; CV – Changbai Volcano; AP – Arabian Peninsula; W – Wrangellian terrain; AT – Aleutian Trench; EPR – East Pacific Rise; CP – Colorado Plateau; K – Kamchatka; A – Anatolia; M – Mexico; B&R – Basin and Range Province; EAR – East African Rift; AAR – American-Antarctic Ridge; MAR – Mid-Atlantic Ridge; ASP – Amsterdam-St Paul Plateau; SWIR – Southwest Indian Ridge; SEIR – Southeast Indian Ridge; KP – Kerguelen Plateau. See Plate 2

As is the case for Hawaii, present-day volcanism is also forming a second flood basalt at Iceland (Fig. 1.10).

Variants of the basic Plume hypothesis can again account for all the unpredicted features. These include lateral flow of material to the mid-Atlantic ridge from a relatively fixed plume that did not pierce the surface at its true location, and pulsing, discontinuous and tilting plumes. The lack of picrite glass samples, which might provide evidence for high temperatures, has been attributed to the inability of dense melts to rise through the thick pile of basaltic lavas. A cool plume under Iceland has also been suggested. Most of these variants, if also applied to Hawaii, would predict observations there at odds with what is found.

[14] http://topex.ucsd.edu/WWW_html/mar_topo.html

Let us back track and consider the additional predictions of the original hypothesis (Morgan, 1971), that have been discarded in the more modern version. These are that plumes are fixed relative to one another, that they transport compositionally distinct, primordial material to the surface from the deep mantle, and that they break continents up and drive plate tectonics.

Relative fixity is difficult to test for the Emperor seamount chain, because there is no other clear, simple, time-progressive chain of the same age (76–50 Ma) on a different plate. The fixity of the Emperor volcanic locus relative to the Earth's geomagnetic pole has been investigated and it has been discovered that Emperor volcanism migrated south by at least 800 km and possibly by as much as 2000 km, across a slow-moving or even stationary Pacific plate during formation of the chain (Sager, 2007; Tarduno and Cottrell, 1997). This is the exact opposite scenario to that of a fixed plume with a plate moving overhead transporting volcanoes away. At ~50 Ma, the time of the "bend" in the Emperor-Hawaiian chain, the behavior appears to have abruptly reversed. The Pacific plate started to move rapidly northward with respect to the geomagnetic pole and the volcanic locus suddenly stopped. In the case of Iceland, volcanism has always been centered on the mid-Atlantic ridge, a location that is not fixed relative to any other melting anomaly.

Again, these features, unpredicted by any version of the Plume hypothesis, have been explained away by *ad hoc* variants. The apparent wandering of the postulated Hawaiian plume has been explained by "mantle wind", i.e., distortion of the plume conduit, and consequential displacement of the surface volcanic locus, by large-scale mantle convection unrelated to plumes (Steinberger et al., 2004). The exact opposite argument has been used to explain the persistence of high-volume volcanism on the mid-Atlantic ridge at Iceland. There, it has been suggested that material flowed laterally from a relatively fixed plume to the northwest, one that was not "blowing in the mantle wind", and which did not penetrate the surface vertically above (Vink, 1984). The volcanism that formed the Emperor-Hawaiian chain is thought to have started on a spreading ridge (Norton, 2007) but it subsequently migrated away and into the deep interior of the thick Pacific plate, penetrating the surface without difficulty, even in the past when its production was much lower than at the present day. This plume material evidently had no tendency to flow laterally and erupt at the nearest ridge.

Let us turn to composition. The prediction that lavas at proposed plume localities are compositionally different from mid-ocean-ridge basalts (MORB) fits the observations well. The enriched basalts found, for example, at oceanic islands, typically contrast geochemically with MORBs in rare-earth- and incompatible-element contents, radiogenic isotopic signatures

and sometimes in their noble gas isotope ratios. The difficulties that arise from this are two-fold. First, the geochemical signatures have been found to arise from the inclusion in the source of near-surface materials such as sediments and subducted oceanic crust. Second, virtually identical geochemical signatures are widespread in basalts scattered throughout the continents and oceans where plumes have not been proposed and are not expected (Fitton, 2007). Such basalts are found on isolated seamounts, in small-volume eruptions in rift valleys, and even in long-lived, low-output volcanic provinces that have erupted similar material for tens of millions of years as the region drifted thousands of kilometers from its original position. Examples of such regions are the Scottish Midland Valley and the rift zones of Europe. A recent plume-compliant explanation for this enigma suggests that the entire asthenosphere is made up of material transported up from the deep mantle in plumes. All surface volcanism, regardless of its cause, then arises from various melt fractions extracted from a global layer of plume material (Yamamoto et al., 2007a; b).

The final prediction, that plumes break continents up and drive plate tectonics is at best only partially consistent with the observations. The North Atlantic Igneous Province is linked to break-up of the Eurasian supercontinent, starting at ~54 Ma, at which time voluminous volcanic margins formed. In contrast, no evidence is available concerning an hypothetical plume head stage for the Emperor-Hawaiian system because the oldest part of the Emperor chain abuts the Aleutian trench. Earlier features may have been subducted, though there is no evidence for this (Shapiro, 2005). The remarkable contrasts between these two huge volcanic provinces, both of which are proposed to be caused by the same geodynamical phenomenon, could hardly be more striking.

Modern plume science, outlined above, thus leaves us in a quandary, both regarding its predictions and the flexibility with which it is applied. On the one hand, clear, specific predictions have been proposed, both early in the history of the hypothesis, and more recently, following decades of research and testing. Some data are consistent with the hypothesis, but in most cases the predictions have not been confirmed. In this respect the Plume hypothesis contrasts starkly with plate tectonics. The predictions of that theory have been spectacularly confirmed over many years, often using methods unconceived when the hypothesis was first proposed. This is not to say that the original plate tectonic concept has survived testing unmodified. Indeed, plates are now known to be not rigid and many details of early suggested plate boundaries have been revised. Nevertheless, observation has tended to strengthen and confirm the basic concept, and discrepancies have been second-order in nature and compliant with the overall, basic model.

In the case of the Plume hypothesis, testing has instead tended to turn up first-order problems requiring major revisions, often different for different proposed plumes. This should scarcely come as a surprise because a striking feature of the volcanic regions proposed to be underlain by plumes is their extreme diversity. They comprise a group much less homogenous than either mid-ocean-ridge spreading segments or subduction-zone volcanoes. No one volcanic system on Earth fulfills all the predictions, and most fulfill very few or even none. A composite of observations drawn from more than one melting anomaly is needed in order to assemble a full set of postulated plume attributes. Notwithstanding this, failure of the predictions is rarely considered to be a threat to the hypothesis. How then, can the Plume hypothesis be tested? Can it ever be rejected, and what is the way forward?

1.8 Moving on: Holism and alternatives

Two possible fresh avenues of approach spring to mind. The first is to take a holistic investiga-

tive approach, both to individual melting anomalies and to the global constellation as a whole. A variant of the Plume hypothesis may indeed be designed to account for an individual discrepancy in a particular data set at a volcanic locality, but is this variant consistent with other, cross-disciplinary observations? Mutually inconsistent variants proposed, for example, to satisfy separate seismic, geochemical and geochronological observations, cannot all be correct. Furthermore, a variant that explains the observations at just one, or very few, proposed plume localities is unconvincing if it is not generally applicable in similar situations elsewhere, that is to say, if it is not predictive.

If the Plume hypothesis does not fulfill minimum reasonable scientific expectations, then what does? The second approach is to develop and test alternatives, and it is therein that the real excitement and potential for progress lies. Interest in alternative models waned during the 1990s. However, the present century has seen a resurgence of interest in response to the growing weight of evidence that the difficulties with the Plume hypothesis are insurmountable. Today, the alternative "Plate" hypothesis encompasses diverse, shallow-based models for melting anomalies that are linked by their common association with the effects of plate tectonics. Describing this hypothesis, and weighing the evidence against both it and the Plume hypothesis, is the subject of the present book.

1.9 The Plate hypothesis

The Plate hypothesis postulates that melting anomalies on the surface of the Earth arise from shallow-based processes related in various ways to plate tectonics (Anderson, 2001a; 2007; Foulger, 2007; Foulger and Natland, 2003).[15] Simply put, it suggests that melting

---

[15] http://www.mantleplumes.org/PTProcesses.html; http://www.mantleplumes.org/PLATE.html

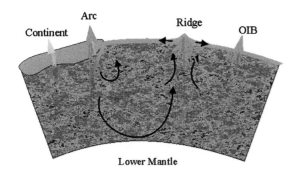

Figure 1.11 The upper mantle is kept inhomogeneous by continuous melt extraction and lithosphere recycling at subduction zones and beneath continents, which makes plate tectonics work (from Meibom and Anderson, 2004). See Plate 3

anomalies arise from permissive volcanism that occurs where the lithosphere is in extension (Favela and Anderson, 1999; Natland and Winterer, 2005). The volume of magma produced is variable, and related to the fusibility of the source beneath and the availability of pre-existing melt (Fig. 1.11). Where extension occurs above a fertile source or regions containing partial melt, enormous volumes of magma may form and be intruded and erupted. Where it occurs above refractory, infertile, sub-solidus mantle, little melt may be produced. For a given temperature, larger volumes of melt will be produced from a source with a low solidus temperature, than from one with a higher solidus temperature (Table 1.1). The eruption rate is controlled by factors such as lithospheric stress and thickness, the quantity of pre-existing melt, and the amount of volatiles.

Plate tectonics constantly maintains a state of variable stress in the lithosphere, with extension in some places and compression in others. This extension may be localized, for example, at continental rifts, mid-ocean ridges, and triple junctions, or it may be distributed in broad regions such as the Basin and Range province in western North America, and in diffuse oceanic plate boundaries. It is in these regions where melting anomalies are most expected and, indeed, where they are most commonly present. The plates are not considered to be absolutely rigid, but they are recognized as being capable of internal deformation, both in continental and oceanic regions. Extension in intraplate oceanic regions is more difficult to study than on land, because most of the sea floor is known only through sensing from space using satellites or from the surface of the ocean. This is comparable to mapping an area on land from several kilometers away, using binoculars. The Plate hypothesis suggests that volcanism itself is an intrinsic indication of where the lithosphere is in extension.

The fertility of source material, often but not necessarily always in the mantle, also varies as a result of processes ultimately related to plate tectonics. New crust is created at spreading ridges by the extraction of melt from the mantle. New mantle lithosphere accumulates beneath the crust as the plates drift away, cool and thicken. The components of the mantle that melt first and form the crust are necessarily the most fusible ones, and the remaining refractory residuum is left behind.

The recycling of slabs comprising crust and mantle lithosphere back into the mantle at subduction zones reintroduces fusible material from the near-surface into the mantle as concentrated packages made up of diverse rock types. Delamination and gravitational instability perform a similar role in recycling continental lithosphere. These are processes by which over-thickened lithosphere, which has densified by mineralogical phase changes, detaches and sinks into the asthenosphere. They are ultimately related to plate tectonics, since it is processes such as continental collision that thicken the lithosphere.

In addition to dehomogenizing the crust and upper mantle compositionally, such processes also dehomogenize it thermally. Melt migrating from one region to another, and the subduction or sinking of cold, dense material, transport heat

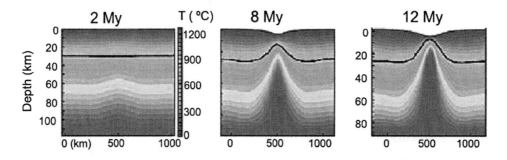

Figure 1.12 Thermal evolution of lithosphere overlying normal-temperature asthensophere during extension leading to continental break-up. Massive volcanism results, building volcanic margins. Black line indicates the Moho (from van Wijk et al., 2001).

as well as compositionally distinct material. In this way, plate-related processes result in variations in temperature from place to place, though these are on the whole expected to be more subdued and less localized than the temperature anomalies predicted by the Plume hypothesis.

These are broad generalizations. Is it possible to be more specific about exactly what is envisaged and predicted? Within the scope of the Plate hypothesis many different phenomena are expected to lead to volcanism. While this may seem untidy and at odds with Occam's Razor – the desirability of maximum simplicity – it is in keeping with the enormous diversity of melting anomalies that is immediately obvious from a mere glance at any list (Tables 1.1–1.5).

Specifically, the main processes that fall under the umbrella of the Plate hypothesis are:

- *Continental break-up:* In the Plate hypothesis, continental break-up, for example, such as formed the Atlantic ocean, is viewed as part of a continual self-reorganization of the plates (Anderson, 2002). Self-reorganization may be brought about by changing plate boundary conditions, for example, the collision of continents, subduction of entire plates, or by local changes in the temperature of the upper mantle brought about by continental insulation. The process of continental break-up is expected to start with extension and to continue through rifting to fragment separation and formation of a new ocean basin. It is expected to be complicated,

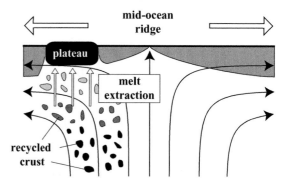

Figure 1.13 Mechanism for the production of excess melt at or near a spreading ridge, as a result of upwelling of fusible material (from Korenaga, 2005).

shedding fragments of continental lithosphere into the new ocean, sometimes large enough to form microcontinents, to cycle continental lithospheric material into the mantle beneath the new ocean, and to be accompanied by massive magmatism that builds volcanic margins (Fig. 1.12).

- *Fertility at mid-ocean ridges:* The most obvious sites of surface extension are mid-ocean ridges. Where these encounter an unusually fertile source region, enhanced magmatism may occur (Fig. 1.13). Many melting anomalies do indeed lie either directly on ridges or close to them, for example, Iceland and Tristan (Figs 1.6 and 1.9). Of the 16 plume localities originally proposed by Morgan (1971), 12 lie on or close to spreading ridges.

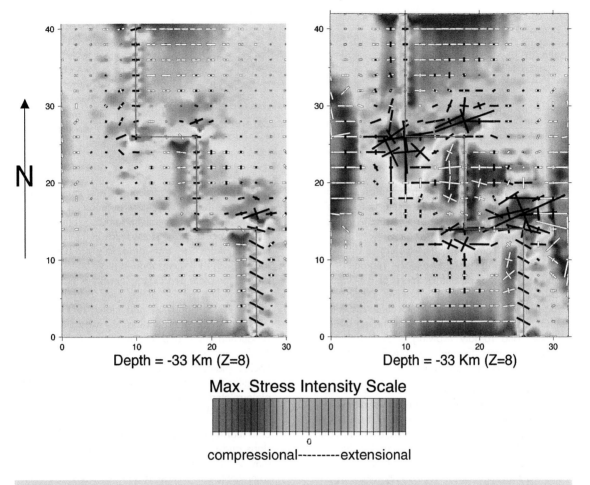

Max. Stress Intensity Scale

compressional---------extensional

Figure 1.14 Predicted variations in stress at ridge-transform intersections: left: a weak transform, right: a strong transform. Lines indicate the type and intensity of maximum stress, black: compressional, white: extensional. Regions of strong lithospheric extension, encouraging magmatism, are predicted on the outer corners of the ridge-transform intersections (from Beutel, 2005).[16] See Plate 4

■ *Enhanced volcanism at plate boundary junctions:* The sites of excess volcanism at or near spreading plate boundaries are often also the sites of complexity in the plate boundary configuration. Excess volcanism commonly develops on the outside of ridge-transform intersection corners. There, extensional stress is predicted to develop when slip on the transform lags behind spreading. The stronger the transform, the larger the extensional stress that builds up (Fig. 1.14) (Beutel, 2005). Examples include Ascension Island and the Amsterdam-St Paul plateau on the South East Indian ridge. Excess volcanism may also occur along oblique or "leaky"

transforms. The Terceira "rift" in the Azores Plateau, the Voring Spur along the trend of the Jan Mayen Fracture zone, and the Iceland-Faeroe Ridge have been proposed as features of this kind (Gernigon et al., 2009).[17]

Excess volcanism also commonly occurs at ridge-ridge–ridge-triple junctions for example, the Azores. There, the flow of upwelling material is expected to

[16] http://www.mantleplumes.org/RTIntersections.html

[17] http://www.mantleplumes.org/JanMayen.html

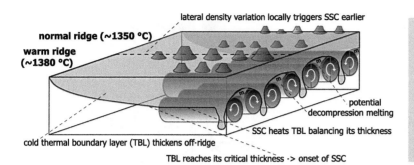

Figure 1.15 Small-scale sublithospheric convection (SSC) spontaneously evolves at the bottom of mature oceanic lithosphere. SSC organizes as rolls aligned by plate motion, and decompression melting occurs above the upwellings (from Ballmer et al., 2009).

be modified by the change in architecture of the base of the lithosphere. As a result, both passive upwelling and temperature are predicted to increase (Georgen and Lin, 2002).

Even more complex is the tectonics of diffuse or rapidly evolving plate boundary regions. Such regions may feature one or more of overlapping parallel ridge segments, microplates, propagating ridges and ridge jumps. At such localities distributed enhanced volcanism is expected to reflect the distributed and rapidly evolving spatial pattern of extensional stress. Examples include Easter, Iceland and the Bouvet triple junction.

■ *Small-scale sublithospheric convection:* Small-scale convection is predicted to onset spontaneously beneath oceanic lithosphere as it is transported away from the ridge where it formed, and cools and thickens (Fig. 1.15). Such convection takes the form of rolls orientated parallel to plate motion (Richter and Parsons, 1975). It successfully explains the corrugated sea-floor fabric on either side of the East Pacific Rise, for example, the Hotu Matua and Sojourn ridges (Fig. 1.16). Individual rolls are predicted to be of the order of 1000 km long and spaced at intervals of 200–300 kilometers. Volcanism may last for a few million years, dwindling as the convection process cools the asthenosphere and shuts itself down. The Marshall and Line islands, the Gilbert ridges and the Cook-Austral and Pukapuka volcanic chains are well explained by this process (Ballmer et al., 2007; 2009; Weeraratne et al., 2007).[18]

■ *Oceanic intraplate extension:* It is a misconception to think of the plates as strong, rigid entities that break only under the influence of some external and

extraordinary *force majeure.* They are in truth weak, and their interiors are maintained in a constant state of stress disequilibrium by plate-tectonic-related processes. The stress fields in the interiors of oceanic plates vary from place to place because of factors such as the variation in plate boundary type around the plate perimeters, thermal effects such as rapid cooling of the plates near ridges (Forsyth et al., 2006; Sandwell and Fialko, 2004), local volcanism and plate bending, for example, near to subduction zones and massive volcanic loads (Fig. 1.17). The widespread diffuse volcanism in the Pacific plate, volcanic ridges, and chains with irregular time progressions have been modeled by plate-related processes (Hieronymus and Bercovici, 1999; 2000).

■ *Slab tearing or break-off:* Disruption of subducting slabs and mantle flow at relatively shallow depths may occur in trenches near to the ends of slabs or when continents collide, subduction stops, and slabs break off. Examples of places where processes of this kind are occurring today are Kamchatka, Anatolia and Mexico (Fig. 1.18).

■ *Shallow mantle convection:* Abrupt lateral changes in the architecture of the base of the lithosphere may induce local convection cells in the asthenosphere. For example, at continental edges, 200-km-thick, cold, cratonic lithosphere may abut oceanic lithosphere only a few tens of kilometers thick. Where warm asthensophere beneath the oceanic lithosphere is juxtaposed against the sub-continental lithosphere, it may cool and sink. This mode of convection has been called EDGE convection (Edge Driven Gyres and Eddies). It can erode the sub-continental lithosphere and enhance upwelling in the interior of ocean basins (Fig. 1.19) (King, 2005; King and Anderson, 1998).

It has also been suggested that convection limited to the upper mantle can result from dense material,

[18] http://www.mantleplumes.org/VolcRidges.html; http://www.mantleplumes.org/Pukapuka.html

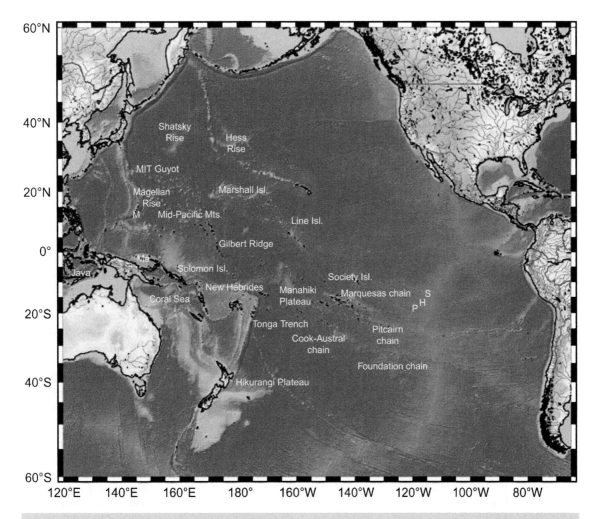

Figure 1.16 Map of the Pacific showing place names: P – Pukapuka Ridge; S – Sojourn Ridge; H – Hotu Matua Ridge; M – Mariana Trench; Ma – Manus Basin. See Plate 5

for example, eclogite blocks, possibly from delaminated continental mantle lithosphere or lower crust, sinking to their level of neutral buoyancy, heating to ambient mantle temperature and then rising again as a result of their newly acquired thermal buoyancy. This process is analogous to the rising and sinking of suspended objects in a Galileo thermometer.[19] It has been termed the "eclogite engine",

[19] http://en.wikipedia.org/wiki/Galileo_thermometer; http://www.mantleplumes.org/TopPages/HotSlabsTop.html

and proposed as an explanation for mid-ocean swells and plateaus such as Cape Verde and the Rio Grand Rise (Anderson, 2007a).

■ *Abrupt lateral changes in stress at structural discontinuities:* The lithospheric stress field may change abruptly where the shapes or structures of continents or plate boundaries change radically, for example, the northern termination of the Tonga trench in the southwest Pacific and the Cameroon region, West Africa (Fig. 1.10).

■ *Continental intraplate extension:* The continental crust can extend both in a localized, focused manner

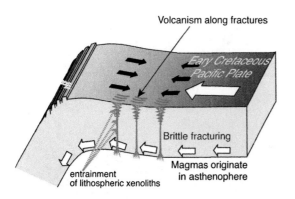

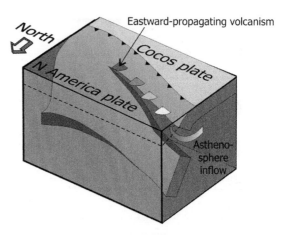

Figure 1.17 Schematic diagram illustrating volcanism resulting from oceanic intraplate extension. Magmas escape from the asthenosphere to the surface as a result of extension of the lower lithosphere and migration through the upper lithosphere via fractures created by flexure of the plate as it approaches a subduction zone (adapted from Hirano et al., 2006). *"Volcanism in Response to Plate Flexure"*, Hirano, N. et al. **313**: 1426–1428, 2006. Reprinted with permission from Science.

Figure 1.18 Schematic diagram illustrating progressive slab tearing and break-off, resulting in migrating volcanism as sub-slab asthenosphere upwells through the gap. This model was proposed to explain migrating Miocene volcanism in Mexico (modified from Ferrari, 2004).[20] *"Slab detachment control on mafic volcanic pulse and mantle heterogeneity in central Mexico"*, Ferrari, L **32**: 77–80, 2004. Reprinted with permission from GSA.

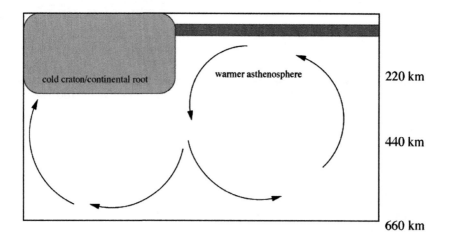

Figure 1.19 Small-scale "EDGE" convection cell at a craton boundary (from King and Anderson, 1998).

and also in a diffuse style throughout broad regions. The Basin and Range province in the western USA (Wernicke, 1981), the continent of Africa, including the East African rift valley (Bailey and Woolley, 2005), and Europe, including the Rhine graben

[20] http://www.mantleplumes.org/Mexico2.html

(Chalot-Prat and Girbacea, 2000; Wilson and Downes, 1991; Ziegler, 1992), are examples of regions that are both in extension and volcanically active. Evidence for extension accompanying the outpouring of continental flood basalts is provided by the widths of feeder dikes, for example, those from which the Columbia River Basalts flowed. The

continental collision

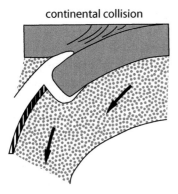

delamination instability

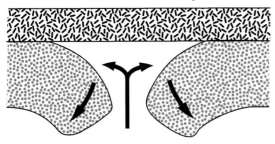

Figure 1.20 Recycling of lithosphere in continental collisions and by delamination[21] (Bird, 1979).

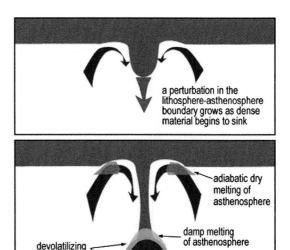

Figure 1.21 Loss of lithosphere by a gravitational instability.[22]

surface extension then amounts to the sum of all the dike widths (Christiansen et al., 2002).

■ *Catastrophic lithospheric thinning:* Where the lithosphere becomes over-thickened, for example in continental collision zones or by deep intrusions, its lowermost parts may become negatively buoyant. This may be aided by sinking to depths and pressures where transformation to eclogite occurs. Eclogite is the dense equivalent of basalt and transformation begins at ~50 km. The high-density lithosphere keel may peel off ("delaminate") (Fig. 1.20) (Bird, 1979) and sink into the asthenosphere, or simply form a bulbous gravitational instability and drop off – a down-going, or inverse plume (Fig. 1.21) (Daly, 1933; Elkins-Tanton, 2005; Kay and Kay, 1993). Part of the lower crust may also be removed in the process. This process can potentially explain flood basalts in intra-continental areas where the lithosphere was initially very thick, and where it was preceded by subsidence and not uplift. It may have played an important role in the Basin and Range province, the Columbia River Basalts, the Colorado Plateau and the Siberian Traps.

■ *Sublithospheric melt ponding and draining:* If the asthenosphere is locally above its solidus, melt may form over a long period of time and large volumes may pond at the base of the sub-continental mantle lithosphere (Fig. 1.22). Later rifting and fracturing of the lithosphere may permit this melt to escape to the surface in a much shorter time than it took to accumulate (Silver et al., 2006). Only relatively narrow feeder dikes are required (Rubin, 1995). This mechanism may explain the rapid formation of flood basalts in the absence of rifting of the magnitude needed for decompression upwelling to account for the melt volumes emplaced. In these cases, flood basalt eruption does not correspond to the melt-production time-scale, but to the reservoir-drainage time-scale.

[21] http://peterbird.name/publications/1979_delamination/1979_delamination.htm

[22] http://www.mantleplumes.org/LithGravInstab.html

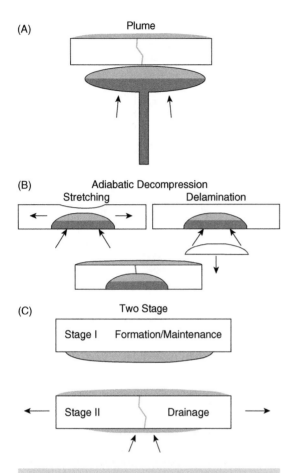

Figure 1.22 Competing models for the formation of continental flood basalts: dark gray indicates regions of elevated temperature or volatile content, light gray indicates zones of partial melt. (A) Plume model; (B) Adiabatic-decompression melting models; (C) Two-stage melt ponding and draining model (from Silver et al., 2006).

In addition to these terrestrial processes, melting anomalies might also be triggered by the impact of bolides arriving from space (Hamilton, 1970; Ingle and Coffin, 2004; Jones et al., 2002; 2003; 2005; Price, 2001).[23] Such impacts can theoretically induce an amount of melting equivalent to a large flood basalt. Melting anom-

[23] http://www.mantleplumes.org/Impacts.html

alies postulated to exist on other planets, and on moons, are also the subject of considerable work. Venus is of particular interest because it features many circular structures variously postulated to be caused by external means – meteorite impacts – or internal plumes (Hamilton, 2005; 2007a; Jurdy and Stoddard, 2007; Matias and Jurdy, 2005; Stofan and Smrekar, 2005; Vita-Finzi et al., 2005). Olympus Mons on Mars is the largest volcanic feature known in the solar system and has been attributed to a giant bolide impact (Reese et al., 2007).[24] Studying features on other planets is more difficult than studying terrestrial examples because the available data are fewer and of poorer quality. Thus, if the origins of melting anomalies on Earth, which can be examined closely and sampled extensively, are disputed, then clearly determining the origins of candidates on remote planets presents an even greater challenge.

The picture painted above is complex. The Plate hypothesis views plate tectonics as a system that is richer and more varied than the simple picture commonly presented in textbooks, of rigid plates floating on a uniform, viscous mantle, with no complications or second-order effects.

A simplifying statement may nevertheless be made. The Plate hypothesis attributes the causative processes of melting anomalies to the Earth's top thermal boundary layer – the surface, in contrast to the Plume hypothesis which attributes them to its bottom one – the core-mantle boundary. The plate view is intrinsically appealing in a number of respects. Some 90% of the heat lost from the Earth's surface is generated internally in the mantle by radioactive decay. Only ~10% is thought to be input into the mantle from the core (Anderson, 2007b). The Earth's top thermal boundary layer has an area four times that of the bottom thermal boundary layer and it is thus a heat transmitter over twice as powerful per unit area. The Plate hypothesis furthermore views mantle convection, plate tectonics and surface

[24] http://www.mantleplumes.org/TopPages/PlanetaryTop.html

volcanism of all kinds as interrelated and interdependent elements of a single planetary dynamic system. It does not suffer from the philosophical awkwardness of the Plume hypothesis, which proposes a convective mode that is separate and independent of normal mantle convection.

1.10 Predictions of the Plate hypothesis

What are the specific predictions of the Plate hypothesis? They are, in general, simply that:

- Melt extraction will occur where the lithosphere is locally in extension. Evidence for extension may include dikes, normal faulting, and continental separation or be found using geodetic surveying, for example, using GPS.
- The volume of magma extracted will be related to the fertility of the source. The primary evidence will be compositional and will comprise mainly petrological and geochemical observations.

The Plate hypothesis is ambivalent regarding several specific predictions of the Plume hypothesis. Vertical motions preceding and accompanying melt extraction are expected to vary according to the dominant process at work. For example, volcanism accompanying the detachment of thickened lithosphere via a gravitational instability is predicted to be preceded by subsidence as the crust is dragged down by initial sinking of the thickened root. The eruption of a flood basalt may or may not be followed by subsequent small-volume volcanism, depending on the ensuing stress state of the lithosphere. Narrow conduits transporting hot material from the core-mantle boundary to the surface beneath currently active melting anomalies such as Hawaii and Iceland are not expected. Volcanic chains may or may not be time-progressive, depending on whether the extending region is localized at one end, or extends along the entire chain. The temperature of the mantle is expected

to vary laterally, but the high, localized anomalies of 200 °C or more that are predicted by the Plume hypothesis to occur beneath active volcanic regions are not expected.

In the Plate hypothesis, long-lived intraplate melting anomalies that lie on the same plate may remain broadly fixed relative to one another or to migrate slowly. This is because they occur in response to large-scale, plate-wide stress fields that are governed by the distant, plate boundary configuration. As long as the boundary configuration remains the same, extending regions and melting anomalies will retain stable relative locations. If the boundary configuration evolves, for example when a ridge is subducted, then the intraplate stress field and relative locations of melting anomalies will change in response. In this way, time-progressive volcanism may develop on a plate that moves rapidly compared with the time-scale of evolution of its boundary. Melting anomalies are not thought to cause continental breakup, but they are thought to form in response to it. They are not thought to drive plate tectonics.

1.11 Testing the Plate hypothesis

How may the Plate hypothesis be tested? As for the Plume hypothesis, almost every discipline in Earth science may be brought to bear. In many cases the same methods and experimental approaches suitable for testing the Plume hypothesis also provide data relevant for testing the Plate hypothesis. These include study of the history of vertical motion and eruption, especially the initial stages, and source temperature and underlying mantle structure.

Additional approaches that may be important target horizontal motions, and the sources of the lavas. The Plate hypothesis predicts that volcanism occurs in extensional regions. Geological techniques can assess past lithospheric extension, and geodetic surveying, for example, using GPS, can measure present-day movements. Geochemistry and petrology may be more important techniques than hitherto appreciated for shedding light on the origin of melting anomalies. They have very little power to determine the depth of origin of magmas, but if the sources and volumes of magma produced are indeed related only to shallow recycling of near-surface materials, then depth is less of an issue. In this case, the nature and origin of the recycled component – issues that can be addressed by petrology and geochemistry – become more important and supercede the search for core-mantle-boundary and deep-mantle tracers, for example, noble gas isotope ratios.

Most importantly, a change to a more open-minded and scientific philosophy is needed. The Plate hypothesis cannot be tested unless there is acceptance that the Plume hypothesis is not proven, and may be falsified at a specific locality. In addition, the concept must be shed that a model may be assumed correct despite serious problems, simply because an alternative perceived to be better has not yet been developed.

How has the Plate hypothesis fared in recent years, and what is its potential for the future? This is for the reader to decide. The purpose of this book is to summarize critically the main facts and arguments that relate to these questions, for consideration alongside contrasting arguments presented elsewhere. To whet the appetite, let us tour Hawaii and Iceland again, and review the observations there from a fresh viewpoint.

1.12 Revisiting Hawaii and Iceland

In the context of the Plate hypothesis, the absence of evidence for a flood basalt and associated uplift at the old end of the Emperor chain is permitted, as is the huge upsurge in volcanic rate over the last 5 Ma that has built the Hawaiian archipelago. Recent conclusions that up to 60% of Hawaiian basalts are derived from mantle peridotite fluxed with melt from recycled crust (Sobolev et al., 2007), are pre-

dicted. A seismic anomaly beneath Hawaii that is continuous from the surface to the core-mantle boundary, currently the target of an ongoing ocean-bottom-seismometer experiment (Laske et al., 2007), is not expected.

A fertile source for Hawaiian lavas is well-documented, but what about the extension predicted by the Plate hypothesis? A major remaining challenge is testing whether Hawaiian chain-normal extension occurs. Numerical modeling suggests that the region is generally in a chain-normal extensional stress field (Stuart et al., 2007), but can extension be measured locally? There are great obstacles in the way. Apart from the few relatively small islands that make up the Hawaiian archipelago, the whole region lies in the deep ocean, presenting huge challenges regarding both technology and cost. Where land is exposed, for example on the Big Island of Hawaii, the vast lava pile blankets all features except those associated with the most superficial layers of the five massive volcanoes that make up the island. How, in such an environment, can we detect lithospheric extensions that might be small, and last only for a few million years at any one locality?

Compared with Hawaii, Iceland is much easier to study. It is surrounded closely by continents and the island itself is relatively large. A much more complete set of observations is thus available than exists for Hawaii. Several observations in the north Atlantic, surprising in the context of the Plume hypothesis, are permitted by the Plate hypothesis. These include the time-scale of the main phase of uplift, which was contemporaneous with continental break-up rather than preceding it (Maclennan and Jones, 2006). The late formation of a second flood basalt, the still-developing Iceland plateau, is also permitted, as is the absence of a time-progressive volcanic chain. The presence of no more than a moderate temperature anomaly in the mantle is expected. The absence of a seismically imaged conduit extending to the core-mantle boundary and the focusing of volcanic production at the extending mid-ocean ridge, are predicted. Geochemical evidence is essentially unequivocal that recycled fusible near-surface material exists in the source of Icelandic lavas (Fitton et al., 1997; Sobolev et al., 2007). The existence of the vast and complex North Atlantic Igneous Province, within which Iceland sits, is considered to be a consequence of Eurasian supercontinent break-up, not a cause of it.

1.13 Questions and problems

The recent explosion of skepticism about the Plume hypothesis has brought with it the realization that many issues that had previously been assumed settled are, in fact, still open questions. The subject has become enriched with new questions. Why do huge volcanic regions erupt without precursory uplift, and perhaps even following subsidence? What process can cause many millions of cubic kilometers of basalt to be erupted in just a few million years or less, even through thick continental lithosphere? Can any process form melt at the huge and sustained rates at which it is sometimes erupted, or must the melt be formed and stored over longer periods? What is the source of the melt, and the ubiquitous enriched geochemical signatures? What is the involvement of recycled near-surface materials such as crust and mantle lithosphere, and how deeply are they circulated? What is the role of the mantle transition zone between 410 and 650 km depth? Why are some volcanic chains time-progressive, whereas others erupted irregularly along much of their lengths for long periods of time? How does the temperature of the mantle vary throughout the Earth, and are elevated temperatures required to produce large volumes of melt, even if fusible source materials are available?

The Plume hypothesis has served scientists well over the last three decades. It has provided a framework within which huge bodies of data have been gathered, and predictions against which the results have been matched. However, a candid appraisal of the hypothesis in the light of the very data whose collection it inspired renders inescapable the conclusion that it is, in many respects, wanting. It must thus lie in facing up to the problems, and posing radical new questions, that the real and important future progress will be made. We must stop sweeping the problems under the rug and deal with them instead.

1.14 Exercises for the student

1 What is a plume?

2 Design an experiment that could conclusively test for the existence of a plume.

3 Which melting anomalies are the most likely to be underlain by plumes?

4 How well do postulated plume localities score, according to the criteria defined in Section 1.4?

5 Develop Plate models for the melting anomalies listed in Table 1.4.

6 How may the Plate hypothesis be tested?

2

Vertical motions

Если на клетке с тигром написано слон, не верь глазам своим[1]

– Kozma Prutkov (1801–1863)

2.1 Introduction

Since Earth science became a subject, geologists have been aware that the thick piles of accumulated sediments on the Earth's surface, and mountain ranges, are evidence for major vertical movements of the crust. The earliest efforts to explain these movements involved geosyncline theory. A geosyncline was envisaged to be an elongated, subsiding basin in which sediments accumulated and which was later compressed and uplifted to form a mountain range. The concept originated in work done during the 19th century by the American geologists James Hall and James Dwight Dana who studied, and sought to explain, the Appalachian Mountains in eastern North America. The concept built on earlier assumptions that the Earth was cooling and contracting. It was retained long after that view had been abandoned, however, and used up to the advent of plate tectonics to explain most mountain ranges (Kay, 1951).

With the passage of time, geosyncline theory had to be much elaborated to encompass an ever-expanding body of diverging observations, and the sub-types into which geosynclines were divided proliferated. Geosyncline theory also had the major disadvantage that it could not explain why vertical motion occurred, in particular how mountains were raised up. Nevertheless, it remained the preferred model for a century. It was not until it was recognized that huge lateral movements of the Earth's crust had occurred, and mobilization, continental drift and plate tectonics were accepted, that the large vertical motions observed were finally understood to result from crustal stretching and continental collisions. As a consequence, geosyncline theory collapsed almost immediately.

From early times, diapirism was also widely considered as a possible explanation for vertical motions (Anderson and Natland, 2005; Glen, 2005). Small-scale examples were known from salt domes, and the dynamics of diapirs was explored in detail in laboratory experiments (Griggs, 1939). A general theory of vertical tectonics based on mantle diapirism was developed by the Russian geologist Vladimir Vladimirovich Belousov, and described in his book published first in Russian in 1956 (Belousov, 1954). This book was translated into English in 1962, the same year as publication of the first paper proposing sea-floor spreading (Beloussov, 1962; Hess, 1962). Mantle diapirism was envisaged to comprise a "lava lamp" style of convection whereby material at depth warms and rises, and

[1] If, on the cage of an elephant, you see a sign reading "buffalo", do not believe your eyes.

Plates vs. Plumes: A Geological Controversy, 1st edition. By Gillian R. Foulger.
Published 2010 by Blackwell Publishing Ltd.

then cools and sinks again. Uplift and subsidence of the crust were explained by the interactions of diapirs with the Earth's surface, and lateral motions were explained as gravity sliding from uplifted regions. These ideas bear close resemblance to features of the Plume hypothesis, which associates uplift with the arrival of plume heads at the base of the lithosphere, and to which continental break-up was also originally attributed (Section 1.4; Morgan, 1971).

The Plate hypothesis offers an alternative suite of candidate models for vertical motions that extend plate-tectonic-related processes beyond the first-order effects of continental break-up and collision (Section 1.9). Processes such as intraplate extension and catastrophic lithospheric thinning predict spatial and temporal patterns of uplift and subsidence that are very different from each other and from the predictions of the Plume hypothesis. The task is then accurate assessment of the detail of motions and testing to determine which model best fits the observations.

Key constraints include the amplitude and sign of the motion, its timing, spatial pattern, and tectonic and volcanic associations. Vertical motions of the sea floor are the most compliant to study as sedimentation is continuous, erosion usually insignificant, and tectonics, in general, simple. Progressive subsidence with ageing and cooling of the oceanic lithosphere can be monitored in sedimentary sequences. On land, the problem is more difficult because sedimentation may be discontinuous and variable and erosion may remove critical evidence. Uplift, an issue particularly important to the Plume hypothesis, may be difficult to assess reliably. Features such as erosion surfaces, conglomerates and river incision are often taken to indicate uplift, but such features may be caused simply by climate change (Bull, 1991) and determining timing may be a challenging problem. For example, if lavas rest unconformably on sub-strata, or overlie erosion surfaces, there may be little evidence with which to date the formation of the surface on which they were emplaced. Using vertical motion observations to match the Plate and Plume hypotheses against one another may be even more difficult. Both predict uplifts and subsidences and the main differences may be timing, spatial patterns and amplitudes of motion. These are second-order issues, however, and may be problematic to address where even the first-order issues are difficult to resolve.

To complicate things still further, it remains the case today that substantial vertical motions of the crust occur for which no ready, reasonable explanation currently presents itself that is not *ad hoc*. Major uplifts, essentially amagmatic, encompassing regions hundreds of kilometers wide such as Southern Africa and the Colorado Plateau, still defy explanation in the context of other tectonic events or any general theory such as plate tectonics (Fleming et al., 1999). Small-scale "egg box" or "piano key" vertical motions around the North Atlantic margins are not obviously related to any larger-scale process (Stoker et al., 2005). Ironically, it may be easiest to find uplifts of the kind predicted by the Plume hypothesis far from flood basalts, in space and time, in locations where no plume has ever been proposed.

2.2 Predictions of the Plume hypothesis

The modern Plume hypothesis predicts that as a rising, bulbous plume head approaches the base of the lithosphere, the surface is uplifted to form a circular dome. Uplift is predicted to start before the onset of flood volcanism, and to be smooth (Campbell, 2006; Farnetani and Richards, 1994; Griffiths and Campbell, 1991a). The amount of time between the onset of uplift and the onset of volcanism is calculated to vary depending on parameters such as the viscosities of the upper mantle, the lower mantle and the plume material, but it is generally expected to start ~10–20 Ma prior to the onset of flood volcanism (Campbell and Griffiths, 1990; Farnetani and Richards, 1994).

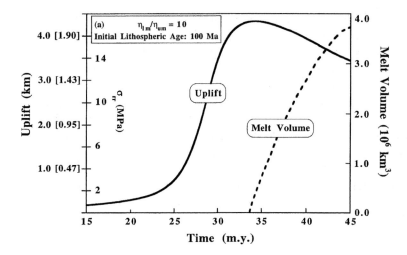

Figure 2.1 Uplift and melt volume predicted to accompany the arrival of a 350°C plume head at the base of the lithosphere. The viscosity contrast between the upper and lower mantle is 10. Bracketed numbers refer to maximum and minimum values (from Farnetani and Richards, 1994).

The amount of uplift depends on the average temperature in the plume head, but for plumes from the core-mantle boundary, and with typical temperatures up to ~350°C higher than the surrounding mantle, uplifts of ~0.5–4 km are predicted at the plume center. The area of greatest uplift is expected to have a radius of ~200 km, and to be surrounded by a zone with a radius of ~400 km within which uplift is less but still measurable (Fig. 2.1).

Uplift is predicted to cease, and subsidence to start, when sublithospheric spreading of the plume head, melt production and eruptions onset (Farnetani and Richards, 1994). On land, a flood basalt subsequently forms, and in the ocean the volcanism creates a basalt plateau elevated above the surrounding sea floor. If the plume rose beneath a rapidly moving plate, for example, the Pacific plate, the site of uplift and the plateau are rapidly transported away from the plume stem that follows. As the plateau drifts away from the residual anomalously hot plume-head asthenosphere, and over cooler mantle, it subsides faster than normal-thickness oceanic crust formed at ridges (Clift, 2005). The higher the temperature of the plume head, the more rapid the anomalous subsidence will be (Fig. 2.2).

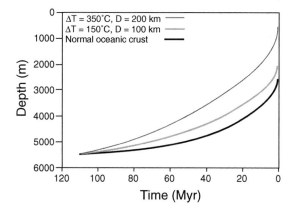

Figure 2.2 Predicted subsidence curves for oceanic lithosphere that drifted away from an anomalously hot plume head (from Clift, 2005).

2.3 Predictions of the Plate hypothesis

The Plate hypothesis attributes melting anomalies to top-driven, shallow processes ultimately related to plate tectonics. Many different processes fall within this scope (Section 1.9) and different patterns of vertical motion are expected to accompany each. In some cases, numerical modeling of the expected motions and the final topographic expression has not yet been done and only qualitative statements can as yet be made.

Plate processes that result in significant vertical motions include extension by rifting, extraction of melt from the subsurface, its emplacement at or near the surface in magmatic and volcanic activity, and catastrophic lithosphere thinning by processes such as Rayleigh-Taylor instability and delamination. The following vertical motions are expected to accompany rifting:

- *Rift shoulder uplift:* Of the order of kilometers of uplift is expected where rifting penetrates the entire lithosphere. It will occur contemporaneously with rifting and form mountain belts running along the flanks of the new rift. Significant uplift will extend for hundreds of kilometers into the interiors of the adjacent continents (Fig. 2.3) (van Balen et al., 1995).

- *Subsidence of neighboring zones:* The new rift valley or ocean basin will subside, as will the hinterland on the continental side of the rift shoulder (Fig. 2.3).

- *Post-rifting erosion:* The shoulder uplift will subsequently erode down, filling the rift and hinterland basins with sediments. The shoulder will rebound isostatically as a result of mass wasting, recouping ~10–20% of the erosional elevation loss. Additional subsidence will occur in the deposition zones as a result of the loading. There will be corresponding flexural responses in adjacent regions.

Where large quantities of melt form in the subsurface, are advected upward and either intruded at shallow depth or erupted onto the surface, several processes give rise to concurrent and subsequent vertical motions. These include:

- *Partial melting of the source:* The volume of melt is less than that of the solid fraction melted, contributing to subsidence.

- *Removal of the melt:* Melt extraction will reduce the source volume, also contributing to subsidence.

- *Chemical depletion of the source, creating residuum:* Removal of partial melt from common mantle materials (e.g., garnet peridotite) will leave residuum that is up to ~1% less dense than the original source rock. This results from the preferential removal of iron (Fe) from olivine and the heavy minerals garnet and orthopyroxene (Schutt and Lesher, 2006). This buoyant residuum may rise and produce uplift, and long-term dispersal will slowly decay that uplift (Phipps-Morgan et al., 1995). The density reduction from this kind of chemical depletion is comparable to that produced by the thermal anomalies typically proposed in plume models. For example, the density reduction that results from the removal of 20% partial melt from spinel lherzolite is similar to that produced by raising the temperature by ~200°C.

- *Heating and expansion of the lithosphere as the melt passes upward through it:* Rising basalt melt – "the bringer of heat" (Daly, 1914) – causes thermal uplift concurrent with magmatism. Subsequent cooling will result in subsidence.

- *Emplacement of the melt at the near surface:* Volcanism and intrusion will result in increased elevation. Basalt has a lower density than the material melted out of the source to form it and the solidified melt will thus occupy a volume of the order

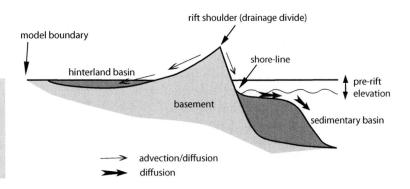

Figure 2.3 Schematic diagram illustrating rift shoulder uplift, hinterland and rift valley subsidence, erosion and sedimentation (from van Balen et al., 1995).

of 10% larger than the melt. The basalt may be underplated onto the bottom of the lithosphere, intruded into the lithospheric mantle or the crust, or erupted onto the surface. If it is retained in the subsurface, the former land surface or seafloor will be uplifted (rock uplift). If part is erupted onto the surface, it will contribute to building a new, elevated land surface or a shallower sea floor, and will depress the old, pre-eruption surface because of the extra loading (rock subsidence). Changes in topography or bathymetry should not be confused with rock uplift or subsidence, i.e., change in the elevation of pre-existing material (England and Molnar, 1990).

- *Weathering of the edifice and redistribution of the sediments:* Where topographic highs are eroded down and sediments deposited elsewhere, isostatic readjustments will cause rock uplift and subsidence, as for post-rifting erosion.

In the case of eruption of a flood basalt, or formation of a volcanic margin, one to tens of kilometers thickness of material is transported as magma from depth and emplaced near and on the surface. In these cases, the net effect of all the above melting-related processes is expected to be an approximately co-magmatic increase in surface elevation of the order of 250 m for every kilometer-thick layer of material transported to the near surface (e.g., Phipps-Morgan et al., 1995). In other words, the integrated volume of the surface elevation increase will amount to some 25% of the volume of the melt transported upward.

The source region from which the melt is extracted may be much wider than the surface area where it is emplaced, so uplift may cover a wider area than the volcanism. Uplift will be followed by erosion causing topographic reduction, which will be partially compensated for by isostatic rebound. Where emplacement of the flood basalt accompanied continental break-up and/or rifting, vertical motions associated with those processes will be superimposed.

With these principles in mind, the vertical motions expected for some mechanisms that fall under the umbrella of the Plate hypothesis may be considered, as follows:

- *Continental break-up:* All vertical motions related to rifting are expected, including subsidence in the rift itself and shoulder uplift of the order of kilometers. The Red Sea escarpment on the Arabian Peninsula and the Transantarctic mountains are clear examples of this. Continental separation is normally expected to be accompanied by massive magmatism and the formation of a volcanic margin (van Wijk et al., 2001). Major contemporaneous uplift, expected as the direct consequence of this magmatism, will be superimposed on the rift-related vertical motions.

- *Excessive melt production from fertility anomalies at mid-ocean ridges:* The main modification to regional, near-ridge topography is expected to be increased elevation, and to result from the excess magmatism itself, for example, Iceland.

- *Enhanced volcanism at plate boundary junctions:* The variable stress fields at these localities is expected to place some regions in extension and others in compression. Subsidence and uplift respectively are expected to result from this, along with the elevation changes expected as a consequence of melt extraction and magmatism. The Easter melting anomaly and the Azores are examples of such localities.

- *Oceanic intraplate extension and volcanism:* Bathymetry will be modified by the effects of extension, melt extraction and near-surface melt emplacement. An example is Hawaii.

- *Continental intraplate extension:* This commonly occurs in association with the formation of rift valleys. Similar motions are expected as for continental break-up, though on a smaller scale. An example is the East African Rift.

- *Catastrophic lithospheric thinning:* The detachment of a Rayleigh-Taylor instability has been modeled in detail (Figs. 1.14 and 2.4) (Elkins-Tanton, 2005). Densification of the lower lithosphere results in subsidence as the surface is weighted down. As the instability develops, necks, detaches and finally sinks, warm asthenosphere rises to replace it and melting, magmatism and volcanism may occur. The surface rebounds, but does not fully regain its former elevation. Vertical motions accompanying the magmatism, described above, are superimposed. A key prediction is that volcanism is preceded by an initial subsidence phase, the opposite of what is expected for a plume. Melting anomalies proposed to be caused by Rayleigh-Taylor instability or

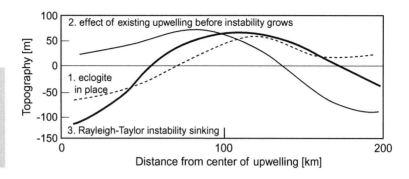

Figure 2.4 Vertical motions predicted to accompany formation and detachment of a Rayleigh-Taylor instability (Fig. 1.14) (from Elkins-Tanton, 2005).

delamination include the Siberian Traps (Elkins-Tanton, 2005; 2007), the Columbia River Basalts (Hales et al., 2005) and elsewhere (Bird, 1979; Gögüs and Pysklywec, 2008; Kay and Kay, 1993; Lustrino, 2005).

2.4 Comparison of the predictions of the Plume and Plate hypotheses

Both the Plume and Plate hypotheses predict substantial uplift, but with contrasting time-scales, amplitudes and spatial patterns. The Plume hypothesis predicts domal uplift, starting well before volcanism. Uplift at the center is expected to be 0.5–4 km. This reduces with distance from the center and is likely to be barely detectable at distances of a few hundred kilometers from the center.

Uplift associated with plate-related processes is mostly expected to be related to rifting and the extraction of melt from depth. In contrast with the Plume hypothesis, uplift is expected to occur more or less contemporaneously with rifting and magmatism, and some will be elongated along rifts. It may be of much greater amplitude than predicted by the Plume hypothesis, possibly amounting to many kilometers locally, and it may be much more widespread if continental break-up is involved.

In the case of subsidence, the Plume hypothesis predicts that the precursory domal uplift will decay away once the plume head region has been transported away from the hot asthenosphere

above the plume tail (Clift, 2005). The Plate hypothesis predicts that:

a) Net subsidence may occur locally, depending on the relative contributions of the various rift and magmatic processes at work, for example, in rift valleys.

b) Post-magmatic processes such as erosion, isostatic adjustment and cooling of intrusives will reduce earlier elevation increases.

Catastrophic lithospheric thinning predicts subsidence precursory to volcanism, the opposite of that predicted by the Plume hypothesis. One might hope that which of these two occurred prior to flood basalt eruption would be easy to distinguish, but a surprising amount of disagreement still exists in many cases (He et al., 2009; Sheth, 2007a; Ukstins Peate and Bryan, 2008; 2009).

2.5 Observations

2.5.1 Classifying melting anomalies

Any attempt to subdivide melting anomalies into groups with similar characteristics is fraught with vexation, as their smooth spectrum of variability immediately becomes obvious (Tables 1.1–1.5). The most spectacular members, for example, Iceland and Hawaii, are essentially unique on the planet. Some involve initial flood basalts but others not, and both these types may or may not have time-progressive volcanic chains. Only a small minority arguably involve both initial flood basalts and time-progressive chains. Just 3 out of the 49 listed in Table 1.4 are indicated as confidently displaying both. Two regions are producing flood basalts at the present day – Iceland and Hawaii.

How does this relate to the vertical motions that may precede or accompany volcanism? Regional domal uplift, followed a few million years later by a flood basalt, is a primary prediction of the Plume hypothesis. The Plate hypothesis predicts that the greatest vertical motions will be quasi-contemporaneous with the greatest rifting and melt extraction events, regardless of when the volcanism occurred in the lifetime of the system.

For the purpose of discussing vertical motions, melting anomalies will be subdivided according to whether or not they involve flood basalts and volcanic chains, as follows.

1 Volcanic chains with initial flood basalts.

2 Volcanic chains without initial flood basalts.

3 Flood basalts currently active, but with no volcanic chain.

4 Extinct flood basalts with no volcanic chain.

Interestingly, there is no known example of an extinct flood basalt plausibly associated with an extinct volcanic chain.

2.5.2 Volcanic chains with initial flood basalts

Of all the melting anomalies on Earth, only the Louisville, Réunion and Tristan systems involve strong spatial and temporal associations between initial flood volcanism and subsequent time-progressive volcanic chains (Table 1.4). The first of these, the Louisville volcanic chain, has been suggested to have started with eruption of the Ontong Java Plateau (Fig. 2.5). This may

immediately be ruled out, however. Detailed plate motion reconstructions show that the Ontong Java Plateau is currently some 2000 km away from where a Louisville plume head would lie, had one formed (Kroenke et al., 2004). Furthermore, there is no geochemical similarity between rocks dredged from the Osbourn trough, an extinct ridge from which the Ontong Java Plateau is thought to have erupted, and Louisville seamounts (Worthington et al., 2006). The Louisville system seems to comprise a volcanic chain only and a plume source is ruled out if the standard criteria are applied (Section 1.4).

(a) Deccan Traps – Réunion

The Deccan Traps have classically been associated, via the Lakshadweep-Chagos ridge, to the currently active Réunion melting anomaly (Fig. 2.6). The Traps erupted between 67 and 60 Ma, shortly after the break-off of the Seychelles microcontinent from the west coast of India.[2] This happened ~30 Ma before India collided with Asia. The Traps are famous, not least because of the celebrated debate over whether their eruption triggered the mass extinction at the end of the Cretaceous or whether it merely coincided with it (Powell, 1998).

Early work on possible vertical motions associated with eruption of the Deccan Traps pointed out that most of the major rivers in India drain from west to east (inset Fig. 2.6) (Cox, 1989). This would be consistent with broad domal uplift centered on the Traps if the drainage pattern developed shortly after uplift. If such uplift occurred a few million years before volcanism started, it would be consistent with a plume origin for the Traps.

The most direct approach to testing this hypothesis is to study sedimentary sequences that were laid down during the few million years prior to trap eruption. These might provide evidence, for example, for progressive shallowing or deepening of the waters in which they were

[2] http://www.mantleplumes.org/Seychelles.html

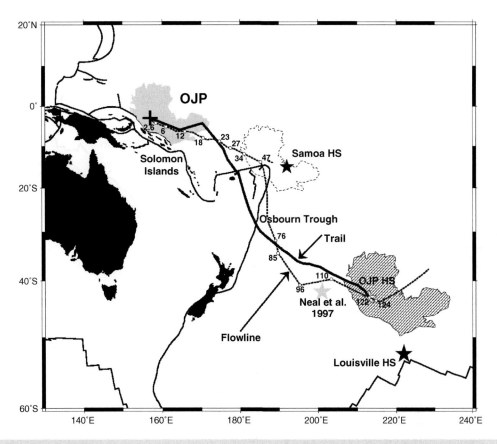

Figure 2.5 Ontong Java flow lines. The present position of the Ontong Java Plateau (gray) and its expected location relative to the Samoa melting anomaly (outline) and the Louisville melting anomaly (hatched). The solid dark line shows the hypothetical Ontong Java plume track (from Kroenke et al., 2004).

deposited, and could be interpreted in terms of uplift or subsidence. Regional uplift is also expected to result in widespread erosion, and evidence for this might be found in conglomerates immediately below the base of the lavas. If uplift occurred before trap eruption, such conglomerates should contain fragments of eroded basement but not trap basalts.

Unfortunately, most of the base of the Deccan Traps is buried beneath the lavas themselves. This horizon can thus only be studied directly in peripheral areas. In some parts of northern India, the early flows overlie marine sedimentary sequences. However, these sequences are due to marine transgression, indicating subsidence, not

to the regression that would be expected in the case of uplift (Sharma, 2007). They also contain clay derived from Deccan lavas, so they must have formed after the eruptions started. Thick conglomerate beds occur beneath the lowest lavas, but they contain both basement and Deccan basalt boulders, so they too must have post-dated the onset of eruptions. It thus seems that sedimentary evidence for vertical motions in the critical few million years prior to the onset of eruptions is lacking.

Similarly, little evidence is to be found in support of uplift in central India. There, both uplifts and subsidences occurred locally prior to trap eruption. However, a systematic pattern

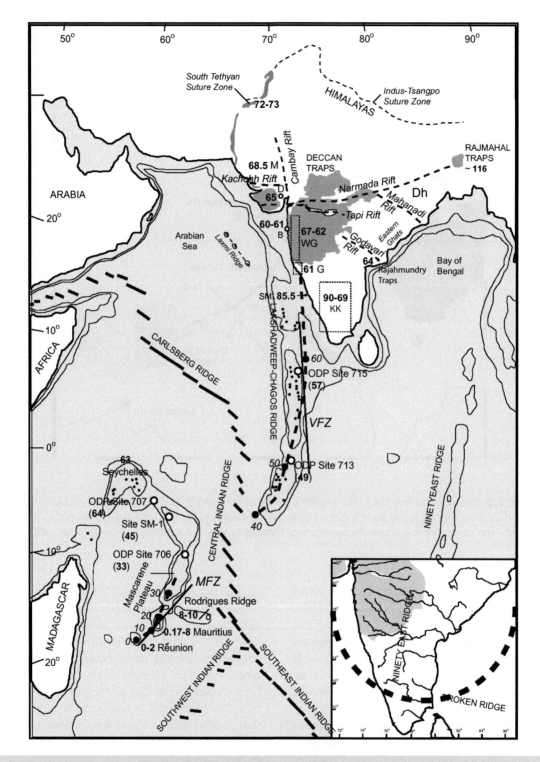

Figure 2.6 The Deccan-Réunion volcanic province. Bold – measured ages, italics – ages predicted for a regularly time-progressive volcanic chain. Dh – Dhanjori Formation. Inset shows rivers and plume head proposed by Cox (1989) (from Sheth, 2005).

of uplift is not observed. Much of the traps erupted onto an extensive, pre-existing, low-lying planation surface – a low-relief surface that cuts across varied rocks and structures – and basement-derived conglomerates beneath the lava pile are absent (Sheth, 2007a). Much of India has uplifted, but subsequent to the eruptions, not prior to it. The western seaboard (the Sahyadri range) uplifted in a margin-parallel, linear pattern during the Neogene (<23 Ma), as did the Eastern Ghats (Fig. 2.6).

Is there evidence that the easterly flowing river-valley drainage pattern developed in response to Deccan-centered domal uplift (Cox, 1989)? On the contrary, there is evidence that it did not. The river valleys are thought to be ancient and to have originally developed in Palaeozoic rifts that formed when Gondwana broke up (Table 2.1). Lavas of the Deccan Traps may have flowed into them, which also suggests that they predate flood basalt eruption (Self et al., 2008). If they formed much earlier than a few million years prior to Trap eruption, they cannot be associated with precursory, plume-head-related uplift.

Post-eruption subsidence is expected for a plume source if the flood basalt or plateau drifted away from the postulated hot plume-derived asthenosphere and onto cooler mantle (Clift, 2005). The Deccan Traps have drifted some 40° of latitude (~4500 km) subsequent to eruption, a distance much greater than the expected radius of a plume head (~1000 km). The post-Deccan, continent-wide, Neogene uplift, including formation of the coast-parallel mountain ranges on the west coast, is therefore the opposite of this prediction. These uplifts might be related to compression in the continental interior following collision with Asia, but the details are unclear. Late (<37 Ma), large-scale (>4 km) subsidence of the rifted volcanic margin in the northern Arabian Sea is also inconsistent with the pattern predicted to accompany drift away from a plume (Calvès et al., 2008). What is clear is that the pattern of vertical motions both before and after eruption of the Deccan Traps

Table 2.1 Geological time scale.

542 Ma to present	Phanerozoic Eon
66 Ma to 0 Ma	Cenozoic Era
2.6 to 0 Ma	Quaternary Period[3]
11,400 a to 0 a	Holocene Epoch
2.6 Ma to 11,400 a	Pleistocene Epoch
23 to 2.6 Ma	Neogene Period
5.3 to 2.6 Ma	Pliocene Epoch
23 to 5.3 Ma	Miocene Epoch
66 to 23 Ma	Paleogene Period
34 to 23 Ma	Oligocene Epoch
56 to 34 Ma	Eocene Epoch
66 to 56 Ma	Paleocene Epoch
245 to 66 Ma	Mesozoic Era
146 to 66 Ma	Cretaceous Period
201 to 146 Ma	Jurassic Period
251 to 200 Ma	Triassic Period
544 to 251 Ma	Paleozoic Era
299 to 251 Ma	Permian Period
318 to 299 Ma	Carboniferous Period (Pennsylvanian)
359 to 318 Ma	Carboniferous Period (Mississippian)
416 to 359 Ma	Devonian Period
444 to 416 Ma	Silurian Period
488 to 444 Ma	Ordovician Period
542 to 488 Ma	Cambrian Period
2500 to 542 Ma	Proterozoic Eon
1000 to 542 Ma	Neoproterozoic Era
1600 to 1000 Ma	Mesoproterozoic Era
2500 to 1600 Ma	Paleoproterozoic Era
3800 to 2500 Ma	Archaean Eon
4570 to 3800 Ma	Hadean Eon

[3] As a result of a review by the ICS of late Cenozoic subdivisions, and subsequent ratification by the IUGS, the Quaternary was formally made a Period co-terminus with the Pleistocene at 2.6 Ma, in 2009. This settles a long-standing confusion regarding terminology and subdivisions in the late Cenozoic (Head et al., 2008).

was complex and varied and does not resemble simple precursory domal uplift and subsequent subsidence.

(b) Paraná/Etendeka – Tristan

The currently active part of the Tristan melting anomaly in the South Atlantic is the basaltic island of Tristan da Cunha, some 500 km east of the Mid-Atlantic Ridge (Fig. 2.7). It is the largest of a group of islands and seamounts spread over an area about 50 km in diameter. It has been proposed that Tristan da Cunha is underlain by a plume that caused continental breakup in this region at ~135 Ma, and has remained near the Mid-Atlantic Ridge subsequently. This model has been invoked to explain the pair of roughly mirror-image ridges that extend from the Mid-Atlantic Ridge west to the coast of Brazil (the Rio Grande Rise) and east to Namibia

on the southwest coast of Africa (the Walvis Ridge).

A large flood-basalt in South America, the Paraná basalts, erupted mostly in the period 139–127.5 Ma (early Cretaceous), at the approximate time when this part of the South Atlantic opened (Comin-Chiaramonti et al., 1999; 2002; Piccirillo and Melfi, 1988). It has been proposed that this flood basalt, along with its smaller counterpart in Namibia, the Etendeka basalts, represents the head of the postulated Tristan plume.

The Paraná basalts overlie a major sedimentary basin ~1500 km wide (Fig. 2.8). This basin developed from Ordovician/Silurian times to the early Cretaceous, during which time it filled to a depth of ~3.5 km (Peate, 1997). The oldest Palaeozoic sediments are marine, but these grade up into subaerial, lacustrine and fluvial deposits. The youngest strata are aeolian sandstones that

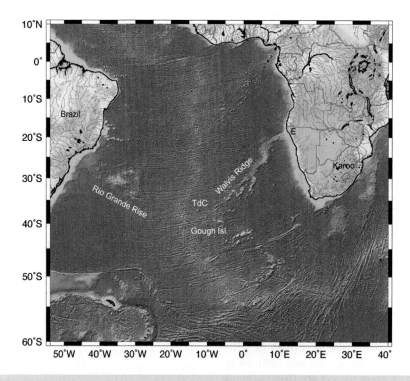

Figure 2.7 Bathymetry and topography of the south Atlantic region. TdC – Tristan da Cunha Island; E – Etendeka Basalts (basemap from Smith and Sandwell, 1997). See Plate 6

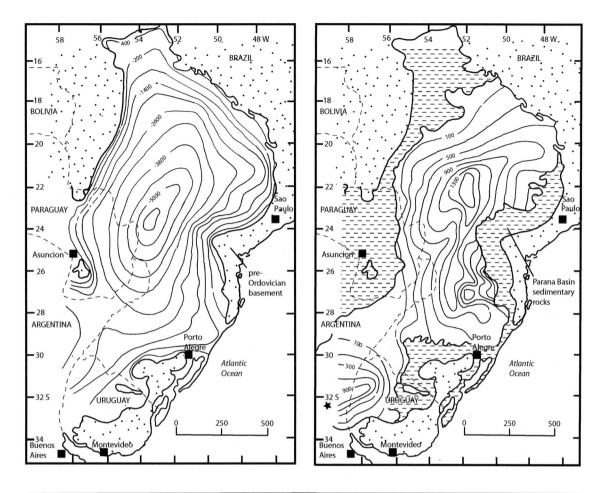

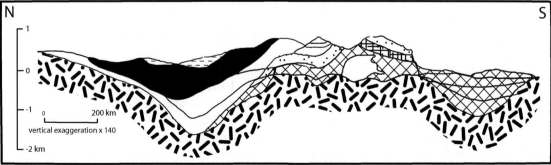

Figure 2.8 Top left: Depth to basement beneath the Paraná sedimentary basin; top right: Isopach map of Paraná lavas; bottom: Cross-section through the Paraná basalts (from Peate et al., 1992).

were laid down in subaerial desert conditions. They directly underlie, are intermingled with, and overlie, Paraná volcanics.

This sedimentary history indicates slow, regional net uplift over a period of several hundred million years. Superimposed on it is the local bowl-shaped subsidence that permitted the deep sedimentary basin to form. Within the sedimentary record, various transgressions and regressions testify to a complex history of local uplifts and subsidences (Fúlfaro, 1995). The Paraná flood basalts erupted directly into this sedimentary basin. The thickest lavas (~1.5 km) coincide with the deepest part of the basin, which lay some 500 km west of the locus of continental break-up (Peate, 1997).

A key question is what vertical movements occurred during the few million years immediately prior to onset of the eruptions. There is little possibility for shedding light on this question by studying macro-features in the aeolian sandstones beneath and within the basalts, because they are uniform, unfossiliferous and provide little evidence for any environmental change during their deposition. Apatite fission track analysis has therefore been used.

Apatite fission track analysis can reveal information on the thermal history of rocks that can be related to uplift and subsidence. Briefly, when uranium (^{238}U) atoms decay, they cause linear zones of radiation damage known as fission tracks in crystal lattices. Fission tracks are typically ~16 μm in length when they form, but they heal and shorten to an extent that is governed by the temperature of the rock. By measuring the lengths of the fission tracks, the thermal history of the rock can be deduced where the temperatures did not exceed ~110 °C. If a formation is uplifted and overburden eroded away, cooling is expected, and if it subsides and is buried, heating is expected (e.g., Donelick et al., 2005).

In the Paraná region, apatite fission track analysis has provided evidence for two periods of basin-wide, kilometer-scale uplift (De Oliveira et al., 2000; Hegarty et al., 1995). The first occurred at 90–80 Ma (Late Cretaceous) and

may have involved several kilometers of differential uplift. The second occurred during the Tertiary, and was smaller. As much as 3 km of material may have been eroded from the Brazilian coastal plain as a consequence. Both uplift episodes significantly post-date eruption of the Paraná basalts, and together they could account for the present-day high elevation of the area (Hegarty et al., 1995). Apatite fission track analysis cannot rule out regional uplift preceding emplacement of the Paraná basalts, but it provides no evidence to support it.

As for the Deccan Traps, rivers in the Paraná region generally drain away from the continental margin where the proposed Tristan plume is postulated to have impacted (Cox, 1989; White and McKenzie, 1989). In common with the Deccan Traps, it has not been demonstrated that this drainage pattern developed in response to uplift in the few million years immediately preceding the eruptions. The later uplifts that are known to have occurred could account for the drainage pattern.

There is no unequivocal evidence for regional, domal, precursory uplift in the few millions of years prior to eruption of the Paraná basalts. Substantial post-eruption uplift is known to have occurred, and can account for the present-day high topography there.

(c) Columbia River Basalts – Yellowstone

The Yellowstone melting anomaly is commonly considered to be the type example of a mantle plume in a continental setting (Fig. 2.9). The eastern Snake River Plain, which contains a time-progressive chain of extinct silicic calderas, is viewed as the trace of the "plume tail". The Columbia River Basalts, a ~1.8-km-thick pile of basalt lavas, which erupted mostly from centers ~400 km north of the proposed volcanic track, are considered to be the flood basalt province representing the "plume head". They erupted at ~17 Ma.

Arguments have been presented both for and against uplift precursory to eruption of the

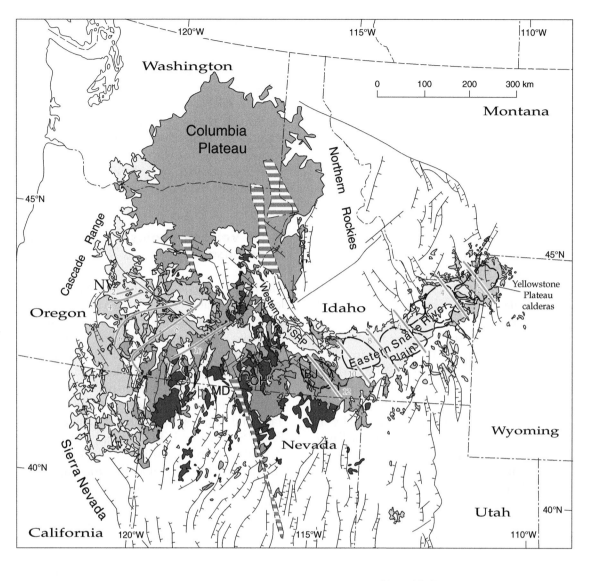

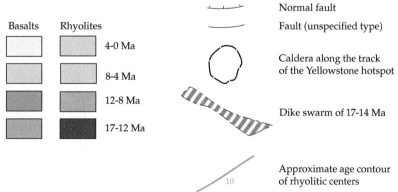

Figure 2.9 Map of the northwestern United States showing features of the Columbia River Basalts, Snake River Plain and Yellowstone. NV – Newberry Volcano (from Christiansen et al., 2002). See Plate 7

Columbia River Basalts (Hales et al., 2005; Hooper et al., 2007; Sheth, 2007a).[4] Early lavas fill deep canyons that must have been formed by erosion before the eruptions started. This has been interpreted as indicating uplift, a view that is supported by palaeobotanical and palaeoclimate altitude estimates (see Hooper et al., 2007 for a discussion). This interpretation has been criticized on the grounds that canyon incision does not necessarily signify absolute uplift of the land surface. It could result simply from climate change (Bull, 1991), or from lowering of the river base level. In the latter case, capture of the lower drainage basin would result in steepening of the gradient and increase in the energy of rivers upstream.

The problem was studied using the topography of individual basalt flows. Extensive lava flows are generally laid down broadly horizontally, so later differential vertical motions are recorded in the deformation of the flow surfaces. Up to 1600 m of differential topography is observed in the flows of the Columbia River Basalts today (Hales et al., 2005).

The results suggest that instead of precursory uplift, mild pre-eruptive subsidence took place near and around the center of eruption. Up to 2 km of post-eruption uplift occurred locally near the center of the flood basalt at the Wallowa Mountains, and lesser amounts, up to a few 100 m, occurred elsewhere. The nature of the pre-eruptive vertical motions is still under debate, but it is not disputed that significant post-eruptive uplift occurred.

The pattern of vertical motions suggested by lava flow topography matches well with that predicted for a Rayleigh-Taylor instability, delamination or other catastrophic lithosphere loss. In that model, initial pre-volcanic subsidence is expected to result from sinking of the instability followed by syn- and post-volcanic uplift as the instability detaches. Long-term residual high

[4] http://www.mantleplumes.org/TopPages/CRBTop. html

elevations are expected as a result of the various processes associated with melt production and eruption (Section 2.3). These can account for the broad, permanent uplift of a few 100 m. Additional, local plutonic processes must be responsible for the very large (~2 km) surface uplift that occurred near the eruptive center at the Wallowa Mountains.

2.5.3 Volcanic chains without initial flood basalts

The largest category of melting anomalies on the longer lists (Tables 1.1–1.5) is small-volume volcanic features lacking initial flood basalts. Examples of islands, seamounts and chains of this sort from the Pacific Ocean include the Bowie, Easter and Cook-Austral chains, the Caroline Islands, the Juan de Fuca/Cobb, Juan Fernandez, Lord Howe, Macdonald and Tasmanid seamounts, Pitcairn, Samoa, San Felix, Socorro and Tahiti (Fig. 1.9). In the Atlantic Ocean, Ascension Island, the Canary and Cape Verde archipelagos, the Discovery, Great Meteor/New England, Meteor and St Helena seamounts also have relatively trivial volumes. Examples from continental areas include Eifel in Europe, Darfur, Hoggar and Tibesti in Africa, and Raton in North America. This includes ~50% of the melting anomalies listed in Table 1.4.

There is essentially no information regarding precursory or contemporaneous vertical motions that may have accompanied the beginnings of these systems. Many lie on bathymetric or topographic highs, but these do not require high temperatures in the mantle to explain them (Stein and Stein, 1993). Processes associated with melt extraction (Section 2.3) contribute to raising the surface elevation, and loading with eruptives will warp the old land surface or sea floor downward. Post-eruptive subsidence is expected as the volcanics cool and the depleted residuum cushion disperses by viscous flow (Phipps-Morgan et al., 1995). High mantle temperature

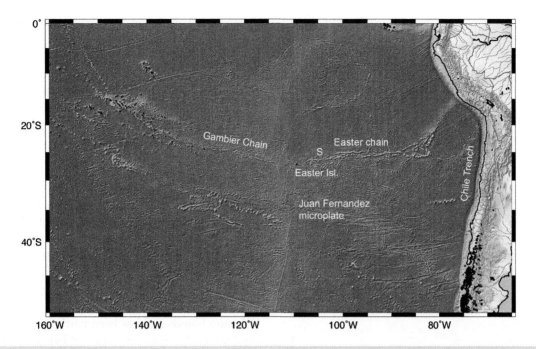

Figure 2.10 Bathymetry and topography in the region of the Easter volcanic chain. S – Sala y Gomez Island (basemap from Smith and Sandwell, 1997). See Plate 8

cannot be concluded simply from high elevation because of the many other contributing effects.

The absence of precursory domal uplift and an initial flood basalt for the many small-volume melting anomalies immediately rules out the Plume model (but see Section 2.6 for a discussion of plume variants). So what are the alternative, plate-related explanations? Many of the small chains in the Pacific have been explained by distributed intraplate extensional stress arising from plate cooling, local loading of the lithosphere by earlier-erupted volcanoes, or distant plate boundary forces (Forsyth et al., 2006; Hieronymus and Bercovici, 1999; 2000; Sandwell and Fialko, 2004). For chains that emanate from plate boundaries, stress perturbations related to complex ridge structure or tectonics have been suggested (Beutel, 2005). The Easter and Foundation chains, for example, emanate from the East Pacific Rise close to the Easter and Juan Fernandez microplates (Fig.

2.10). The Cameroon volcanic chain mirrors the Benue Trough, a rift valley ~250 km to the northwest. Also, it almost exactly bisects the ~100° angle in the continental margin of West Africa (Fig. 2.11). It has been suggested that it is related to the mechanical weakness of an object with an edge the shape of the west coast of Africa, but this has yet to be modeled numerically.

(a) The Emperor-Hawaiian system

This system is the most remarkable volcanic chain lacking an initial flood basalt. It is the type example plume locality, but offers little opportunity to assess vertical motions precursory to or contemporaneous with its initiation. It comprises a ~6000-km-long chain of islands and seamounts stretching from the central Pacific, northwest to terminate in the acute angle formed by the junction of the Kamchatka and Aleutian

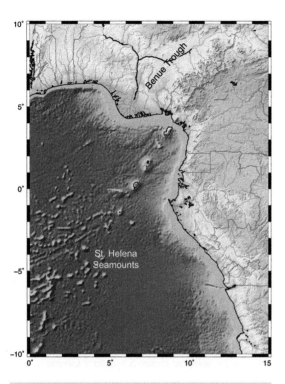

Figure 2.11 Bathymetry and topography in the region of the Cameroon volcanic line (basemap from Smith and Sandwell, 1997). See Plate 9

subduction zones Fig. 2.12). Is this a curious coincidence or does it have some more fundamental meaning?

The oldest end of the Emperor chain lacks a flood basalt. However, because it abuts a subduction zone the possibility exists that older volcanoes, and perhaps an initial flood basalt, once existed but have been either obducted onto land or subducted. Whether such older features once existed, or whether Meiji seamount, the most northerly of the Emperors, is the first seamount formed, is an important issue because it could potentially discriminate between the Plume and Plate hypotheses for the Emperor chain (Fig. 2.13). Plate motion reconstructions and geochemical analyses suggest that Meiji seamount formed at or near a recently subducted spreading ridge (Keller et al., 2000; Norton, 2007; Regelous

et al., 2003). It is not clear what this curious observation means as melting anomalies near ridges tend to maintain their proximities rather than to migrate away. This is predicted by the Plate hypothesis, and the common observation that it occurs is often explained in the context of the Plume hypothesis as lateral flow of plume material from a distal source (Section 2.6).

There is no evidence in the geology of Kamchatka for an obducted oceanic plateau or seamounts (Shapiro, 2005).[5] Thus, if a plateau once existed, it must have been subducted. It is not known whether plateaus can be subducted. When they collide with continents they are often obducted onto land (Cloos, 1993). That process has formed the Wrangellia terraine in the Pacific Northwest, and the Ontong Java Plateau, which is currently colliding with the Solomon islands and has been partly pushed up onto the land to form exposures of basalt. There is no evidence for recycled subducted Emperor chain material in the lavas of Kamchatka.[6]

It has been suggested that Meiji seamount is indeed the first formed volcano of the Emperor chain, and that it erupted near a spreading ridge as a consequence of a complex plate boundary reorganization (Norton, 2007). During the Cretaceous three plates, the Pacific, Farallon and Kula/Izanagi, met at a ridge-ridge-ridge triple junction there. A tectonic reorganization is thought to have occurred that involved a ~2000-km jump of the triple junction to the southeast, leaving two branches of the old triple junction to continue life as a straight ridge segment. Meiji seamount formed at the newly extinguished triple junction. Volcanism is postulated to have continued, and ultimately formed the Emperor chain, though why it should have done so has not yet been explained. It is known from seafloor magnetic isochrons that the mother ridge was very recently subducted at the Kamchatka and Aleutian trenches.

[5] http://www.mantleplumes.org/Kamchatka2.html
[6] http://www.mantleplumes.org/Shapiro2006.html

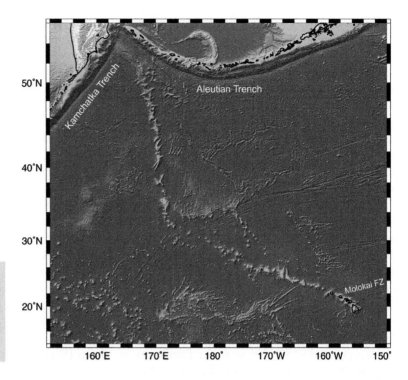

Figure 2.12 Bathymetry and topography in the region of the Emperor and Hawaii volcanic chains (basemap from Smith and Sandwell, 1997). See Plate 10

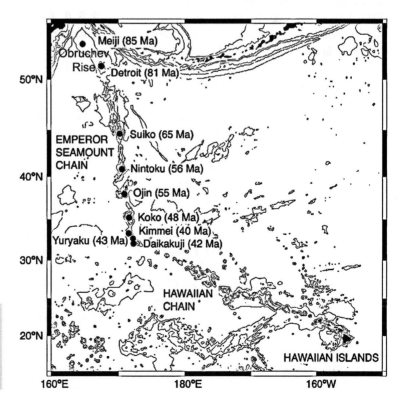

Figure 2.13 Bathymetric map of the North Pacific showing the Emperor and Hawaiian volcano chains (from Regelous et al., 2003).

There are other examples of excess volcanism at ridge-ridge-ridge triple junctions from this part of the northwest Pacific Ocean. Shatsky Rise, a large plateau, formed along a migrating ridge-ridge-ridge triple junction, and Hess Rise is thought to have been formed by anomalous spreading-ridge volcanism (Fig. 2.12) (Sager et al., 1999).[7] It appears, however, that long-lived volcanism did not emanate from those plateaus, to form long chains, as it did from Meiji seamount.

The Emperor-Hawaiian system is not entirely lacking a flood basalt, however. In a curious reversal of the predictions of the plume model, the system is currently forming a flood basalt at its young end. The volcanic archipelago there is so large that it fulfils even the upper-end criterion for a large igneous province (Bryan and Ernst, 2008; Sheth, 2007b).[8] It is currently producing magma at the massive rate of $0.25\,km^3/a$ and has covered ~150,000 km^2 of the Pacific sea floor with tholeiitic lavas in the last 5 Ma (Robinson and Eakins, 2006; Van Ark and Lin, 2004). These lavas cap the Hawaiian swell – an ~800-km-wide region of sea floor that is elevated by up to ~1 km above the normal Pacific sea floor level (Fig. 2.12).

How can the swell topography be explained? Could it be caused simply by the magmatism itself? A remarkable observation in support of this is that the volume of the swell along the chain is almost everywhere approximately proportional to the volume of the magma produced (Phipps-Morgan et al., 1995). If the swell results from the formation, extraction, and near-surface emplacement of magma, the ratio of swell volume to magma volume would be expected to be ~0.25 (Section 2.3). However, at the young end of the chain, where magma volumes are known best (Robinson and Eakins, 2006), this ratio is ~0.5. About half the swell topography can thus be attributed simply to melt extraction, but the rest may require additional explanation.

The plume model predicts that the magmatism results from anomalously high source temperatures, and that the swell is thermal in origin. However, measurements on the sea floor have failed to detect the high heat flow predicted (Chapter 6; von Herzen et al., 1989).[9] It has been countered that shallow water circulation would mask the heat flow expected from a plume (Harris and McNutt, 2007).[10] Nevertheless, it remains the case that a quantitative hypothesis has yet to be proposed to explain the Hawaiian swell whose predictions have been confirmed by observation. More detailed work is needed to explore whether melt extraction can account for the entire swell topography, for example by taking into account the unusual mantle petrology indicated by the geochemistry of Hawaiian lavas (Section 7.6.2).

2.5.4 Active flood basalt provinces lacking volcanic chains

Two flood basalt provinces that are not associated with time-progressive chains but which are still presently active are Iceland and Afar. In addition, widespread, ongoing volcanism is presently occurring in a number of continental regions, such as eastern Anatolia.

(a) The North Atlantic Igneous Province and Iceland

Of the melting anomalies presently active today, the North Atlantic Igneous Province and Iceland comprise the system that is the most clearly and uncontroversially associated with substantial, regional uplift early in its history (Fig. 2.14). This uplift is still apparent, for

---

[7] http://www.mantleplumes.org/Shatsky.html
[8] http://www.mantleplumes.org/TopPages/LIPClassTop.html

[9] http://www.mantleplumes.org/Heatflow.html
[10] http://www.mantleplumes.org/Heatflow2.html

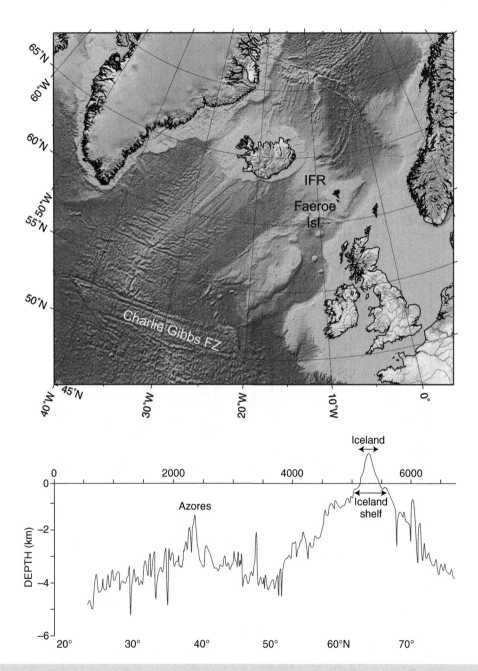

Figure 2.14 Bathymetry and topography of the Iceland region. IFR – Iceland-Faeroe Ridge (basemap from Smith and Sandwell, 1997). Inset shows bathymetry along a profile along the Mid-Atlantic Ridge. Iceland is not an isolated topographic anomaly but simply the summit of a vast bathymetric high ~3000 km wide that fills much of the North Atlantic (from Vogt and Jung, 2005). See Plate 11

example in the highlands of Scotland, Greenland and Norway, which still stand hundreds of meters above the level at which they stood prior to continental break-up. For this reason, the North Atlantic is frequently cited as providing a clear example of plume-related uplift (e.g., Campbell, 2006).

The first phase of eruptions that contributed to the North Atlantic Igneous Province occurred during the 3-Ma period 61.6–58.7 Ma. They are traditionally considered to be the earliest associated with the postulated Iceland plume. They mostly erupted in a northwest-trending

zone traversing Britain, east Greenland, west Greenland and Baffin Island (Fig. 2.15).

An hiatus of 2–5 Ma followed, after which a second volcanic phase began, contemporaneous with break-up of the Eurasian supercontinent and opening of the North Atlantic Ocean. Vast volcanic margins formed, comprising a ~2600-km-long belt flanking the new ocean and lying orthogonal to the volcanic zone formed in the first phase (Fig. 2.15). As the new ocean basin widened, in most places the volcanic rate quickly dwindled to levels that produced oceanic crust of fairly typical thickness only. However, the

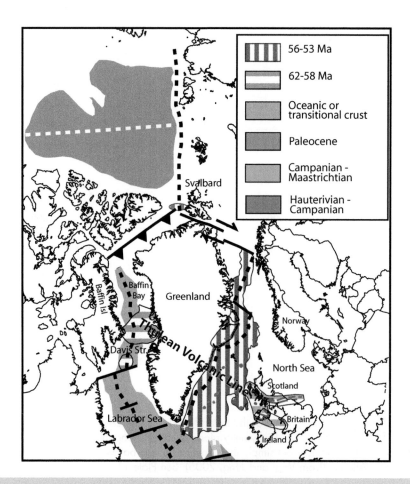

Figure 2.15 Volcanics in the North Atlantic Igneous Province. The volcanism associated with the first phase of volcanism, at ~62–58 Ma, and the second phase associated with continental break-up at ~56–53 Ma, are distributed very differently (courtesy of Erik Lundin).

rate remained high at the present-day latitude of Iceland.

The magmatic volume produced in the first phase, and the accompanying vertical motions, were relatively minor. Uplift occurred only locally near igneous centers (Dam et al., 1998; Maclennan and Jones, 2006; Ukstins Peate et al., 2003). In contrast, the volume produced in the second phase was an order of magnitude greater and possibly exceeded $6 \times 10^6 km^3$ (Eldholm and Grue, 1994). Eruption was accompanied by widespread uplift amounting to several hundred meters, even approaching 1 km in some places.

Unusually, it has been possible to study this uplift in great detail in the sedimentary basins surrounding Britain and Ireland, because they have been intensively researched for decades for hydrocarbon exploration (Maclennan and Jones, 2006). Uplift there is up to ~900 m. It occurred in several short, discrete episodes from ~59–55 Ma. About half the total occurred in a single, particularly intense period that started at ~55.8 Ma. It lasted no more than 1 Ma, but sedimentological and palaeontological studies suggest that it occurred even more quickly than this. A period even as short as 30,000 years cannot be ruled out, and the rate of uplift might then have been as great as 30 mm/yr (Fig. 2.16). The spatial pattern of uplift also fits a circular, domal pattern poorly. The best-fit model of this kind is centered in the Atlantic, off the coast of east Greenland, over 1000 km from the postulated plume impact position (Jones, 2005).

Several Deep Sea Drilling Project wells around the North Atlantic margins provide data that show regional subsidence faster than would be expected for normal sea floor. This is consistent with temperature anomalies of up to ~100 °C throughout a shallow layer up to ~100 km thick, but it precludes very high temperatures or very thick thermal sources (Clift, 2005) (Section 6.5). Curiously, a drill hole on the Iceland-Faeroe ridge, which flanks Iceland and is postulated to be a plume track, subsided at a normal rate.

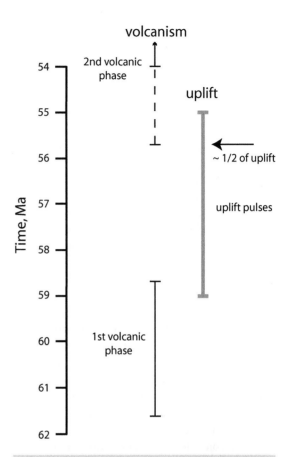

Figure 2.16 Relative chronologies of volcanism and uplift in the period preceding and accompanying break-up of the North Atlantic.

Superimposed on this broad picture is a complex post-volcanic history of local vertical motions widespread in and around Britain and occurring throughout much of the Cenozoic (Holford et al., 2009; Japsen et al., 2006; Praeg et al., 2005; Stoker et al., 2005). This has involved subsidences, uplifts, and tilting and formed a complex of basins and anticlines – so-called "egg box" or "piano key" tectonics. Motions were often rapid, local, and sometimes of the order of a kilometer in amplitude. Much, if not all of the ~2.5-km rock uplift observed in southern Britain occurred in the Neogene (23–2.6 Ma) and has been attributed to compressional stresses from the Mid-Atlantic Ridge and the Alpine collision (Hillis et al., 2008).

The first-order observations, of uplift of the order of 1 km over an area in excess of 1000 km wide, followed by subsidence more rapid than usual, are consistent with the predictions of the plume model. However, any more detailed scrutiny immediately reveals problems. The Plume hypothesis predicts that uplift occurred gradually over 10–20 Ma before the onset of the first phase of volcanism. However, at Iceland it occurred much later, some of it in rapid bursts. The limited, local uplifts that preceded the first phase of volcanism are consistent with local diapirs within the lithosphere rather than an asthenospheric upheaval on a thousand-kilometer-wide scale.

The major, widespread uplift along the North Atlantic margins that is commonly attributed to an arriving plume (Campbell, 2006; White and McKenzie, 1989) occurred much later, and was quasi-simultaneous with continental rifting and the second and most voluminous phase of volcanism (Maclennan and Jones, 2006). The ~100 °C temperature anomaly consistent with the post-volcanic subsidence history of the North Atlantic margins is dependent on the assumption that none of the topographic decay resulted from dispersal of the residuum left by volcanic margin formation (Clift, 1996; Clift and Turner, 1995). That effect is so large that it can explain much of the decay of the ~1 km Hawaiian swell (Section 2.5.3). It is thus unlikely to be negligible in the North Atlantic where the volcanic margins cover an area ten times larger than the Hawaiian archipelago plateau. A plume model would predict that the subsidence rate of the Iceland-Faeroe ridge would be fastest of all, since it is the closest location studied to the proposed plume center, but on the contrary, it is slowest there.

The integrated volume of uplift expected from melt extraction and emplacement at the near surface is ~25% of the volume of melt extracted (Section 2.3). Thus, for $6 \times 10^6 \, \text{km}^3$ of melt, an uplift $~1.5 \times 10^6 \, \text{km}^3$ in volume is expected. This is equivalent to ~1 km of uplift along the entire 2600-km volcanic margin, extending to over 250 km inland on both sides of the Atlantic. This uplift would be expected to decay slowly with time as the lithosphere cooled and the residuum root dispersed. Major rift-related shoulder uplift up to several kilometers in amplitude and hundreds of kilometers wide would also be expected to accompany the rifting of the continental lithosphere. The timing of the major uplifts, near-sychronous with volcanism and rifting, is qualitatively consistent with the observations from the North Atlantic. No quantitative modeling has yet been attempted.

Present-day Iceland lies on the summit of a broad topographic rise that is global in scale. It fills the entire width of the North Atlantic, extends from the Charlie-Gibbs fracture zone in the south to Svalbard in the north, a distance of ~3000 km, and has an amplitude of up to ~2 km (Fig. 2.14). In the context of this broad rise, Iceland itself comprises a topographic anomaly of no more than ~1 km, the highest peak rising to no more than ~2 km above sea level. It is interesting to compare this to Hawaii. There, the topographic anomaly is only ~800 km wide and ~1 km high, but the highest peak rises to ~8 km above the level of the surrounding sea floor.

Multiple seismic experiments show that the layer with crust-like seismic velocities (the "seismic crust") beneath Iceland and the surrounding shelf is ~15–40 km thick. If this layer were basaltic crust with normal densities, Iceland would be expected to rise in general to ~4 km above sea level, a full 3 km higher than its actual general ~1 km elevation. Iceland must stand so low because the difference in density between the lower part of the seismic crust and the upper mantle is much smaller than normal (Menke, 1999). The reasons for this are not understood and no reasonable petrology has been found that can explain it (Gudmundsson, 2003). It is unlikely that Iceland will be understood until this problem is solved. Thus, we stand at present in the bizarre situation of puzzling over why the elevation of this primary plume candidate, the highest point anywhere along the oceanic spreading ridge system, is as extraordinarily low as it is.

(b) Afar

The Afar melting anomaly has certain things in common with Iceland, but also contrasts with it in a number of ways. Its primary feature is an elongated flood basalt that extends for ~3200 km from Yemen, southerly along the Ethiopian rift. It is ~1,300 km broad at its widest point and covers an area of ~500,000 km^2 (Fig. 2.17) (Baker et al., 1996). Many of the volcanics were erupted in the period 31–28 Ma, but volcanism continues to the present day in the rift and in the Afar depression. This includes diking associ-

ated with crustal extension in a similar style to that which occurs in Iceland, and probably on the mid-ocean ridge (Wright et al., 2006). The basalt pile is locally over 2 km thick and an additional 1–2 km have been removed by erosion. This flood basalt has been attributed to the impingement of a plume (Morgan, 1971; White and McKenzie, 1989).

The region is unique tectonically. Afar lies at the center of a triple junction, the three branches of which are the Red Sea and the Gulf of Aden spreading centers, and the intra-continental East African Rift. These zones extend up to ~2500 km from Afar. Like the North Atlantic margins, much of the region was uplifted in association with the magmatism and subsequent rifting (Bohannon et al., 1989; Menzies et al., 1992). The uplift extends along all three

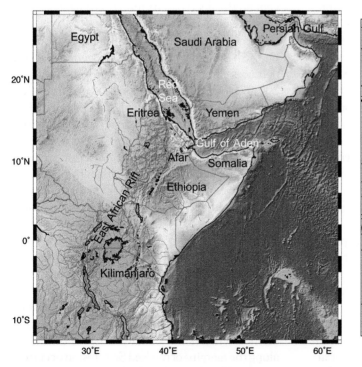
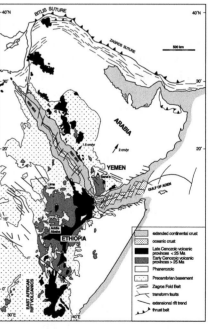

Figure 2.17 Left: Bathymetry and topography of the Afar region (basemap from Smith and Sandwell, 1997); right: Geological map showing the distribution of Cenozoic volcanism of different ages (modified from Baker et al., 1996). See Plate 12

branches of the triple junction, and laterally away from them and into the flanking plates for ~500–1000 km. Total basement uplift has been as much as 3–4 km locally. The region affected includes parts of Saudi Arabia, Yemen, Egypt, Ethiopia, Eritrea, and Somali (Fig. 2.17).

The start of flood basalt development at ~31 Ma was not the earliest volcanism in the region. It was a continuation of volcanism, albeit at an accelerated rate, that had already been ongoing in Ethiopia and Eritrea from ~38 Ma or even earlier. There is no unambiguous evidence for simultaneous extension. Rotational block faulting and detachment faulting started some time after flood volcanism dwindled, and was underway by ~25 Ma. This led to rifting and further magmatism and volcanism, ultimately forming the Gulf of Aden and the Red Sea. Sea-floor spreading onset in the Gulf of Aden at ~20 Ma and in the Red Sea at ~5 Ma (Drury et al., 1994).

There is little evidence for major uplift preceding the onset of flood volcanism at ~31 Ma (e.g., Menzies et al., 1992). As for the North Atlantic, uplift was concurrent with the magmatism and the subsequent rifting. The entire Afro-Arabian continent lay at low elevations or below sea level for 45 Ma before rifting precursory to Red Sea opening started at ~25 Ma. Throughout Yemen, Saudi Arabia and Ethiopia, the oldest flood basalts overlie disconformably (i.e., without any angular unconformity) shallow marine sediments and lateritic palaeosols. The laterites can be traced for thousands of kilometers throughout the Arabian peninsula and East Africa, and provide an important datum. In some places, uneroded marine sediments overlie the laterites. There is no evidence for any significant erosional periods in the sedimentary sequence, and no angular unconformities between the sediments and the overlying basalts. This is consistent with lack of major pre-volcanic uplift (Menzies et al., 1997; 1992; Ukstins Peate et al., 2005). In Saudi Arabia to the north, complete sedimentary sections dating from late Cretaceous to early Oligocene provide no evidence for doming (Bohannon et al., 1989). Palaeocurrent data provide no evidence for a radiating drainage pattern as would be expected for domal uplift (Cox, 1989), and the extensional basins that matured to form the Red Sea formed near sea level.

Some arguments in support of uplift have been presented. The palaeoenvironment of the sediments underlying the oldest basalt flows changes from shallow marine to fluvial continental and then to subaerial, at which stage the laterite palaeosols were deposited. This sequence indicates a long-term (perhaps tens of millions of years) base level rise of the order of tens of meters preceding the initiation of magmatism at ~31 Ma. (Menzies et al., 1997). Parts of north Egypt may have uplifted at ~27 Ma, prior to the 20-Ma peak of magmatism there. This locality is ~2000 km from the proposed plume center at Afar, however. In Eritrea, the laterites are absent locally, suggesting uplift of ~1 km. However, in general the flood basalts rest conformably on the lateritic palaeosols.

A case has been made for broad domal uplift on the basis of deposition of Oligocene-Miocene deltaic sediments in the Persian Gulf (Sengor, 2001). However, inference of a dome from this observation requires extrapolating a small region of dipping sediments for some 2000 km across the entire Arabian peninsula. There is no evidence that such an extrapolation is justified, nor is there evidence from sedimentological or fission track data for such a regional dome.

Major surface uplift in the region occurred after the onset of flood volcanism, and took two forms. First, uplift occurred simultaneously with the volcanism, as expected as a direct consequence of melt extraction from depth and emplacement at the near surface. Second, very large rift shoulder uplifts and massive erosion accompanied continental rifting and basin subsidence. Fission track analysis has shown that along the margins of the Red Sea erosion occurred at 21 ± 2 Ma, coeval with the beginning of the main phase of basin subsidence and extension (Omar and Steckler, 1995). In Saudi Arabia, uplift began at ~20 Ma and totaled 2.5–4 km. In

Yemen, the rift flank is up to 3.6 km high at present, over 3 km of material were removed after ~20 Ma, and up to 4 km of subsidence may have accompanied basin formation. This suggests over 10 km of differential vertical motion there (Menzies et al., 2001). Near the base of the flood volcanics, marine sediments occur at a height of 2.4 km, indicating this amount of base-level uplift since 28 Ma. If this were solely a consequence of the melt extraction process, the 25% rule of thumb would require ~10 km of melt to be emplaced. If 3–4 km of flood volcanics were erupted, a typical estimate, a further ~6–7 km of intrusives is thus predicted.

The history of vertical motions associated with emplacement of the Afar flood basalt province is thus inconsistent with the predictions of the Plume hypothesis. It is, however, qualitatively consistent with what would be expected from melt extraction and rifting associated with creation of the new Red Sea and Gulf of Aden ocean basins. Again, quantitative modeling of this and formulation of predictions, for example, regarding the amount of intrusions expected, has not yet been done.

It remains to speculate what might have triggered formation of the Afar province. The flood basalt, the Red Sea and Gulf of Aden spreading centers, and the East African Rift are clearly all interrelated in some sense and form part of the global plate boundary system. The Plate hypothesis predicts that the whole tectonomagmatic phenomenon formed in response to the constantly evolving global plate configuration. The continent-continent collision between India and Eurasia at ~34 Ma is an example of a boundary reconfiguration that would have needed to have been accommodated by changes in the plate boundary elsewhere (Bohannon et al., 1989). The formation of the Afar triple junction province was most probably a passive reaction to the need to reconfigure plate motions within the closed system that is the surface of the Earth. However, the sequence of events, with eruption of the flood basalt substantially preceding rifting and sea floor spreading, remains to be fully explained.

(c) Anatolia

The eastern Anatolian volcanic region lies on a high plateau in the Alpine-Himalaya mountain belt that currently stands at an elevation of up to ~2 km above sea level (Fig. 2.18). Two-thirds

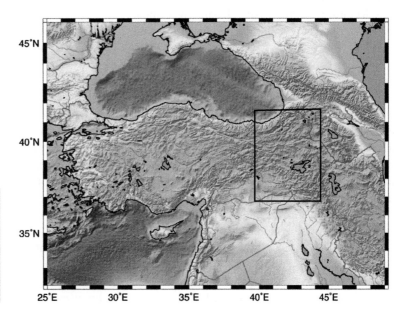

Figure 2.18 Bathymetry and topography of the Anatolia region (basemap from Smith and Sandwell, 1997). Box indicates the location of recent volcanism. See Plate 13

of this plateau, an area of some ~43,000 km², is covered by young volcanics that may originally have had a volume of ~15,000 km³. The shape of the topography is domal, with an extent and amplitude similar to what is predicted by the Plume hypothesis (Sengor et al., 2003). This has led to the suggestion that the province was formed by a plume (Ershov and Nikishi, 2004).

The volcanics range in age from 11–0 Ma, and are locally up to ~1 km thick (Keskin et al., 1998). The entire region extends for a distance of ~1000 km from eastern Turkey to southern Russia along the Alpine-Himalaya collision belt. In addition to the plume model, numerous other models have been proposed to explain the volcanics, most of them related to collision-belt tectonic scenarios (see Keskin, 2007 for a detailed review).

A plume model is implausible in a setting of this sort. Until recently a slab was subducting northward beneath the area, consuming the last remnants of the Tethys sea floor. A plume would have had to have penetrated this slab. Plumes have nevertheless been proposed to penetrate slabs at several locations to explain volcanism behind subduction zones, including Mexico, where no uplift occurred at the postulated plume impact site (Ferrari and Rosas-Elguera, 1999),[11] the Manus back-arc basin, Japan (Mashima, 2005),[12] the Basin and Range Province and Italy.

The slab beneath Anatolia started to break off at ~11 Ma, when the last of the Tethys oceanic crust had been consumed and subduction gave way to continent-continent collision (Sengor et al., 2003). Volcanism and uplift both began almost simultaneously, starting in the north and migrating southward with systematic changes in lava chemistry (Keskin, 2003). A radial pattern is not observed. This sequence of events suggests that the volcanism was related to local restructuring of the plate boundary.

Seismology has shown that the lithospheric mantle is exceptionally thin or absent beneath the region, and the asthenosphere lies at the shallow depth of 40–50 km (Al-Lazki et al., 2003; Gök et al., 2000; 2003). In view of these observations, the hypothesis of least astonishment is that both the uplift and the magmatism resulted from progressive slab steepening, break off and removal of the mantle lithosphere. Asthenosphere flowed passively in to replace the lithosphere, and melted as a result of adiabatic decompression (Fig. 2.19) (Gögüs and Pysklywec, 2008; Keskin, 2003).

The case of Anatolia is important because in this tectonic setting it is virtually inconceivable that the plume model can be correct. On the other hand, the present-day topography is almost identical to what is predicted for a plume. Anatolia thus provides an example of how the residual topography of a volcanic province produced by plate-related processes can resemble closely that predicted by the Plume hypothesis (Keskin, 2007). It therefore highlights the dangers inherent in interpreting old uplifted areas and drawing conclusions where the timing of events and the tectonic setting may be much less clear than for modern examples.

2.5.5 Extinct flood basalt provinces lacking volcanic chains

While few flood basalt provinces are forming at the present day, some 300 formed in the past and are now extinct (Ernst and Buchan, 2001). Examples include the Hess and Shatsky rises and the Manahiki and Hikurangi plateaus, all in the Pacific Ocean. The Dhanjori Formation, India, and parts of the Karoo province are examples of Proterozoic flood basalts (Jourdan et al., 2004; Mazumder and Sarkar, 2004).[13] The Siberian Traps and the Ontong Java Plateau are the largest continental and oceanic flood basalts,

[11] http://www.mantleplumes.org/Mexico1.html; http://www.mantleplumes.org/Mexico2.html
[12] http://www.mantleplumes.org/Japan1.html

[13] http://www.mantleplumes.org/Dhanjori.html; http://www.mantleplumes.org/Karoo.html

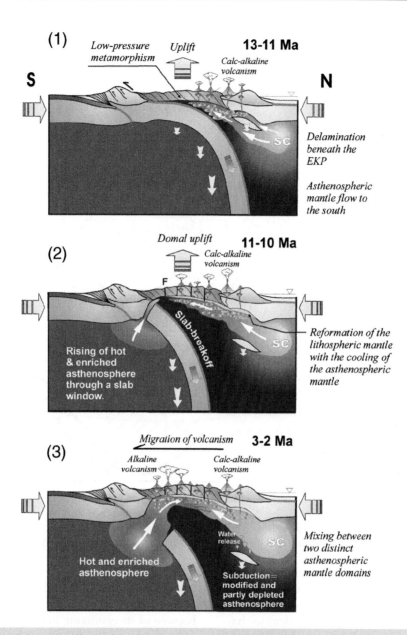

Figure 2.19 Slab break-off model for the generation of volcanism in eastern Anatolia (from Keskin, 2007).

respectively. Whereas three or so extinct flood basalts are plausibly linked to presently active volcanic chains (Section 2.5.2), no extinct flood basalt is confidently associated with an extinct volcanic chain.

(a) The Siberian Traps

The Siberian Traps erupted in the largest sub-aerial volcanic event ever known (Fig. 2.20). They formed in the period ~225–265 Ma, but

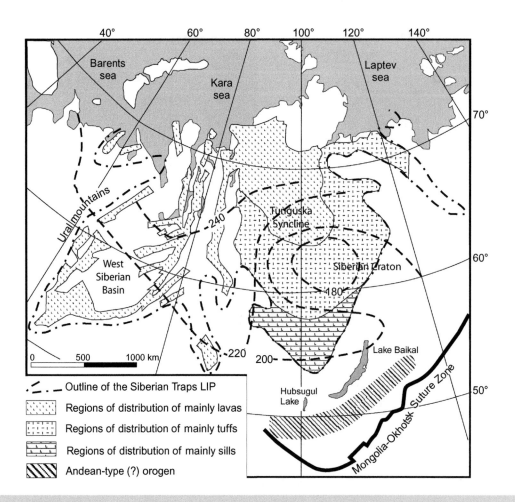

Figure 2.20 The Siberian Traps, showing the western and the better-exposed eastern lobes. Dashed contours show depths in kilometers to the base of the lithosphere determined using seismic data from nuclear explosions (Egorkin, 2001; Pavlenkova and Pavlenkova, 2006) (adapted from Ivanov, 2007).

the major peak of production was at ~251 Ma at the Permian-Jurassic boundary, coincident with a mass extinction. Much of the volcanics has now been eroded away, but the basalt pile may locally have attained a thickness of ~6.5 km. It originally extended over an area of $\sim 7 \times 10^6\,km^2$ and had a total volume of $\sim 4 \times 10^6\,km^3$ (Masaitis, 1983). It comprises two lobes, overlying the West Siberian Basin and the Siberian Craton to the east.

It is spectacularly clear that the eastern lobe, the thickest part, erupted onto lithosphere that subsided continually prior to and during early volcanism (Czamanske et al., 1998)[14]. The Traps overlie the largest coal resource in the world and because of its economic importance it has been researched intensively for several decades, including the drilling of hundreds of boreholes. The geology and stratigraphy of the region are thus known in great detail.

The subsidence was substantial and widespread. The uppermost coal-bearing layers, whose sedimentary facies indicate shallow-water lakes or lagoons, are interbedded with the

[14] http://www.mantleplumes.org/Siberia.html

earliest volcanics (tuffs), showing that sub-sidence was ongoing when volcanism started. Subaerial conditions prevailed during later stages of volcanism, indicating co-magmatic uplift, along with some local areas of extremely fast subsidence that might have resulted from drainage of large magma chambers.

The situation in the West Siberian Basin is somewhat different. The Traps are much less well exposed, there is much Mesozoic sediment cover, and the extent of the volcanics is less well known. They were erupted onto Proterozoic to Upper Palaeozoic rocks, not Archean craton as for the eastern lobe. Much of the area was subaerial before the Traps erupted, with elevations possibly exceeding 2 km locally. These elevations have been attributed variously to Hercynian (Late Palaeozoic) orogenic uplift, and to the arrival of a hot plume shortly before the Traps were erupted.

It has been suggested that subsidence of the Siberian Craton prior to trap eruption does not rule out a plume, because one could have arrived beneath the West Siberian Basin and fed the eastern lobe by lateral flow (Saunders et al., 2005) (see Section 2.6 for a general discussion of plume variants). However, the pattern of ver-tical motion in the West Siberian Basin had been ongoing and uniform since at least the Early Permian, several tens of millions of years before eruption of the Siberian Traps. Also, such a plume model would require that most of the magmas flowed laterally 2000 km or more to erupt beneath the much thicker Siberian Craton, where they formed by far the thickest succession. Such flow is contrived, and at odds with the "upside down drainage" concept of lateral flow, which suggests that laterally flowing buoyant material drains upward and toward areas of thin lithosphere (Sleep, 1996) (Section 2.6).

In an interesting contrast with the Afar flood basalt, the East Siberian Traps erupted simulta-neously with major rifting and extension along horst-graben structures, accompanied by erosion and sedimentation. The vertical motions associ-ated with those processes would thus be expected to be co-magmatic. An association between

rifting and the flood basalt volcanism thus sug-gests itself in this case.

Because of the precursory subsidence of the eastern lobe, a Rayleigh-Taylor instability model has been tested to see if it fits the observations for the Siberian Traps (Elkins-Tanton, 2005; Elkins-Tanton and Hager, 2000). The model that fits best involves weakening of the lower lithosphere by heating, and injection of melt which freezes to form dense eclogite. Onset of the Rayleigh-Taylor instability might also be triggered by hydration weakening (Ivanov, 2007). The lower lithosphere then sinks, causing subsidence at the surface, followed by necking and detachment, creating space into which asthenosphere flows from deeper levels (Figs 1.14 and 2.4). Decompression melting follows, and uplift, once volcanism is underway.

This model predicts that the lithosphere beneath the Siberian Craton was thick prior to the eruptions, and thinner afterwards. There is debate over the present thickness of the lithos-phere there. Values of >300 km (Czamanske et al., 1998; Saunders et al., 2005) and 180 km (Ivanov, 2007) have been suggested on the basis of seismic tomography, thermal modeling and nuclear explosion seismology. The latter work suggests that the thinnest lithosphere underlies the thickest part of the Traps, the Tunguska syncline on the Siberian Craton. Confirmation of this would provide important support for the Rayleigh-Taylor instability model and other lithosphere removal models, although it is diffi-cult to estimate the effects of processes during the ~251 Ma following Trap eruption. However, this model can explain the chronology of subsid-ence and uplift associated with the Siberian Traps, which other models cannot do.

(b) Ontong Java Plateau

Being the largest oceanic plateau on Earth, the Ontong Java Plateau is the oceanic counterpart of the Siberian Traps. It is the largest magmatic event ever known to have occurred on Earth. It formed at ~122 Ma, in the Early Cretaceous, covers an area of some $2 \times 10^6\,km^2$ and has a volume of ~$50 \times 10^6\,km^3$ (Fitton et al., 2004a, b). It has recently been suggested that the Manihiki and Hikurangi plateaus were formed together with it. If so, the composite plateau would have had the astounding total volume of ~$100 \times 10^6\,km^3$ (Taylor, 2006). The Ontong Java Plateau currently abuts the Solomon trench and is obducting onto land on the Solomon islands (Fig. 2.5).

Since the vast majority of the plateau lies beneath the ocean, it is difficult to study in detail. Its crest is currently ~1.7 km below sea level, and elsewhere it is ~2–3 km deep. It is, however, one of the better sampled oceanic plateaus because seven Ocean Drilling Project holes have sampled it, compared with the one or two that are typical for most oceanic plateaus. In addition, sections up to 3.5 km thick are available for study on the Solomon islands (Tejada et al., 2004). The information that does exist is consistent with the rapid emplacement of a body that is mostly homogeneous in structure and chemistry.

The plateau is remarkable for the small amount of uplift and subsidence that accompanied and post-dated its formation (Korenaga, 2005; Roberge et al., 2005).[15] Plume models for such a vast magmatic event would predict several kilometers of pre-eruption uplift (Farnetani and Richards, 1994; Ito and Clift, 1998). This, coupled with the increased surface elevation expected from the extraction and near-surface emplacement of a >30-km-thick layer of melt would predict that during late stages of formation the elevation of the top of the plateau should have lain in the range from approximately sea level, up to several kilometers above sea level (Ingle and Coffin, 2004; Roberge et al., 2005). Extensive subaerial volcanism would have occurred. Nevertheless, a large subaerial plateau never formed and all the lavas recovered by drilling, both on the plateau and on land in the Solomon Islands, were erupted well below

[15] http://www.mantleplumes.org/OJ_Puzzle.html

sea level. Only a few subaerial rocks have been found, and these are in the far southeast of the plateau, well away from the area where the greatest uplift is predicted. Eruption elevations obtained from the H_2O and CO_2 contents of basalt glass samples recovered from drill cores suggest eruption depths of ~1–3 km below sea level (Roberge et al., 2005).

In addition to insufficient uplift, much less post-eruptive subsidence occurred than predicted by a model involving a hot mantle source. For the Ontong Java Plateau, a total subsidence of 2.7–4.1 km would be expected since its formation. However, the overlying sedimentary strata suggest subsidence of only ~1–2 km with an average of ~1.5 km over much of the plateau (Clift, 2005; Ingle and Coffin, 2004). This is consistent with post-emplacement drift over warmer, not cooler mantle.

The co-emplacement plateau low-stand is comparable to the present-day situation of Iceland. There, the 15–40-km seismic thickness of the crust suggests that the general elevation of the island should be ~4 km, some 3 km higher than it actually is. The discrepancy requires an exceptionally small density contrast between the lower crust and the upper mantle (Section 2.5.4). The unexpectedly small post-formation subsidence at the Ontong Java Plateau is analogous to that of the Iceland-Faeroe ridge (Clift, 2005). The crustal thickness of the Ontong Java Plateau, and the structure of the underlying mantle, are similar to those of Iceland.

The Ontong Java Plateau is not unique in its lack of post-formation subsidence. Plateaus thought to represent anomalously thick crust, that have subsided more slowly than normal cooling oceanic crust, include the Magellan rise, the Mannihiki plateau, the Ninetyeast ridge, the Kerguelen plateau and the Rio Grande rise (Clift, 2005). Globally, the rapid subsidence predicted by plume models to follow the formation of an oceanic plateau is the exception rather than the rule.

It has been suggested that plateaus may be buoyed up by underlying low-density residuum

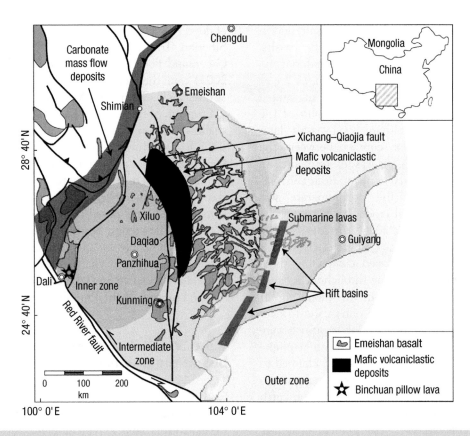

Figure 2.21 Geological map of the Emeishan basalts, southwest China, showing the distribution of mafic volcaniclastic deposits originally mis-identified as terrestrial conglomerates. Inset shows the location of the Emeishan area in China (from Ukstins Peate and Bryan, 2008).

left after melt extraction. In the case of the Ontong Java plateau, this would require that the residuum cushion traveled with the plateau for the last 122 Ma, a distance of ~7000 km relative to the geomagnetic pole. Such a process would conflict with residuum models for the Hawaiian swell, which predict that the cushion disperses by viscous flow on a time-scale of a few tens of millions of years (Phipps-Morgan et al., 1995) (Section 2.5.3).

(c) The Emeishan basalts

The Emeishan basalts in southwest China provide an ironic cautionary tale. This flood basalt erupted in the Permian at ~260 Ma (Zhou et al., 2002). It is estimated to have covered an area of ~2.5 × 10⁵ km² and has a preserved extrusive volume of ~0.3 × 10⁶ km³ (Xu et al., 2001) (Fig. 2.21). The volcanics are petrologically diverse, but dominated by basaltic flows, and with a total thickness up to ~5 km.

The main basaltic succession is underlain by a massive fossiliferous limestone formation with a thickness of ~250–600 m. This is part of a shallow marine carbonate platform that extends over much of southeast China. It was initially reported that in the Emeishan area its thickness varies and it is eroded in some places. In addition, a conglomerate layer with gravel-to-boulder-sized volcanic and limestone clasts and abundant fossils was described to lie between the limestone and the overlying flood basalt, and in some places to be interbedded in the basalts. This was considered to be erosional material and alluvial fan deposits.

The entire set of observations was interpreted as indicating pre-eruptive, domal uplift of the area where the Emeishan flood basalt later formed. The irregular contact between the limestones and the basalts was interpreted as an erosional unconformity resulting from this pre-volcanic doming and the "conglomerate" was interpreted as material eroded from the domal high (He et al., 2003; 2006). On the basis of detailed composite stratigraphic sections, an uplift model was developed comprising three concentric zones extending throughout a circular region ~800 km wide. In this model, uplift varied from 0–250 m in the outer zone to over 1 km in the inner zone. The observations were interpreted as being consistent with a plume origin for the Emeishan basalts, and claimed to be the best-documented case of regional, precursory domal uplift to be found anywhere on Earth (Campbell, 2006; 2007).

The observations were subsequently re-interpreted by other workers (Ukstins Peate and Bryan, 2008).[16] The variable thickness of the limestone formation was attributed to fault motion, carbonate mass flow, and deposition in rift basins, possibly associated with Himalaya-related deformation, most influential in the west of the region. Drawing an analogy with modern reefs, the uneven top surface of the limestone was interpreted as natural reef topography and not erosion or karst formation.

Most significantly, re-examination of the "conglomerate" revealed features such as accretionary lapilli, free marine fossils, pillow lavas, volcanic bombs, bomb sags and ductile deformation of basaltic clasts. These unequivocally identify it, not as a conglomerate, but as a widespread hydromagmatic deposit produced in a shallow marine environment by the initial Emeishan eruptions. Such deposits are common at the base of flood basalts and are seen also underneath the Siberian Traps (Ross et al., 2005), the East Greenland flood basalts (Ukstins Peate et al., 2003) and the Ferrar flood basalts (Ross and White, 2005; White and McClintock, 2001).

Eruption of the earliest Emeishan basalts in a shallow sea, in the shallow marine regional setting of southwest China that existed in the Permian, suggests magma injection and eruption through an active carbonate platform. This precludes kilometer-scale, pre-volcanic uplift, which would have raised the region well above sea level. The modest positive relief that developed later probably resulted from accumulation of the volcanic pile itself. This re-interpretation negates all evidence for significant uplift precursory to eruption of the Emeishan basalts.

2.5.6 Vertical motions without flood basalt magmatism

Continental regions commonly experience major vertical movements, and our understanding of why this happens is incomplete. Substantial uplift has occurred in many regions, some in the deep continental interiors and far from plate boundaries, without subsequent flood basalt volcanism or other characteristics of plumes, for example, in Mongolia (Petit et al., 2002).[17] The causes of these uplifts would ideally be understood because those associated with tectonism and volcanism are attributed to mantle-scale thermal diapiric upwellings.

(a) The Colorado Plateau

The Rocky Mountain orogenic plateau in western North America includes the southern and central Rockies, the adjacent western Great Plains and the Colorado Plateau (Fig. 2.22). It forms part of a vast belt of high elevation that extends from Alaska, through the western United States and

[16] http://www.mantleplumes.org/Emeishan.html

[17] http://www.mantleplumes.org/MongoliaGravity.html

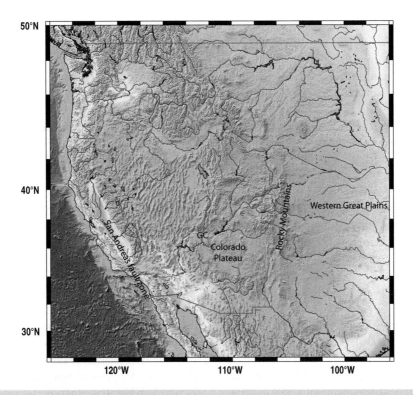

Figure 2.22 Topographic map of Western North America. GC: Grand Canyon (basemap from Smith and Sandwell, 1997).

into Mexico. The region was close to sea level in the Late Cretaceous, when shallow marine sediments were deposited. Today these sediments are uplifted to >2 km above sea level. The Colorado Plateau is a distinct sub-unit that comprises a deeply eroded land surface lying at a somewhat lower elevation than the surrounding mountains. Its most famous deep erosion feature is the Grand Canyon. Volcanic rocks are rare in the interior of the plateau, of small volume and typically confined to the margins. Flood basalt eruption did not occur anywhere in the Rocky Mountain orogenic plateau.

The exact timing and cause of Rocky Mountain and Colorado Plateau uplift is the subject of a longstanding debate. The problem is complicated by several ambiguities. Contributions to changes in surface elevation in a region include true rock uplift and subsidence,

erosion and deposition. Isostatic response to erosion and deposition will contribute to true rock elevation changes and must be estimated and accounted for if the contributions to uplift from processes such as tectonism or mantle convection, are to be separated out.

This problem has been addressed for the Colorado Plateau using Geographical Information Systems analysis of topography, erosion and sedimentation (Pederson et al., 2002b). Palaeoelevations have also been estimated from the vesicularity of basaltic lava flows, a relatively new method that is being pioneered to estimate palaeoatmospheric pressure and from it, elevation (Sahagian et al., 2002). Since the Late Cretaceous there has been ~2 km of rock uplift, a net loss of ~400 m of elevation as a result of faster erosion than deposition, and ~300 m of isostatic uplift.

Some of the uplift may have occurred in a rapid resurgence in the last few million years. This notion is appealing because of the recent intense erosion that formed features such as the Grand Canyon. However, such a resurgence may not be required by the data. Erosion is not a reliable indicator of true rock uplift since it can also occur as a result of climate change or the lowering of the base level further downstream (Bull, 1991). In this case, the Grand Canyon may have formed as part of basin and range extension and subsidence, and not as a result of accelerated uplift of the Colorado Plateau itself (Pederson et al., 2002a).

Regardless of the details, the results are consistent with ~2 km of largely amagmatic Cenozoic true rock uplift. What caused this? Uplift from non-isostatic sources might be explained by the Laramide orogeny, which occurred at ~80–40 Ma (McMillan et al., 2006). That orogeny is presumed to have been linked to reorganization of the plate boundary to the west, which at that time comprised a subduction zone but is now represented by the San Andreas transform fault system. Other processes that have been suggested include flat subduction of the eastward-subducting Farallon slab and delamination of the continental lithosphere, though it is not clear why this did not result in major volcanism. Intriguingly, these plate-related processes predict more uplift than is observed and so, as at Iceland, the problem at hand is insufficient observed uplift rather than the absence of candidate explanations for it.

(b) Bermuda

A remarkable example of regional uplift in the absence of major volcanism in the ocean is Bermuda (Vogt and Jung, 2007). The Bermuda islands comprise small subaerial carbonate reefs that cap the eroded summits of four large seamounts. They formed on the Bermuda Rise, a bathymetric swell in the central Atlantic Ocean ~600–800 km east of the North American continent–ocean transition to the west. The rise

is ~1500 km long, 500–1000 km wide, and stands up to ~1 km above the level of surrounding sea floor of the same age (Fig. 2.23).

Seismic experiments show that, unlike oceanic plateaus such as Ontong Java, the Bermuda Rise is underlain by slightly thin or normal-thickness ocean crust – it is not a flood basalt. This crust formed at the Mid-Atlantic Ridge in the period ~100–140 Ma, and began to uplift at ~47–40 Ma (Middle Eocene), continuing into the Miocene. The seamounts beneath the islands started to erupt at about the same time as the uplift, forming on crust that was, at the time, ~76–84 Ma and is now ~123–124 Ma. There has been little or no subsidence of the Rise since its formation.

About 900 km southeast of the Bermuda Rise is the Southeast Bermuda Deep. There, the sea floor lies up to ~600 m below normal ocean floor depth for its age. The Deep thus comprises a bathymetric anomaly comparable to the Bermuda Rise but of opposite sign. In contrast to the Bermuda Rise, it is mirrored by a similar bathymetric low the opposite side of the Mid-Atlantic Ridge, on the African plate. This suggests that this anomalously deep oceanic crust formed at the ridge axis, post-dating and unrelated to the Bermuda Rise.

A satisfactory explanation for the Bermuda Rise is currently elusive. It does not stand high as a result of crust thickened either during formation or by later eruption of a flood basalt. There is no evidence for adjacent surface loading that might be responsible for flexural upwarping, and no heat flow anomaly that might support a thermal model. Low-density residuum or other igneous processes might offer a solution but there is no evidence for large-volume magmatism. Nevertheless, the Rise summit is associated with a geoid high of ~5–10 m which, together with the depth anomaly, implies compensation by relatively low-density material at a depth of ~50 km. This would place it in the lithosphere and/or upper asthenosphere. Whatever the compensating body is, it must have been carried along with the plate for the ~47–40 Ma since

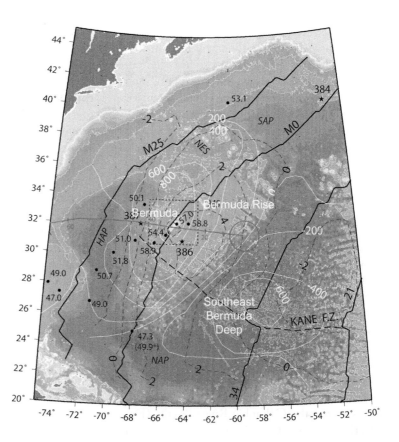

Figure 2.23 Bathymetry and topography of the Bermuda region (basemap from Smith and Sandwell, 1997) with key magnetic lineations, DSDP leg 43 drill site locations (stars), heatflow (W/m²), residual depth anomaly (white contours at 200 m intervals), residual geoid anomaly (in meters, dashed black lines) and predicted track of the North America plate over a fixed Bermuda "hot spot" (Duncan, 1984), with the predicted present "hot spot" location shown by the large gray circle. NES – New England seamounts; MS – Muir seamount (from Vogt and Jung, 2007).

swell formation began, a distance of the order of the swell diameter. Bermuda is not unique, but shares many features with other oceanic swells such as the Cape Verde swell.

2.6 Plume variants

It is challenging to make observations of the required sensitivity, accuracy and distribution to test for the kilometer-scale, regional domal uplift predicted by the Plume hypothesis. However, the issue of model variants presents a more subtle difficulty. Variants on the simple thermal diapir model have been designed that can account for almost any pattern of vertical motion observable, or its absence. If such variants are accepted, uplift cannot be used to test the hypothesis.

The problem of too little uplift being observed has been dealt with by proposing models that downsize the amount predicted. Scaling down the size of the proposed plume, for example, to a feature that occupies the upper mantle only, reduces its height from the ~3000 km necessary to traverse the entire mantle to a mere ~650 km. The diameter of the plume head and the amplitude and lateral extent of the predicted vertical motions then scale down proportionately and the predicted uplift might be too small to be measurable as a practical matter. Such a short plume has been suggested for the Iceland region, to explain the confinement of the seismic low-wave-speed anomaly there to the upper mantle (Section 5.5.1). Ironically, however, the North Atlantic and the Iceland region is one of few places on Earth where kilometer-scale uplift influencing a region thousands of kilometers wide is associated with flood volcanism. Small thermal diapirs limited to the upper mantle only are not generally sufficient to explain the features that plumes were invoked to explain, such as vast flood basalts, persistence of volcanism for tens of millions of years, and relative fixity between melt extraction loci.

Variants have also been suggested that redistribute the predicted uplift. In these situations, where precursory uplift did not occur where the flood basalt was erupted, cases have been made to associate the volcanism with uplift elsewhere. This is achieved by invoking lateral flow of plume material along the underside of the lithosphere, which can be used to explain both volcanism without uplift and uplift without volcanism. Topography on the underside of the lithosphere is postulated to guide the flow of buoyant plume material ("upside-down drainage"; Sleep, 1996). Such a model could potentially account for the formation of the Yellowstone volcano trail along the eastern Snake River Plain, ~300 km south of the eruptive center of the Columbia River Basalts (Fig. 2.9), the eruption of the Paraná basalts ~500 km west of the proposed Tristan plume impact site (Fig. 2.8), and the formation of the Iceland volcanic system ~900 km southeast of the postulated plume arrival site under Greenland (Vink, 1984). In the latter two cases, it is not explained why the respective supercontinents split hundreds of kilometers away from the postulated plume impact sites.

A few plume scenarios have also been suggested that require no uplift at all. Laboratory experiments and numerical models suggest that if a rising plume has a higher viscosity than the mantle around it, then a bulbous head is not predicted. In such cases, uplift and a flood basalt are not predicted (Davies, 1999, p. 303). Recently started plumes that have not yet reached the surface have also been postulated (Montelli et al., 2006). It is not clear how such models can be tested.

Lastly, more complicated patterns of vertical motion than simple domal uplift have also been attributed to plumes. Continental lithosphere rheologies more realistic than a single, homogeneous layer with high viscosity and a rigid top have been modeled numerically.[18] Real continental lithosphere is more realistically likened to a stack of layers with different rheologies that produce elastic, viscous and plastic deformation. Layers including the surface may move laterally with respect to one another in response to an upward force on the base of the stack. Continental lithosphere may also vary from being relatively young, warm, and soft to old, cold and stiff. Modeling these factors suggests that the prediction of regional domal uplift prior to the arrival of a plume is over-simplified, and a more complex pattern is expected that involves both uplift and subsidence on various spatial scales (Fig. 2.24). It has been suggested that the vertical motions observed in the East African Rift system, the French Central Massif, and the Pannonian-Carpathian region in eastern Europe, are consistent with such models.

2.7 Discussion

Both the Plume and the Plate hypotheses predict that uplift is associated with melting anomalies. The classical Plume hypothesis predicts that major precursory uplift precedes flood volcanism by millions or tens of millions of years. This is followed by subsidence as the plume head spreads beneath the lithosphere, melt forms and is drained from the subsurface, and the region is transported away from the hot source by plate motion. The Plate hypothesis predicts different chronologies and patterns of motion, depending on the process at work. The formation of a large volume of melt at depth and its extraction and emplacement at the near surface in itself results in substantial vertical motions, both transient and permanent, along with creation of a new, higher-elevation surface simply from the addition of eruptive material. Continental break-up

---

[18] http://www.mantleplumes.org/LithUplift.html

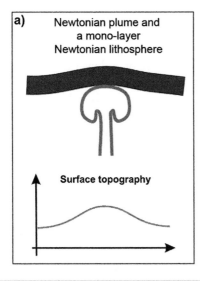
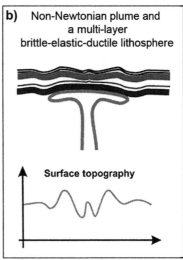

Figure 2.24 Patterns of uplift predicted for impingement of a plume beneath: (a) a simple lithosphere structure, and (b) lithosphere with realistic rheology (Burov and Guillou-Frottier, 2005).[19]

and catastrophic loss of the mantle lithosphere in particular are predicted to result in large vertical motions. The motions predicted to accompany plate processes are, however, expected to occur quasi-contemporaneously with volcanism. It is not the occurrence of uplift, but its pattern and timing that distinguishes the predictions of the two hypotheses.

Only a minority of currently active melting anomalies are plausibly linked to initial flood basalts. Where major regional precursory uplift followed by flood basalt eruption did not occur, the classical plume model fails. Flood basalts claimed to have been accompanied by uplift consistent with a plume origin include the Deccan Traps, the Paraná basalts, the North Atlantic Igneous Province, Anatolia, the Columbia River Basalts, Afar and the Emeishan Basalts. For these regions, the critical questions are first whether uplift did indeed occur, and second whether the chronology and pattern of uplift are consistent with the predictions.

There is no unequivocal evidence for widespread regional precursory domal uplift of the sort predicted by the Plume hypothesis at any of the localities listed above. The drainage patterns suggested as evidence for uplift domes in India and South America post-date eruption of the flood basalts. Large elevation increases occurred in the North Atlantic Igneous Province, but it was quasi-contemporaneous with flood volcanism and does not have a domal shape centered beneath Greenland, as predicted. The vertical motions and volcanism are concentrated along the margins of the new ocean basin and are more consistent simply with the continental rifting process. Uplift in Anatolia is of the magnitude and pattern predicted by the Plume hypothesis, but a plume is unreasonable behind a subducting slab, and a mechanism for volcanism related to slab break-off fits the holistic observations better.

Evidence for significant regional uplift prior to eruption of the Columbia River Basalts or the Afar flood basalt is lacking. Major flood basalts, where there is positive evidence that uplift did not occur, include the Emeishan Basalts and the Ontong Java plateau. There is clear evidence for major subsidence prior to the eruption of the

[19] http://www.mantleplumes.org/burov2005.html

Siberian Traps, consistent with catastrophic lithosphere loss, a mechanism that has also been suggested for the Columbia River Basalts. The post-emplacement subsidence expected to occur during transport of oceanic plateaus away from hot source areas is generally not observed.

There is thus an extensive body of evidence, from the largest and best-studied flood basalts and plateaus on Earth, that precursory uplift of the kind predicted by the Plume hypothesis does not occur (Sheth, 1999a; Ukstins Peate and Bryan, 2009). In fact, there is not a single unequivocal example of such uplift in a setting where a plume model would be reasonable. Can this be accounted for by adapting the Plume hypothesis? Scientific models are expected to evolve in the light of new observations and increased understanding – they are, after all, merely frameworks within which ongoing research can be organized. However, in order for a model adaptation to be a serious candidate, it has to fulfill a number of criteria, listed below.

1 Be physically reasonable.

2 Fit the observations from the melting anomaly in question consistently. For example, a small plume confined to the upper mantle is not a reasonable explanation for the absence of uplift preceding the largest flood basalts on Earth.

3 Improve the Plume hypothesis by increasing its predictive powers, including at other localities. Model adaptations are not scientifically useful if they are invoked on an *ad hoc* basis to explain observations where required, and not applied successfully to other localities.

Minor uplift would be expected for a small plume confined to the upper mantle. However, in addition to a proportionately small flood basalt being then predicted, an explanation is required for why a hot plume should rise from the base of the upper mantle, which is a mineralogical phase change and not a thermal boundary layer. A local heat source just beneath the upper mantle would be rapidly exhausted if not continually replenished. The lifetime of the surface melting

anomaly would then be short and a long time-progressive volcanic chain would not be expected. If an upper-mantle plume rose from the top of a lower-mantle plume (Courtillot et al., 2003), a structure traversing the entire mantle would then be expected to be imaged using seismology. Such structures have not been reliably detected (Chapter 5).

If lateral flow governed by an "upside-down drainage pattern" is postulated, then precursory uplift is predicted in the true source region and lithosphere basal topography should be able to explain the distribution of volcanics at all proposed plume localities. Lateral flow has been suggested for melting anomalies where perceived to be required (Vink, 1984), but a global analysis of model compliance has not yet been done. Such global compliance is also required of the prediction of complex patterns of vertical motion as a result of non-uniform continental lithosphere structure (Burov and Guillou-Frottier, 2005) but again this has not yet been investigated.

Model variants are not useful if applied only where required, as in this case they simply serve to make the hypothesis infallible and unscientific. At the same time, a great deal of work remains to be done to compare the quantitative predictions of the Plate hypothesis to the rapidly growing body of observational data that is being collected at many volcanic regions throughout the world.

2.8 Exercises for the student

1 Classify melting anomalies.

2 Model numerically the vertical motions expected to accompany continental break-up.

3 Can the elevated bathymetry at small-volume melting anomalies be explained by the buoyancy of the residuum left after melt extraction?

4 Did kilometer-scale, domal precursory uplift precede any flood basalt by a few million years, as predicted by the Plume hypothesis?

5 What density reduction occurs when partial melt is removed from eclogite?

6 Does the Caribbean contain a flood basalt province, and if so was its emplacement preceded by uplift?

7 Develop a plate model for the Emeishan basalts, China.

8 What causes the volcanism at the Canary Islands and Cape Verde?

9 Does the bathymetry of the Hawaiian swell require high temperatures?

10 Can the Cameroon volcanic line be explained by a mechanical weakness resulting from the shape of the African continent?

11 What caused the Bermuda Rise?

12 What caused uplift of the Colorado plateau?

13 What caused uplift of South America and Southern Africa?

14 What elevates oceanic swells, for example, the Cape Verde Swell?

15 Can the vertical motions in the Red Sea–Afar-Gulf of Arabia region be accounted for by continental break-up?

16 What is the timing of uplift relative to volcanism and rifting in the Afar region?

17 Can processes associated with continental break-up account for vertical motion in the North Atlantic during the early Cenozoic?

18 What vertical motions and bathymetry are expected at ridge-ridge-ridge triple junctions?

19 What are the systematics of vertical motions prior to the eruption of flood basalts?

3

Volcanism

*La verdadera ciencia enseña, sobre todo, a dudar
y ser ignorante.*[1]

Miguel de Unamuno (1864–1936)

3.1 Introduction

How does Earth make melt, how much can be made and stored, and how is it extracted from the interior and emplaced on and near the surface? The primary observation that has given rise to the Plume hypothesis is the observation of volcanism on the Earth's surface that seems to be anomalous in the context of plate tectonics. But what exactly does "anomalous" mean? It could refer to quantity, distribution, emplacement rate, or history of eruption, including the presence or absence of a time-progressive volcanic chain.

The question that immediately follows then concerns degree. For volcanism to be labeled "anomalous", how large a volume is required? How far does it have to be from a plate boundary where "normal" magmatism is expected, and what does the rate of emplacement have to be? Before these questions can be answered, it is first necessary to reflect on what is considered to be "normal".

[1] True science teaches, above all, to doubt and be ignorant.

Magmatism traditionally viewed as "normal" in a plate-tectonic context includes much that occurs at spreading plate boundaries and subduction zones. At the former, largely peridotitic mantle is conventionally thought to rise and melt to a degree of ~10–20% through isentropic decompression. This melt forms the basaltic oceanic crust. If this simple picture is correct, the thickness of the crust and the spreading rate then provide a measure of the magmatic rate.

The crust is defined, and its thickness measured, using seismology. It is typically assumed that the interface between the basaltic melt layer and underlying mantle peridotite is the Mohorovočić discontinuity, or "Moho", where the compressional seismic wave-speed increases abruptly from ~7.3 km s^{-1} to ~7.8–8.2 km s^{-1} (Mohorovičić, 1909). This assumption is a simplification. The lowermost part of the crust and the mantle immediately beneath is probably a transition interval where largely basaltic crust above grades down into largely peridotitic mantle beneath (Cannat, 1996). Where the crust is exceptionally thick, the composition of the lower part may be ambiguous (Section 3.5.2).

Crustal thickness varies extremely. At some ultra-slow spreading ridges, mantle peridotite is exposed at the surface (Dick et al., 2003). There, the production rate of basaltic magma is low. At the other extreme, the crustal thickness beneath Iceland on the Mid-Atlantic Ridge is up to ~42 km (Foulger et al., 2003). In between these

Plates vs. Plumes: A Geological Controversy, 1st edition. By Gillian R. Foulger.
Published 2010 by Blackwell Publishing Ltd.

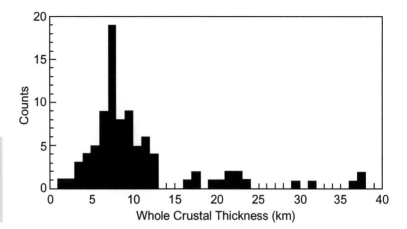

Figure 3.1 Histogram of oceanic crustal thicknesses from a compilation of 89 measurements (Mutter and Mutter, 1993).

two extremes, crustal thicknesses are variable, with the most common value being 7–8 km (Fig. 3.1).

The plates move apart at full rates that vary from ~1 cm a^{-1} at "ultra-slow" spreading ridges, for example, the Gakkel Ridge in the Arctic Ocean, to ~18 cm a^{-1} at the "ultra-fast" spreading East Pacific Rise. The minimum melt production rate along the spreading ridge is zero. At the fast-spreading East Pacific Rise, where the crustal thickness is 5–6 km and the full spreading rate is ~15 cm a^{-1}, the magmatic rate of a 100-km length of plate boundary is ~0.08 km^3 a^{-1} (Fig. 3.2). The full spreading rate at Iceland is ~2 cm a^{-1}. If the 42-km-thick seismically defined crust there (Section 5.5.1) is all basaltic, then the magmatic production rate is similar to the East Pacific Rise. This value is also similar to the production rate of the Hawaiian chain for most of its history.

At subduction zones, magmatism occurs in spectacular arcs of stratovolcanoes. These occasionally generate "super-volcano" eruptions that may produce volumes of up to ~3000 km^3 instantaneously on a geological time-scale. Such eruptions are exceedingly rare, however, and the long-term magmatic rate along arcs, including intrusions, is a relatively modest ~0.01–0.001 km^3 a^{-1} per 100 km length of arc (Fig. 3.2) (Clift and Vannucchi, 2004; Hildreth, 2007; Jicha et al., 2006). This magmatism is thought to arise from fluxing of the mantle wedge with volatiles, for example, water, which rise from the dehydrating, down-going slab. These volatiles reduce the solidus of mantle wedge material, which partially melts and rises to the surface. Arc volcanoes demonstrate that fluxing the mantle with volatiles can induce magmatism, even in a compressional regional stress field.

Volcanism near subduction zones also occurs in back-arc basins. These are thought to result from extensional stresses induced by slab rollback. They were not predicted by the plate tectonic hypothesis, but they are consistent with it and do not require awkward modifications. Back-arc basins widen by normal sea-floor spreading at rates approaching those of fast-spreading ridges, and their magmatic rates are similar. Their products are also similar compositionally, except they contain up to five times more water, probably derived from the dehydrating down-going slab.

Magmatism is also widespread throughout the interiors of plates. It is largely this activity that gave rise to the concept that a type of volcanism exists that is unrelated to plate tectonics and requires a separate theory to explain it. However, an immediate difficulty with such a view is that intraplate volcanism does not form a single generic type. It varies greatly in volume, spatial distribution and time history of eruption. It includes huge flood basalts and tiny seamounts,

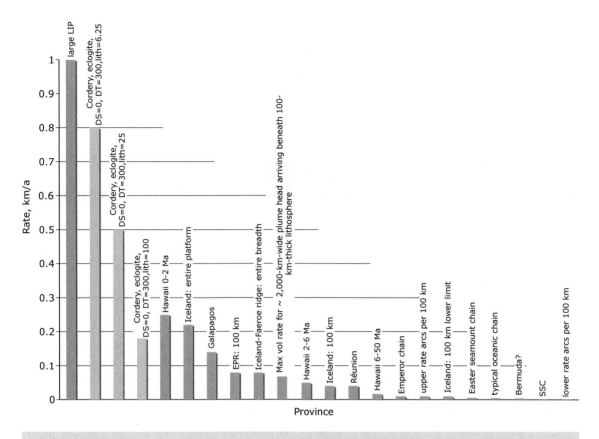

Figure 3.2 Melt production rates for some volcanic regions.

vast regions of scattered volcanism and compact, localized volcanoes, both brief and prolonged histories of volcanism, and volcanic systems with or without regularly time-progressive chains (Chapter 4).

3.1.1 Flood basalts and oceanic plateaus

The term "large igneous province" (LIP), was originally proposed for volcanic areas larger than 100,000 km^2 in size and comprising predominately tholeiitic basalt which "originated via processes other than normal seafloor spreading" (Fig. 3.3) (Coffin and Eldholm, 1994). Well-known flood basalts traditionally placed in this

category include the Columbia River Basalts, the Deccan Traps, the North Atlantic Igneous Province, the Ontong Java Plateau and the Siberian Traps (Table 3.1).

This new terminology encourages the perception that all volcanic provinces that fall into this category are emplaced rapidly and that this, along with their large volumes, makes them distinct and separate from other volcanic regions. This is a misleading simplification that is not supported by observation, as is readily appreciated from even a brief glance at the standard global LIP map (Fig. 3.3).

Magmatic provinces that fit the definition of a LIP form a continuum as regards size, emplacement rate and longevity. LIPs range from the smallest the definition allows up to the vast

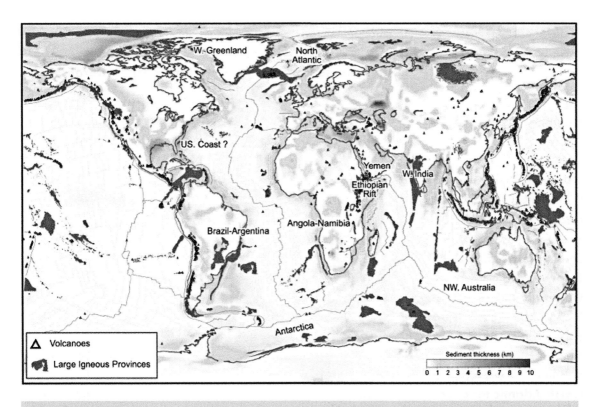

Figure 3.3 Sedimentary basins, large igneous provinces, volcanic rifts and volcanic rifted margins (from Yamasaki and Gernigon, 2009).

Siberian Traps, which cover 4,000,000 km², an area over half the size of Australia (Fig. 2.20). Both large and small volcanic provinces may be localized or diffuse and scattered across areas thousands of kilometers wide, for example, in Africa, central and east Asia, and the southwest Pacific (Bailey and Woolley, 2005; Barry et al., 2003; Moore et al., 2008; Natland and Winterer, 2005).[2] They may comprise oceanic ridges or linear chains that may be short or long. The broader the region the less clear it may be whether the whole province should be considered a single entity or not, or, indeed, what that fundamentally means.

Eruption rates range from small, for example, a modest 0.006 km³ a⁻¹ for the Easter seamount

chain, to large, exceeding 1 km³ a⁻¹ for periods of 1 Ma or more, for example, the Siberian Traps (Fig. 3.2). It is difficult to estimate volume rate. The amount of material intruded as dikes, sills and plutons is generally not known, but may account for as much as 90% of the total volume (Crisp, 1984). In addition, emplacement may span many millions of years during which the magmatic rate varies extremely. The North Atlantic Igneous Province has been erupting for ~54 Ma, but the Columbia River Basalts are thought to have been emplaced largely within ~1 Ma. Because deeper igneous layers are inaccessible to sampling, however, it may be impossible to be certain of how long activity lasted. The volume rate both between and along volcanic chains may vary by several orders of magnitude. Along the Emperor-Hawaiian system, it

[2] http://www.mantleplumes.org/SWPacific.html

Table 3.1 Volumes, areas and emplacement durations of volcanic provinces.

	Total volume $km^3 \times 10^6$	area $km^2 \times 10^6$	Duration Ma
Afar	0.35	0.5	4
Anatolia	0.015	0.043	11
Azores	1.4	0.4	20
CAMP	2.4	10	1–2
Columbia River Basalts	0.15	0.16	~1
Continental flood basalts	0.1–10	0.05/0.1–4	various
Deccan Traps	0.5	0.5	8
Eifel	2×10^{-5}	7×10^{-5}	0.2
Emeishan Basalts	0.3	0.25	?
Emperor chain	0.35	–	30
Galapagos	3.2	–	23
Hawaiian chain 6–50 Ma	0.77	–	45
Hawaiian chain 2–6 Ma	0.7	–	5
Iceland plateau	3.5	0.4	26
Karoo	3	2	8
North Atlantic Igneous Province	10	–	62
Oceanic plateaus	10–60	up to 100	unknown
Ontong Java Plateau	60–100	2	unknown
Paraná Basalts	1	1.2	11.5
Shatsky Rise	3.7	0.48	25
Siberian Traps	4	4	40
Yellowstone (last 2 Ma)	7×10^{-3} (rhyolite) 10^{-4} (basalt)	–	2

has varied by a factor of 250, from 0.001–0.25 $km^3 a^{-1}$ (Fig. 3.4). Volcanism may also be continuous or intermittent. In the Scottish Midland Valley, for example, during the Carboniferous and Permian, diffuse volcanism was intermittent but persistent at a low rate for over 70 Ma, while the plate drifted through ~15° of latitude (~1700 km).

LIPs and melting anomalies are also diverse regarding tectonic context and spatial distribution. They may occur on or near spreading plate boundaries, behind or in front of subduction zones, in continental rift valleys, for example, the East African rift, or in continental extensional zones, for example, the Basin and Range province in the western USA. They are found in regions associated with continental collision, for example, Eifel, at the sites of continental break-up, such as the North Atlantic Igneous Province, and in the interiors of plates, apparently distant from any active tectonic features.

There is no clear separation between the largest and fastest-erupted volcanic provinces and others. Volcanic provinces range continu-

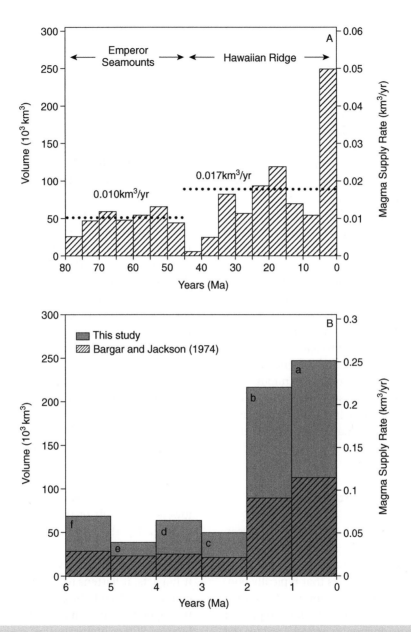

Figure 3.4 Magmatic rates for the Emperor and Hawaiian chains. (A) Histogram along the Emperor and Hawaiian volcanic chains at 5-Ma intervals (Bargar and Jackson, 1974). Dotted lines indicate long-term average rates. (B) Histogram of magmatic production of the Hawaiian Islands for 1-Ma intervals. Gray bars show revised volumes and magma supply rates for the last 6 Ma, that take into account subsidence of the Cretaceous sea floor (Robinson and Eakins, 2006). The hatched areas indicate the estimates of Bargar and Jackson (1974).

ously in size. The lack of a clear separation between major and minor volcanic regions, based on a distinct set of characteristics, recently led to a re-examination of the definition of a LIP. The term "silicic large igneous province" (SLIP) was suggested for LIPs dominated by silicic volcanic rocks (Bryan et al., 2000). However, many extensive magmatic provinces were still excluded, for example, broad areas of basalt magmatism such as Indochina, and large, layered, mafic intrusions such as the Bushveld complex (Sheth, 2007b). This led to suggestions that the minimum size for LIPs should be reduced to $50,000 \, km^2$ and a complex hierarchical classification scheme introduced involving up to 10 sub-categories based on intrusive/extrusive nature, petrology and tectonic context (Bryan and Ernst, 2008; Sheth, 2007b).[3]

The perceived need for a much more complex classification system is symptomatic of an unsuccessful attempt to pigeonhole members of a group that in reality forms a continuous spectrum. Even below the suggested $50,000 \, km^2$ threshold, the sizes of melting anomalies range smoothly down to very small seamounts (Hillier, 2007).

3.1.2 Normal or anomalous?

Where, then, can a line be drawn between "normal" and "anomalous" volcanism? It is human nature to focus on extremes rather than the more numerous members that typically populate the middle of a continuity. It is then natural to assume that the extremes represent entirely separate phenomena, and to miss the point that they are merely the end members of a continuous spectrum. Many phenomena in nature are self-similar, or governed by power laws, with numerous small members, few large ones, and a total range that spans many orders of magnitude. Earthquakes are an example, and

[3] http://www.mantleplumes.org/TopPages/ LIPClassTop.html

yet seismologists do not attribute different fundamental genesis processes to earthquakes, merely on the basis of size. Despite their vast range in size, the smallest earthquakes are linked to the largest by a basic, common causality, and volcanic provinces are likely to be the same.

3.2 Predictions of the Plume hypothesis

The primary observation that motivated the Plume hypothesis was anomalous magmatism. Everything else is secondary. Ten years after his original proposal, Morgan (1981) pointed out that some volcanic chains apparently emanate from flood basalts, for example, the Lakshadweep-Chagos-Réunion ridge and the Deccan Traps (Fig. 2.6). He therefore suggested that plumes produce an initial large burst of magmatism followed by long-lived, small-volume, activity.

This extension has been incorporated into the modern hypothesis, which currently envisages plumes as nucleating at the core-mantle boundary to form first a bulbous head. This head rises through the mantle, growing continually as it entrains some of the mantle through which it rises, and is inflated by injection of the faster-moving plume tail beneath it. Its temperature lowers as a result of the entrainment of cooler mantle. By the time it reaches the base of the lithosphere it has attained a diameter of ~500 km (Griffiths and Campbell, 1990). When it impacts, it flattens to form a disc with a diameter approximately twice that of the rising head (Campbell, 2006; Campbell and Griffiths, 1990). Large-volume, high-rate volcanism ensues over an area ~2000 km in diameter and typically lasts ~10–20 Ma (Cordery et al., 1997; Farnetani and Richards, 1994). When the head has exhausted itself, low-volume, long-lived volcanism associated with the plume tail takes over, forming a time-progressive volcanic chain as the moving surface plate transports older volcanoes away.

Can this model match quantitatively the geological observations, including the features popularly associated with flood basalts, i.e., high eruption rates and large volumes, produced on very short time-scales (Table 3.1)? This question has been explored in detail using finite-element analysis (Cordery et al., 1997; Farnetani and Richards, 1994; Leitch et al., 1997). A critical factor is lithospheric thickness. Rising, decompressing magma does not begin to melt until it reaches depths of ~100 km, and it does not melt substantially until it reaches depths of a few tens of kilometers. However, continental lithosphere is typically 100–200 km thick, and oceanic lithosphere has cooled and thickened to 50–100 km by the time it is ~100 Ma old. The thickness of the Cretaceous lithosphere near Hawaii is, for example, of this order (Fig. 3.5). There is little difficulty in simulating large eruptive rates and volumes for hot plumes impinging on the base of thin lithosphere. However, it cannot be done for large flood basalts forming over thick lithosphere.

A peridotite plume rising beneath thick continental lithosphere will produce no melt at all unless temperature anomalies exceeding 300 °C are assumed (Farnetani and Richards, 1994). There is little evidence for such temperatures, however (Chapter 6). It has been suggested that impinging plumes thermally erode the lithosphere, enabling them to rise to shallower depths (Crough, 1983), but numerical modeling has failed to simulate a significant effect of this kind (Ribe and Christensen, 1994). Furthermore, large precursory surface uplifts of 2–4 km are predicted by such hot plumes, but not observed (Chapter 2). Because of these difficulties, plumes containing eclogite have been modeled (Cordery et al., 1997). Eclogite, the high-pressure form of basalt, forms where oceanic crust is subducted to depths greater than ~60 km. It is fusible and has a solidus temperature several hundred degrees lower than peridotite. There is also geochemical evidence that it exists in the

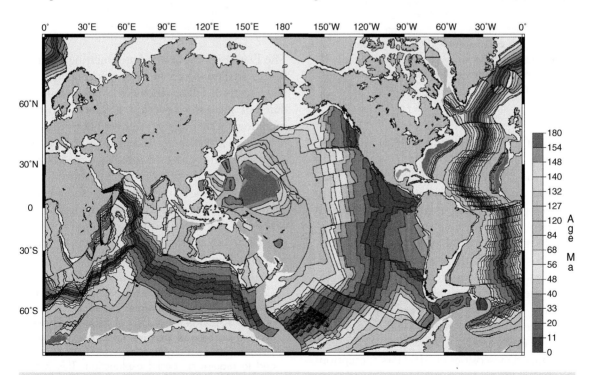

Figure 3.5 Sea-floor age based on identified magnetic anomalies and relative plate reconstructions. The age of the Cretaceous "quiet zone" is ~64–127 Ma (from Sandwell et al., 2005). See Plate 14

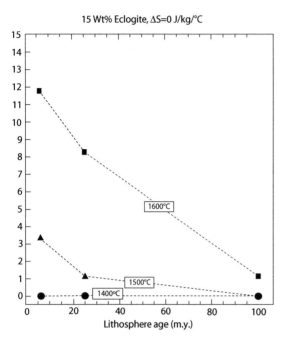

15 Wt% Eclogite, ΔS=0 J/kg/°C

Figure 3.6 Total melt volumes predicted for plumes with various temperature anomalies, rising through mantle with a temperature of 1300°C. It is assumed that 100% of the melt comes from eclogite in the plume, and that the latent heat of fusion required may be ignored (from Cordery et al., 1997).

sources of basalts erupted at melting anomalies (Chapter 7).

The most successful modeling results were obtained assuming that rising plumes contain 15% eclogite (Fig. 3.6). It was further assumed that there was no entrainment cooling, only the eclogite melted, and that it was distributed as blocks no more than a kilometer or so in thickness. Under these circumstances the blocks could thermally re-equilibrate with the surrounding mantle and thus the latent heat of fusion needed to melt them could be neglected (Cordery et al., 1997).

For mantle viscosities of $\sim 10^{21}\,Pa\,s^{-1}$, eruption rates up to several cubic kilometers per year could be simulated. However, the maximum volume that could be simulated was $12 \times 10^6\,km^3$,

for a plume head with a temperature anomaly of 300°C impacting beneath lithosphere only 50 km thick. For lithosphere 200 km thick, only $\sim 1 \times 10^6\,km^3$ of magma could be simulated. Huge flood basalts such as the Siberian Traps, which erupted on lithosphere $\sim$200 km thick, and the Ontong Java Plateau, which erupted on thin lithosphere but has the vast volume of $60–100 \times 10^6\,km^3$, could not be simulated.

The melt volumes calculated can be increased by assuming higher plume temperatures, lower plume melting temperatures and thinner lithosphere (Farnetani and Richards, 1994; Leitch et al., 1997). However, even the most extreme assumptions cannot simulate the largest melt volumes observed. It is commonly claimed that only plumes can account for the large volumes and eruption rates associated with flood basalts, but this is not so – they cannot.

3.3 Predictions of the Plate hypothesis

The Plate hypothesis attributes melting anomalies to shallow-based processes associated with plate tectonics. This includes a broad suite of different processes (Section 1.9) and a correspondingly wide range of melt volumes, eruption rates, spatial distributions and time-histories of eruption. Some processes predict large-volume flood basalts and others minor volcanism only. Activity may be either long- or short-lived, depending on the process. A current limitation of the Plate hypothesis is that rigorous quantitative modeling of the volumes of melt expected has only been done for a few cases:

■ *Continental break-up:* Massive volcanism and the formation of volcanic margins are expected to be an intrinsic part of the process of continental break-up. The absence of massive volcanism, i.e., non-volcanic passive margins, is expected to be unusual and to require a special explanation. This is the reverse of the view that it is volcanic margins that are exceptional and in need of special explanation. It is, however, consistent with the observation that worldwide, more than 50% of rifted passive margins are volcanic.

Continental break-up typically involves the separation of lithospheric blocks 100–200 km thick. As they separate, asthenosphere rises to fill the gap. Because complete break-up permits asthenosphere to rise to very shallow depths, large-scale decompression and melting occurs. The volumes and rapid eruption rates typically observed at volcanic margins, for example, in the North Atlantic, have been successfully simulated using finite-element modeling at normal asthenospheric temperatures (Fig. 1.12) (van Wijk et al., 2001; 2004).[4] This result is intuitive. Plume modeling shows that melt volumes are much enhanced if a plume can rise to shallow depths. Continental break-up modeling simply shows that if asthenosphere rises to very shallow depths, it is not necessary to have anomalously high temperatures as well.

■ *Fertility at mid-ocean ridges:* The magmatic productivity of mid-ocean ridges is related to factors that include the local plate boundary configuration, source composition including volatiles, and temperature. A more fertile source will have a lower solidus and thus yield more melt at a given temperature. This will result in thicker oceanic crust. If magmatic enhancement is major and relatively brief, the area of thickened crust formed will be restricted and may be classified as an oceanic plateau.

Normal basaltic oceanic crust is traditionally assumed to be formed largely from 10–20% partial melting of mantle peridotite upwelling beneath mid-ocean ridges. This basalt thus represents the most fusible fraction of the peridotite source and is completely molten at a temperature below the solidus of the residuum. It transforms to eclogite at ~60 km depth as it sinks in subduction zones, and it may be recycled in the shallow mantle. If it later contributes to the source of mid-ocean ridge basalts, it may melt completely before the surrounding peridotite has melted at all.

The details of this process are complicated in nature by incremental melting, extraction of melt fractions, and reaction with the host peridotite through which the melt rises. However, this does not alter the fact that productivity is increased by enhanced source fertility (Fig. 3.7). Simple calculations suggest that melt volume is at least twice as

[4] http://www.mantleplumes.org/TopPages/VMTop.html

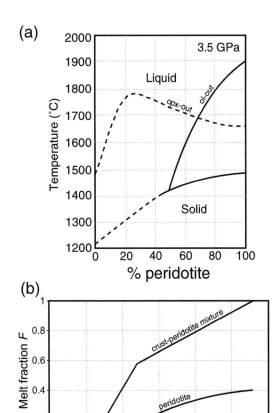

Figure 3.7 (a) Solidus and liquidus for fertile peridotite containing varying percentages of recycled oceanic crust. opx – orthopyroxene; ol – olivine (adapted from Yaxley, 2000). (b) Melt fraction F vs. temperature for fertile peridotite and a mixture of 30% oceanic crust and 70% fertile peridotite. The higher average dF/dT and lower solidus temperature of the crust-peridotite mixture results in enhanced melt productivity at a given temperature (from Foulger and Anderson, 2005).

large for eclogite upwelling. Unfortunately, precise calculations are not at present possible, because the relevant physical constants at the required temperatures and pressures are not known.

Ridges are not fixed relative to one another, and thus must migrate laterally over the underlying mantle at speeds of the same order as spreading rates. They thus sample lateral mantle compositional heterogeneities where these exist, and variations with time in the thickness of crust produced will result. Source compositional heterogeneities in the North Atlantic may be responsible for variations in the thickness of crust there (Presnall and Gudfinnsson, 2008). Upwelling at a spreading ridge of a source with fusibility enhanced by entrained eclogite fragments is at present the only model that can explain the melt volumes in the vast Ontong Java Plateau (Korenaga, 2005).

■ *Enhanced volcanism at plate boundary junctions:* Excess melt is predicted in local areas of extension associated with complexities in the plate boundary. At ridge-transform intersections, excess volcanism is expected on the outsides of the corners (Beutel, 2005).[5]

For ridge-ridge-ridge triple junctions such as the Azores, both mantle upwelling rate and temperature are predicted to increase along the slowest-spreading ridge branch as the triple junction is approached. The effect is greatest for low spreading rates, where the temperatures may rise by ~150 °C, thickening the crust by as much as 6 km (Georgen, 2008; Georgen and Lin, 2002).

In the case of more complex plate boundary configurations involving microplates, propagating ridges and diffuse oceanic plate boundaries, the excess melt production is expected to be related to the pattern of extension. The simultaneous occurrence of fertile mantle at such localities will enhance melt production still further.

Some of the most remarkable oceanic melting anomalies can be explained by such models. These include the Shatsky rise, which formed at a migrating triple junction (Sager, 2005), the Easter volcanic chain, which emanates from the southern boundary of the Easter microplate, Iceland, which is a diffuse spreading plate boundary involving two microplates, and the Bouvet triple junction.

■ *Small-scale sublithospheric convection:* The time-averaged magmatic rates predicted for small-scale sublithospheric convection (Section 1.9), which can simulate the orthogonal zones of thickened crust on either side of the East Pacific Rise, are up to ~0.002 km^3 a^{-1}. Convection vigor is dependent on

asthenosphere viscosity, density heterogeneity and temperature (Ballmer et al., 2007).[6]

■ *Oceanic lithospheric extension and cracking:* Intraplate volcanism is common without major, sustained extension. It is particularly widespread in the Pacific Ocean, where observations suggest permissive leaking of melt up from the asthenosphere via cracks that penetrate the entire lithosphere (Dana, 1849; Jackson and Shaw, 1975; Natland and Winterer, 2005). Melt is thought to pre-exist in the asthenosphere, as suggested by depressed seismic wave-speeds in the "low velocity zone" that is widespread in the Earth beneath the lithosphere (Chapter 5). An example of a region where all other models can seemingly be ruled out is in the ocean off shore of Japan. There, small-volume volcanism appears to be associated with warping of the sea floor as it approaches the Japan Trench. That volcanism can only be explained by trans-lithospheric cracks, opening in response to the warping, that permit melt to rise from the asthenosphere to form small volcanoes – so-called "petit spots" (Hirano et al., 2006).

Traces of melt are expected to exist in the mantle below ~60 km because CO_2, which is ubiquitous in the mantle, lowers the solidus of mantle rocks below the geotherm (Presnall, 2005). Phase relations suggest that the composition of this melt changes in the depth range ~60–200 km such that at shallower and deeper levels it is alkalic in composition, and at intermediate depths it is tholeiitic (Section 6.1.3). Lithosphere-traversing cracks will tap these melts, and alkalic or tholeiitic lavas will be erupted, depending on the depth from which they are extracted (Presnall, 2010).

Support for this model may be found in the sizes and compositions of Pacific volcanoes. There, the sizes of the largest seamounts are related to the age of the sea floor on which they formed. Large seamounts can form on thick lithosphere but only small seamounts form on thin lithosphere (Fig. 3.8) (Hillier, 2007; Wessel, 1997). This suggests that the maximum volume and the rate at which magma can be tapped from the asthenosphere increases with the depth to which cracks must penetrate to reach the melt. The geochemistry of lavas also varies systematically with lithosphere thickness, suggesting that different compositions are extracted from different depths (Humphreys and Niu, 2009).

[5] http://www.mantleplumes.org/TopPages/RTIntersectionsTop.html

[6] http://www.mantleplumes.org/SSC.html

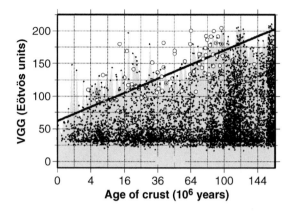

Figure 3.8 Sizes of seamounts vs. age of the sea floor on which they rest. The maximum size of seamounts is greater for older crust, and the largest tend to be the youngest, i.e., they formed after the lithosphere on which they rest had thickened (adapted from Wessel, 1997). "*Sizes and ages of seamounts using remote sensing: Implications for intraplate volcanism*", Wessel, P. **277**: 802–805, 1997. Reprinted with permission from Science.

This observation is the exact opposite of what is predicted by the Plume hypothesis. That model predicts that more melt should form the thinner the lithosphere and the shallower the melting. However, the largest volcanic edifices in the Pacific, for example, Hawaii, Tahiti and Savai'i/Upolu in Samoa, all erupted on old, thick lithosphere and the former two contain tholeiites, in contrast to the alkalic rocks that characterize most small seamounts. Such observations support the lithospheric cracking model.[7]

In the lithospheric cracking model, the volume of melt produced is not expected to be related to the amount of extension because pre-existing melt is tapped. However, the cracks are expected to be orientated normal to the direction of extensional stress. The intraplate stress field in the Pacific is expected to arise from sources such as basal plate traction, ridge-push forces from the East Pacific rise, slab pull in the subduction zones to the north, west and south and thermal contraction of the cooling plates (Forsyth et al., 2006; Lithgow-Bertelloni and Guynn, 2004; Sandwell and Fialko, 2004; Stuart et al., 2007). The latter may be the most important

effect. Globally, it is estimated to be comparable in total magnitude to an unrecognized slow-spreading plate boundary, and to largely account for the failure of plate motion circuits to close (Gordon et al., 2005a, b). Loading of the lithosphere by adjacent volcanoes is also important locally (Hieronymus and Bercovici, 1999; 2000).

■ *Shallow mantle convection:* If EDGE convection[8] (Fig. 1.12) onsets when continents break up, it is expected to be most vigorous at its onset and melt production is expected to dwindle with time. This has been confirmed by two-dimensional finite-element simulations. The magmatic volumes commonly observed at volcanic margins can be reproduced, but the time history of magmatism is sensitive to mantle viscosity. An increase in mantle viscosity along with melt extraction is necessary in order to simulate the observed reduction in magmatism with time, as volcanic margin formation gives way to formation of oceanic crust with normal thicknesses. Such a viscosity increase might come about through dehydration (Boutilier and Keen, 1999).

■ *Continental intraplate extension:* At continental rifts, the magmatic rate is expected to be related to extension rate, as decompression upwelling of the mantle will permit rocks to melt. Exact spatial and temporal correspondence between rifting and volcanism is not expected, however. Melt will rise where extension is greatest and this may be different at depth from at the surface – the pattern of extension of the lower crust and mantle lithosphere may be different from the upper crust. Depth-dependent stretching, for example, may result in more extension at depth than at the surface (Kusznir and Karner, 2007).

Normal faults induce extension at depth in their footwalls, even though at the surface those regions are topographically high and contain minor compressional features (Ellis and King, 1991). Thus, volcanism is expected some kilometers or tens of kilometers from the surface traces of normal faults bounding continental rifts rather than within the rifts themselves. This is commonly observed. For example Mt Kilimanjaro lies on the flank of the East African Rift, and volcanism occurs outside the Baikal rift in Siberia.

Continents also extend by diking. Dikes intruded shortly before eruption of the Columbia River

[7] http://www.mantleplumes.org/TopPages/SamoaTop.html

[8] http://www.mantleplumes.org/EDGE.html

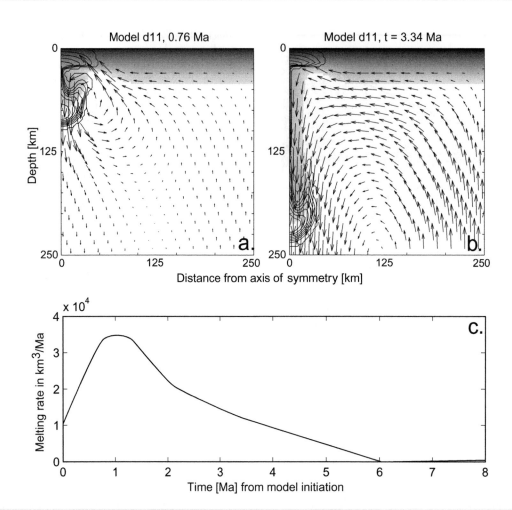

Figure 3.9 Mantle flow field and melt volumes calculated to accompany formation of a small gravitational instability (from Elkins-Tanton, 2005).

Basalts, for example, had a total integrated width of 17 km.

■ *Catastrophic lithospheric thinning:* Melt formation associated with catastrophic loss of the lithosphere by gravitational instability[9] (Figs 1.20 and 1.21) has been modeled numerically (Fig. 3.9) (Elkins-Tanton, 2005; Elkins-Tanton and Hager, 2000). Removal of the mantle lithosphere can induce the magmatic volumes and eruption rates observed for small flood basalts with volumes of 50,000–100,000 km³ in the presence of normal mantle temperatures. Larger flood basalts, with volumes of the order of one or more million cubic kilometers, require higher mantle temperatures. The largest magmatic rates are predicted to occur early, as the instability develops and falls. High rates last for 1–2 Ma, and the entire magmatic episode is predicted to last for 5–10 Ma.

■ *Sublithospheric melt ponding and draining:* This model can potentially explain the rapid emplacement of large flood basalts over thick lithosphere, for example, cratons, where lithospheric thickness may exceed 200 km. Such events are known to occur where there is no evidence for major lithospheric extension or thinning, and they apparently require pre-existing melt. The asthenosphere imme-

[9] http://www.mantleplumes.org/LithGravInstab.html

diately below the lithosphere is envisaged to be above its solidus. Melt accumulates over a long period of time, perhaps even hundreds of millions of years. Melting may be aided by the fluxing of the asthenosphere with volatiles, water or CO_2, which lowers the solidus of the asthenosphere below the geotherm, but not that of the less fertile, more refractory lithosphere. Fracturing traversing the entire lithosphere, perhaps associated with rifting during an orogen, then provides a pathway for the melt to drain to the surface over a much shorter time period than it took to accumulate (Silver et al., 2006).

The strength of this model is that it does not require the melt to be formed on the same short time-scale as flood basalt eruption. It is thus free of a major problem that besets all other models for flood basalt formation – that of accounting for how melt can be produced as quickly as it is erupted.

3.4 Comparison of the predictions of the Plate and Plume hypotheses

The predictions of the Plate and Plume hypotheses regarding the volumes, pattern and time-history of volcanism differ in three fundamental ways. First, the Plume hypothesis predicts a large degree of uniformity among melting anomalies whereas the Plate hypothesis, involving as it does many different sub-processes, predicts extreme diversity. The Plume hypothesis requires that essentially all melting anomalies start with a flood basalt (the plume head), followed by persistent, small-volume volcanism from a relatively localized focus that forms a time-progressive volcano chain on relatively moving plates (the plume tail). The Plate hypothesis does not require such a head-tail sequence of volcanism. On the contrary, this is expected to be unusual, as is observed to be the case (Table 1.4). Some plate processes are expected to produce flood basalts, for example, continental break-up and sinking of gravitational instabilities. Others, for example, permissive volcanism at complex spreading plate boundaries, and intraplate extension, permit but do not require large-volume magmatism. The melt volumes are expected to range from

extremely large down to very small. Depending on the process, a time-progressive volcanic chain may or may not be expected (Chapter 4). A wide range of longevities of volcanism are anticipated. Persistent, ongoing small-volume magmatism may follow formation of a flood basalt, but it is not required, not necessarily expected to erupt from a restricted focus, and it is expected to be unusual.

The second contrast concerns the temporal relationship between melt production and eruption. The Plume hypothesis assumes a "melt as erupted" scenario – the mantle is assumed to be subsolidus, and melt forms in a rising plume on a similar time-scale to that on which it is erupted. It is this feature of the Plume hypothesis that means that high temperatures and shallow depths of plume penetration are needed to simulate the largest and most rapid eruptions.

The Plate hypothesis assumes that in some cases the majority of the melt extracted pre-existed in the mantle. In these cases pre-existing melt, either ponded at the base of the lithosphere or distributed throughout the asthenosphere, is drained when a crack or conduit opens a pathway traversing the entire lithosphere. The existence of ponded or interstitial melt may be controlled by solidus-lowering volatiles such as water and CO_2 and the mere provision of a pathway to the surface leads to volcanism. In some cases, millions of cubic kilometers of melt are extracted – sufficient to form flood basalts.

The third fundamental difference between the Plate and Plume hypotheses relates to the distribution of melting anomalies on the surface of the Earth. Plumes are predicted to impinge randomly on the Earth's surface. In contrast, the Plate hypothesis predicts that clear associations between extensional tectonic features and anomalous volcanism will be common. Of the melting anomalies on the list of "classic hot spots" of Courtillot et al. (2003; Table 1.4), 15 out of the total 49 (30%) lie on or close to ridges. In addition, many occur at unusually complex parts of the ridge, for example, the Azores, Bouvet and Afar triple junctions, Iceland, and the Easter microplate.

3.5 Observations

3.5.1 Classifying melting anomalies

Subdividing melting anomalies into discrete groups for the purpose of structured discussion is problematic because they display a huge range in almost every aspect. In the following discussion, they will be subdivided according to their volume, longevity, and proximity to spreading ridges. Such subdivisions must necessarily be loose because of their continuous variation in these characteristics.

3.5.2 Large-volume, sustained volcanism

(a) On-ridge volcanism

Iceland, Afar and the Azores are examples of relatively large-volume, long-lived melting anomalies currently centered on spreading ridges. Of these, the North Atlantic Igneous Province is by far the largest in every respect. It has a total volume of some $5–10 \times 10^6 \, km^3$, and has been active for ~62 Ma. Its current expression is Iceland, an enigmatic ridge-centered melting anomaly which, despite many decades of study, still poses many bewildering and unanswered questions.

The earliest volcanism in the North Atlantic region was distributed in a northwest-trending zone stretching from Britain to west Greenland (Fig. 2.14). This zone has been called the "Thulean volcanic line". Magmatism lasted for ~3 Ma during which time relatively modest volumes were erupted roughly synchronously in Britain and Greenland (Fig. 2.16). Following a hiatus of 2–5 Ma, volcanism restarted at ~56 Ma

in a zone trending northeast, cross-cutting the earlier trend approximately orthogonally, and running along the line of the frontal thrust of the 400-Ma-old Caledonian suture. This second phase of volcanism accompanied the onset of continental break-up. A total of several million cubic kilometers of excess magma were produced, with the magmatic rate gradually decreasing over a period of ~6 Ma (Holbrook et al., 2001). This activity built the volcanic margins that flank the present-day North Atlantic Ocean (Fig. 3.10).

This complex space-time pattern fits a plume head model poorly and has inspired many suggested plume variants (Section 3.6). Not only is the distribution of volcanism and its two-phase emplacement style unexpected, but so is the absence of volcanism at the predicted plume center. Spreading in the Labrador Sea occurred in the period ~61–54 Ma, when the proposed plume is predicted to have underlain western Greenland (Fig. 3.10), and yet volcanic margins and thickened oceanic crust did not develop there. Meanwhile, magmatism occurred over 2000 km away, as diking in the North Sea.

A model that fits the observations more naturally attributes the volcanism to the complex history of break-up of the North Atlantic. Prior to ~54 Ma, sea-floor spreading occurred in the Labrador Sea and Baffin Bay. At ~62 Ma, changes in the regional stress field encouraged a new rift to form sub-parallel to the Labrador Sea rift and ~500 km to the northeast – the Thulean rift. Such a rift would have formed a straight extension to the Baffin Bay rift, bypassing the Davis Strait transform (Fig. 2.15) (Dewey and Windley, 1988; Lundin and Doré, 2005).[10] Evidence for such an extensional rift includes northwest-trending graben structures, dikes, fissures and fjords in Britain, the Faeroe Islands and east and west Greenland (Lundin and Doré, 2005). The nascent Thulean continental rift did not develop to full continental break-up, but gave way to the later orthogonal rifting that culminated in the opening of the North Atlantic.

[10] http://www.mantleplumes.org/Iceland2.html

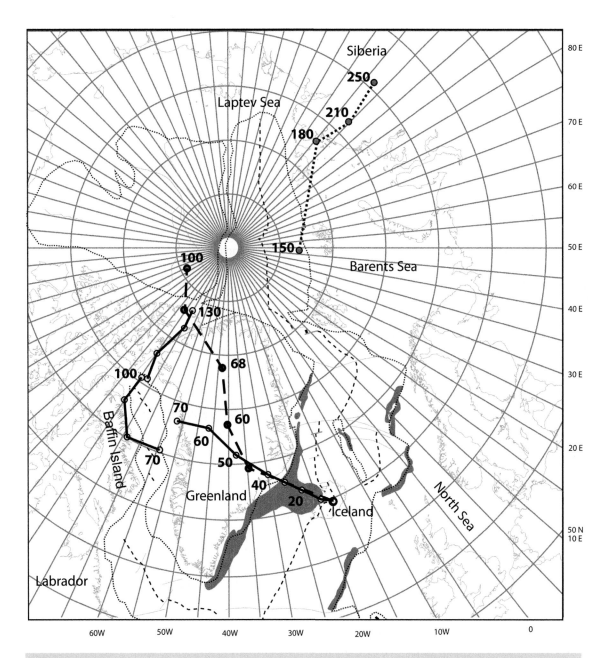

Figure 3.10 Tracks calculated for an hypothetical Icelandic plume assumed to be fixed relative to others postulated to lie on the African, Indian, North and South American and Australian plates (dots, labeled in Ma). Dashed track 0–100 Ma from Forsyth et al. (1986); solid tracks 0–70 Ma and 70–130 Ma from Lawver and Muller (1994); dotted track 150–250 Ma from Lawver et al. (2002); thin dashed lines are spreading ridges. Gray regions are seaward-dipping reflector sequences. (adapted from Lundin and Doré, 2005).

This final change in tectonics coincided with an abrupt change in deformation style in Europe. At this time there was a pause in convergence between Europe and Africa, which resulted in compressional tectonics being replaced by stress-relaxation features. Numerical modeling of these stress changes has shown that they are consistent with continental break-up occurring in response (Nielsen et al., 2007). Passive decompression upwelling accompanying continental break-up can explain the melt volumes, assuming normal mantle temperatures (Fig. 1.12) (van Wijk et al., 2004).

Following continental break-up, the bathymetric feature known as the Icelandic transverse ridge developed. It currently extends from east Greenland, through Iceland to the Faeroe islands (Fig. 2.14). It is likely that one part or another of this ridge was subaerially exposed for much, if not all, of the time since the Atlantic opened, since palaeosols have been drilled on the Iceland-Faeroe ridge (Nilsen, 1978). Ironically, this may have been essentially the only place where a "land bridge" existed, of the sort rejected by Wegener (1915) as a viable explanation for the simultaneous existence of old species on continents now separated by wide oceans.

Formation of the transverse ridge took place in two phases. First, the Greenland-Iceland and Iceland-Faeroe ridges were built by volcanism along a section of Mid-Atlantic Ridge ~200 km in north-south extent. At ~25 Ma, this expanded to ~600 km and the much wider Icelandic platform, including Iceland itself, was built.

How much excess melt is produced on this part of the Mid-Atlantic Ridge? Despite the fact that present-day Iceland happens to protrude above sea level, it actually represents rather a minor topographic anomaly in a regional context. It is the tip of a vast bathymetric swell ~3000 km in north-south extent (Fig. 2.14). The island itself stands, in general, only ~1 km higher than a smooth continuation of the swell. Such a small topographic excess could be supported isostatically by just a few kilometers of extra crustal thickness. It is also conceivable that some of the topographic anomaly results from the low-density tuffs and hyaloclastites that make up much of the near-surface volcanics in Iceland. These have formed as a result of the perpetual subaerial and subglacial eruption environment.

This line of reasoning ignores the seismological evidence, however. The crust underlying the Greenland-Iceland-Faeroes transverse ridge has been extensively investigated for over four decades using both earthquakes and explosion seismology (Section 5.5.1; Bott and Gunnarsson, 1980; Darbyshire et al., 1998; Flovenz, 1980; Foulger et al., 2003; Palmason, 1971). The layer with crust-like seismic wave speeds is ~30 km thick beneath the Greenland-Iceland and Iceland-Faeroe ridges, but is locally as much as 42 km beneath Iceland (Fig. 3.11). Elsewhere in the North Atlantic, the oceanic crust typically has a thickness of only 5–10 km (Foulger and Anderson, 2005).

The enigma is that it is unclear exactly how much of the excess crustal thickness detected using seismology corresponds to excess melt. It is commonly assumed that the oceanic crust corresponds essentially entirely to melt. However, this ignores the low topography of Iceland. The thick crust cannot be entirely made up of rocks with basalt- or gabbro-like densities, because then Iceland would rise to over 4 km above sea level instead of the ~1 km elevation of most of its broad dome (Gudmundsson, 2003; Menke, 1999). Furthermore, mineralogical phase relations require that Icelandic lavas are extracted from the depth range ~15–45 km (~0.5–1.3 GPa) (Sections 6.2.2 and 7.1), which is largely within the seismically defined crust itself. This serious problem is usually ignored, but that approach runs the risk that all subsequent reasoning that depends on it may be wrong.

Some of the excess thickness must be attributable to a submerged microplate, which is required by the kinematics and evolution of Iceland (Fig. 3.12) (Foulger, 2006; Foulger and Anderson, 2005). Palinspastic reconstructions require that a block of older crust ~150 km in east-west extent underlies Iceland, submerged beneath older lavas. This crust may be wholly oceanic and resemble the Easter microplate, or it may be

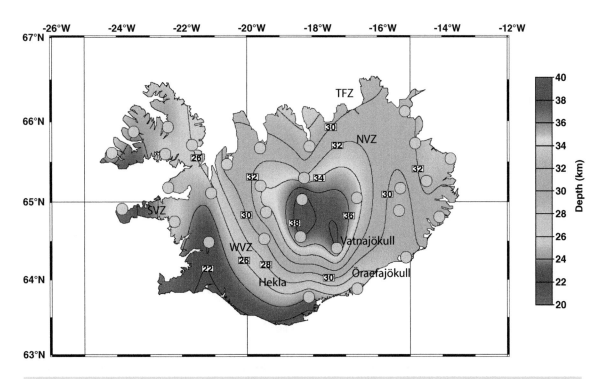

Figure 3.11 Thickness of the layer with crust-like seismic wave speeds, defined as $V_S \leq 4.1\,\mathrm{km\,s^{-1}}$. It is commonly assumed that this approximately indicates crustal thickness, but density considerations are incompatible with the entire layer corresponding to basaltic melt (Section 5.5.1). WVZ, NVZ, SVZ – Western, Northern and Snaefellsnes Volcanic Zones; TFZ – Tjörnes Fracture Zone (from Foulger et al., 2003).

wholly or partly continental, a southerly extension of the Jan Mayen microcontinent that lies just northeast of the Icelandic plateau. There is no correlation between crustal thickness and distance from the volcanic zones – the crust is not systematically thinner beneath the active volcanic zones compared with neighboring regions (Darbyshire et al., 2000; Foulger et al., 2003). Furthermore, there is no band of thick crust traversing Iceland from west to east as might be expected if excess melt had been produced at a point source beneath central Iceland for a long time (Fig. 3.11).

Some of the excess seismic crustal thickness may well represent excess magmatism. The spreading plate boundary at Iceland is diffuse. Since at least ~15 Ma and possibly much longer, spreading has persistently occurred along an unstable, parallel pair of spreading ridges with one or more deforming microplates between them (Fig. 3.12) (Foulger and Anderson, 2005).[11] The ridges propagate longitudinally, jump laterally to both the east and west, and the plate boundaries in north and south Iceland evolve independently. On top of this, the lava dome of Iceland seems to be collapsing outward, as suggested by a radial component in the directions of surface motion measured in Iceland using GPS (Foulger and Hofton, 1998; Perlt et al., 2008). Under such circumstances an inhomogenous stress field, including areas of enhanced extension permitting excess volcanism, is inevitable. The dome collapse in particular might account for the culmination of volcanism in south-central Iceland, beneath the icecap Vatnajökull.

[11] http://www.mantleplumes.org/Iceland1.html

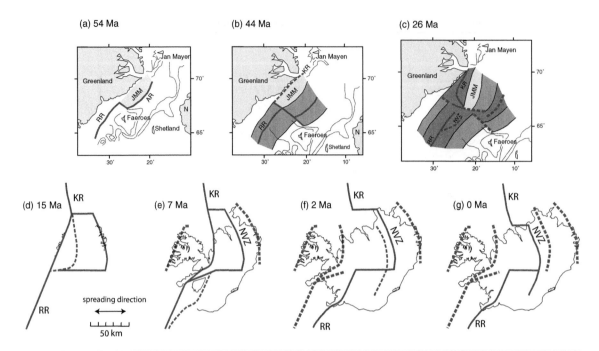

Figure 3.12 Tectonic evolution of the Iceland region during the past 54 Ma. Solid lines – currently active plate boundaries; dashed lines – imminent and extinct plate boundaries; thin lines – bathymetric contours; JMM – Jan Mayen microcontinent; KR – Kolbeinsey Ridge; N – Norway; NVZ – Northern Volcanic Zone; RR – Reykjanes Ridge, AR – Aegir Ridge (adapted from Foulger et al., 2005a).

Why is mid-ocean ridge magmatism so complex at the latitude of Iceland? The Icelandic transverse ridge formed where the new Mid-Atlantic Ridge crossed the westernmost frontal thrust of the Caledonian suture. It also coincided with the Thulean volcanic line (Fig. 2.15) and at this latitude the trend of the Mid-Atlantic Ridge abruptly changes by ~35° (Fig. 2.14). This may have resulted in complications in tectonics at this locality from the onset (Nunns, 1983). The geochemistry of Icelandic lavas is consistent with a source partly made up of recycled, fusible subducted slab material of Caledonian age. These slabs might be late-subducted Iapetus slabs trapped in the Caledonian suture. Such a source in itself would yield larger melt volumes than common mantle peridotite at the same temperature (Foulger and Anderson, 2005).

Exactly how much excess magmatism occurs at Iceland? This question cannot be answered with confidence. The topography of Iceland requires no more than a few kilometers at most, which would suggest an excess magmatic rate of no more than ~$0.01\,km^3\,a^{-1}$ per 100 km of ridge. On the other hand, if the whole layer with crust-like seismic wave speeds corresponds to melt, then for a 550-km-long section of Mid-Atlantic Ridge, a total excess magmatic rate is implied of ~$0.22\,km^3\,a^{-1}$, or $0.04\,km^3\,a^{-1}$ per 100 km of ridge. When the Iceland-Faeroe ridge, which is only ~200 km wide, was built, the total excess magmatic rate would have been ~$0.08\,km^3\,a^{-1}$.

A question of no lesser interest is where the volcanism occurred through time. The entire Icelandic transverse ridge was formed by magmatism persistently centered at the Mid-Atlantic

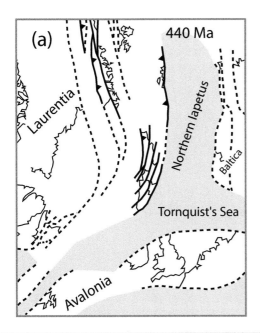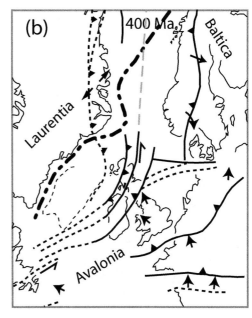

Figure 3.13 Caledonian collision zone associated with the closure of the Iapetus ocean at (a) 440 and (b) 400 Ma, formed by convergence of Laurentia, Baltica and Avalonia. Arrows – convergence directions; thick lines – faults and orogenic fronts. Black triangles indicate sense of thrust faults. Gray dashed line – inferred position of the Caledonian suture. Slabs were subducted beneath Greenland, Baltica and Britain. Bold dashed line indicates position of MAR that formed at ~54 Ma (from Foulger and Anderson, 2005).

Ridge – volcanism did not migrate systematically east with respect to the ridge as is often claimed (Section 4.5.1). Time progressions in the volcanism that built the Icelandic transverse ridge are essentially the same as across any other expanse of oceanic crust, i.e., systematic ageing away from the spreading ridge where it was formed. A time-progressive volcanic trail, such as that which adorns Hawaii, is clearly absent at Iceland.

In summary, the Greenland-Iceland-Faeroes ridge, including Iceland itself, formed at a section of the mid-ocean ridge that is diffuse, lies at the outer boundary of the Caledonian suture, and has persistently apparently produced excess melt. The geochemistry of Icelandic rocks indicates that recycled subducted Caledonian slab material contributes to the source (Chapter 7; Breddam, 2002; Chauvel and Hemond, 2000;

Korenaga and Kelemen, 2000; McKenzie et al., 2004). These facts suggest that the Iceland melting anomaly results from complex tectonics coupled with a source made more fusible by the inclusion of recycled, subducted near-surface materials.

An unusual feature of the North Atlantic is the chevron bathymetric ridges that adorn the Reykjanes ridge between Iceland and the Charlie Gibbs fracture zone at ~53 °N (Fig. 3.14) (Vogt, 1971). They are a few hundred meters high, and most clearly visible in the gravity field. Such ridges are also weakly developed to the east of the Kolbeinsey ridge north of Iceland (Jones et al., 2002). These unusual ridges probably represent variations in crustal thickness of ~2 km. They formed at intervals of 5–6 Ma, and their chevron shapes are a consequence of the locus of enhanced melt extraction on the

Figure 3.14 Satellite gravity field of the North Atlantic with the long-wavelength component corresponding to plate cooling removed. The chevron ridges are clearly seen. JMM – Jan Mayen Microcontinent (from Jones et al., 2002). See Plate 15

Mid-Atlantic Ridge propagating south (or north) at ~20–25 cm a^{-1}.

The chevron ridges have popularly been attributed to fluctuations in a postulated plume stem beneath Iceland that channels melt along the Reykjanes and Kolbeinsey ridges. Several scenarios have been suggested, including different plume geometries, either vertical or horizontal flow, and pulses of either temperature, composition or flux (e.g., Fitton and Hardarson, 2008; Hardarson et al., 1997; Ito, 2001; White et al., 1995).

Such models can be adjusted to fit the amplitudes and propagation rates of the enhanced melt extraction loci. However, they require the *ad hoc* assumption of quasi-regular plume pulses of some sort, which lacks independent confirmation. They also imply a sphere of influence for what is supposedly a 60-Ma-old (at least) plume stem that rivals that of the largest postulated plume heads and which is without parallel among other ridge-centered melting anomalies. Lastly, pulsing Plume hypotheses fail to explain three remarkable features. First, the region of sea floor occupied by chevron ridges is broader west of the Reykjanes ridge (up to ~300 km) than east of it (up to ~100 km). Second, it is triangular in shape, ~400 km wide close to Iceland, and it pinches out where the Reykjanes ridge changes direction at ~56 °N. Third, the point south of which the chevron ridges do not extend represents no more than a minor (~28 °) change in strike of the plate boundary. It is not clear how such a feature could be a serious barrier to the flow of plume material.

These features are predicted, however, if the chevron ridges result from minor enhancement of volcanism at the tips of small-offset rift propagator systems that repeatedly migrate south along the Reykjanes ridge (Hey et al., 2007). That process predicts a wider expanse of crust on one side of the spreading ridge than on the other, because it transfers lithosphere from one plate to the other (Fig. 3.15) (Hey et al., 1989a,

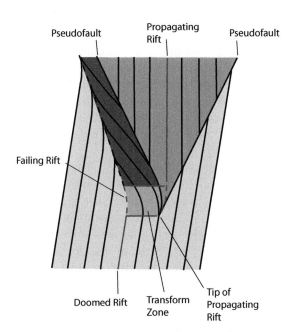

Figure 3.15 Idealized oceanic propagator system. Mid-gray – crust formed at propagating rift; light gray – crust formed at dying rift; dark gray – lithosphere transferred from one plate to the other (Hey et al., 1989a, b).

b). It also predicts the triangular shape of the sea floor occupied by chevron ridges. This develops because the boundary between the crust formed at the propagating and dying spreading segments (the "pseudofault") lies at a high angle to the spreading ridge as a whole. The reason why chevron ridges do not form south of ~56 °N, where the Reykjanes ridge changes from being oblique to the spreading direction to being more nearly parallel to it, then lies in a change in ridge tectonic and spreading style. Evidence for such propagators has recently been found by bathymetric and gravity surveying (Hey et al., 2007).

(b) Near-ridge volcanism

Similar features can be seen in the sea-floor bathymetry on either side of the Cocos-Nazca spreading axis in the Pacific Ocean (Fig. 3.16). There, several ridge propagators strongly influence the local morphology. Eastward and westward families of propagators are separated by a ~100-km-long transform fault known as the 91 ° transform (Hey et al., 1992). The most northerly of the Galapagos islands lies just 25 km south of this transform.[12]

The entire Galapagos region is highly complex tectonically. Oceanic crust formed at the East Pacific Rise north of its junction with the Cocos-Nazca spreading axis subducts beneath Central America in a direction of N 40 °E. Crust formed at the East Pacific Rise south of the junction subducts beneath South America in an easterly direction. The resulting 50 ° difference in plate motion direction has required a triangular swathe of ocean floor to form in between, and this has been produced at the Cocos-Nazca spreading axis (Hey et al., 1972). This swathe of crust is ~1000 km in north-south extent at its eastern end, progressively narrowing to the west to pinch out at a triple junction at the East Pacific Rise.

[12] http://www.mantleplumes.org/TopPages/GalapagosTop.html

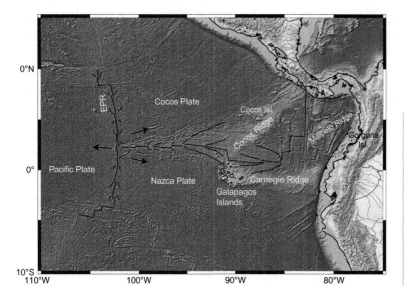

Figure 3.16 Seafloor bathymetry in the region of the Cocos-Nazca spreading axis and the Galapagos islands, with the main plate boundary features (from Hey et al., 1992) superimposed. The 91° transform lies immediately north of the Galapagos Islands (basemap from Smith and Sandwell, 1997). See Plate 16

The Cocos-Nazca spreading axis has been persistently unstable and complex throughout its ~23-Ma history. The difference in subduction direction between the Cocos and Nazca plates requires the spreading axis to migrate geographically north if spreading were symmetric. However, the axis has repeatedly jumped south, resulting in asymmetry of the sea floor formed about it (Wilson and Hey, 1995). A pair of bathymetrically shallow ridges, the Cocos and Carnegie ridges, emanate from the 91° transform in the central part of the Cocos-Nazca spreading axis. This situation has much in common with that at Iceland.

The Carnegie and Cocos ridges rise to ~2 km above the surrounding sea floor, considerably higher than the ~1-km-high Iceland-Faeroe Ridge. The amount of excess melt produced can be estimated from seismic measurements of crustal thickness. This is greatest beneath the oldest, easternmost parts of the ridges, where it is ~19 km thick. It reduces to ~15 km beneath the Galapagos platform, and to ~13 km beneath the western Carnegie Ridge. Assuming all the excess crustal thickness represents melt, the magma production rate over and above that required to produce normal oceanic crust, is ~0.14 km^{3}a^{-1} (Sallares and Charvis, 2003). The present most active locus of volcanism is at the Galapagos islands, some distance from the spreading axis. This is a recent situation that has probably existed for only ~5 Ma, however. For the last ~20 Ma, excess melt extraction has predominately occurred at the spreading axis itself, near the 91° transform, and along the Cocos and Carnegie ridges.

The Galapagos islands have a land area of ~8000 km^2 and form the westernmost part of the Carnegie Ridge. Despite their great volcanic activity, seismic measurements of crustal thickness suggest that magmatic rates were higher in the past, early in the life of the Cocos and Carnegie ridges. It is postulated that a plume underlies the Galapagos islands and channels melt to the nearby Cocos-Nazca spreading axis, enabling the Cocos and Carnegie ridges to form, and that these ridges are time-progressive (Geist et al., 1995; White and McBirney, 1993).

Precise dating has shown that, on the contrary, the Cocos and Carnegie ridges are not simply time-progressive (O'Connor et al.,

2007).[13] Volcanism has persistently been widespread and long-lived throughout the region. High-quality $^{40}Ar/^{39}Ar$ radiometric ages show that in the last ~3 Ma almost the entire ~1000-km length of the Cocos Ridge and 600 km of the Carnegie Ridge have been volcanically active. Some distal parts of these ridges have been volcanically active for ~10 Ma of their ~15 Ma lifetime. Other islands, for example, the 24 km^2 Cocos island, are maintained above sea level by active volcanism.

A model involving dispersal of melt from a localized, plume-fed source (Harpp and Geist, 2002) fits the observations poorly. Such a model requires that melt flows laterally and preferentially leaks from the 91° transform, not from the nearby ridge. Melt is also required to flow along the Carnegie and Cocos ridges, to fuel ongoing volcanism along them. In the Plume hypothesis, the location of the Galapagos islands immediately to the south of the 91° transform is a coincidence. Also a coincidence is the location of the Carnegie Ridge along the mega-pseudofault boundary between crust formed at the Cocos-Nazca spreading axis and the East Pacific Rise.

The observations are suggestive of widespread leakage of melt to the surface throughout a broad region of chronically unstable and evolving tectonics. The divergent directions of motion of the Cocos and Nazca plates impose extension throughout the region. The unstable Cocos-Nazca spreading axis, with its propagator systems, lateral ridge jumps, and extreme along-axis variability must induce a heterogeneous stress field. The 91° transform not only separates westward-from eastward-migrating families of propagating systems, but it strikes at 15° oblique to the direction of spreading about the Cocos-Nazca spreading axis and is thus in transtension (Harpp and Geist, 2002). A continuation of it bisects the Galapagos archipelago, so the western and eastern islands are underlain by lithosphere of differing thickness. The more active and voluminous western islands are underlain by the thicker lithosphere and the petrology and volcanic morphology differ between the western and eastern volcanoes. Northwest- and northeast-trending volcanic alignments are similar to the propagator pseudofault orientations and to volcanic lineaments associated with transtension at the 91° transform.

Several processes may be operating to encourage enhanced volcanism in a region of such complexity as this. Extensional stress is predicted to occur at plate boundary junctions, including ridge-transform intersections (Beutel, 2005). The Galapagos islands lie in such a locality. Magmatism there is expected to be enhanced by the transtension of the 91° transform. This same process is also likely to be responsible for the building of the Cocos and Carnegie ridges. Further down these ridges, persistent volcanism testifies that stress field complexities extend into the plates for up to ~500 km from the spreading axis. Small-scale sublithospheric convection, which can explain minor volcanism near to ridges (Ballmer et al., 2007), may also contribute to this distributed magmatism. Comprehensive modeling of the stress field of the entire region is needed to test whether regions predicted to be in extension are those where active magmatism is observed to occur.

(c) Off-ridge volcanism

Hawaii is the most remarkable currently erupting melting anomaly anywhere on Earth that is not associated with an active spreading ridge.[14] There are major challenges to studying it, however. Not only do most of the Emperor and Hawaiian volcanic chains lie beneath deep ocean, but they are also emplaced on sea floor that formed during the "Cretaceous quiet zone", a ~32-Ma-period when the Earth's magnetic field did not reverse (Fig. 3.5). Thus, any relationships to variations in age of the oceanic crust

[13] http://www.mantleplumes.org/GalapagosDating.html

[14] http://www.mantleplumes.org/TopPages/HawaiiTop.html

on which they are emplaced are elusive. The islands that rise above sea level, the southern part of the Hawaiian chain, are small in extent, limiting experiments. The largest, the "Big Island", is only ~100 km in diameter, less than a quarter the width of Iceland. Most of the Big Island is blanketed by very young basalt flows, rendering much of the volcanic edifices out of reach of direct sampling. In addition there are few high-quality petrological samples from the submarine parts of the chain. Dredged samples are altered, they may also have experienced sub-aerial weathering before they subsided beneath sea level, and little drilling has been done.

There is no evidence that flood volcanism occurred when the Emperor seamount chain began to form, but there is evidence that the chain started life as enhanced volcanism at a ridge-ridge-ridge triple junction (Norton, 2007).[15] The oldest seamount of the Emperor chain, Meiji, has an age of ~82 Ma and forms part of the Obruchev Rise. It has been questioned whether Meiji truly forms part of the Emperor chain, raising the question of what exactly is meant by "the Emperor chain".

The magmatic volume rate varied extremely as the chains formed. It was on average ~0.01 km³ a⁻¹ when the Emperor chain formed (Fig. 3.4). After the "bend" at ~50 Ma, the rate fluctuated greatly. For the first ~10 Ma it was only ~0.001 km³ a⁻¹, but subsequently it increased and averaged ~0.017 km³ a⁻¹. In the last ~2 Ma it suddenly surged by an order of magnitude. Its present unprecedented rate of 0.25 km³ a⁻¹ (Fig. 3.4) (Robinson and Eakins, 2006) has constructed a volcanic plateau so big that is technically a LIP.

The spatial distribution of volcanism is remarkable for its exceptional narrowness. In this, it contrasts with its two sister oceanic tholeiitic melting anomalies, Iceland and the Galapagos. Almost all the volcanism in the Emperor and Hawaiian chains occurred within a narrow corridor often considerably less than

100 km wide. Volcanism has occurred on the surrounding sea floor but much of this pre-dates the oldest Emperor seamount.

The trajectories of the chains have varied with time. To begin with, the orientation of the Emperor chain was about N 135 °E. This changed at ~60 Ma to approximately north-south, and at the famous bend it changed from ~N 170 °E to ~N 120 °E. It became more easterly between ~30–20 Ma, but during the last ~1 Ma it has hooked around strongly to the south. Loihi, popularly considered to be the newly forming future volcanic center, is directly south of Kilauea, currently the most active volcanic center.

Hawaii is the type locality of the Plume hypothesis (Morgan, 1971; Wilson, 1963). The key observations that inspired the hypothesis include the perceived regular time progression along the chain, the match with the present northward rate of motion of the Pacific sea floor (i.e., the belief that the locus of melt extraction was fixed relative to the geomagnetic pole), the high volcanic rate, and the restricted area of active volcanism. However, later, detailed work has tended to undermine these perceptions rather than strengthen them. The time progression is now known to have varied by more than a factor of three, and there is no correlation between the time progression of the Emperor chain and the rate of motion of the Pacific sea floor when it was emplaced (Section 4.5.4). The current high volcanic rate is a recent surge without precedent. Only the unidirectional age progression and the restricted area of active volcanism remain unchallenged. The latter is, however, a relatively unusual feature among proposed plume locations.

Plume models have difficulty in explaining large volumes of melt erupted rapidly through thick lithosphere. The volumes currently being erupted at Hawaii, supposedly from a ~100-Ma-old plume tail, can only be simulated by models that involve either thinning of the lithosphere, for example, by thermal rejuvenation (Crough, 1983), raising the plume temperature, or lowering the source solidus. Such models require inde-

---

[15] http://www.mantleplumes.org/Hawaii2.html

pendent observational support, which has not tended to be forthcoming. Variation in the volume rate with time can only be explained on an *ad hoc* basis.

Although the Emperor chain may have begun life at an oceanic triple junction, most of it, and also the Hawaiian chain, formed in the interior of the ocean far from any plate boundary. Present-day Hawaii is almost in the geometric center of the Pacific plate, the largest of all the Earth's tectonic plates and almost as far from any plate boundary as it is possible to get anywhere on Earth. Of the possible melt-forming processes discussed in Section 3.3, those dependent on proximity to a plate boundary, or which predict only ephemeral volcanism, therefore do not apply. Only oceanic lithospheric extension and cracking is a candidate. Such an hypothesis was essentially proposed first for Hawaii as long ago as 1849 by James Dwight Dana, who suggested that the islands formed over a "great fissure" caused by thermal contraction cracking of the Earth's surface (Dana, 1849).

There is qualitative support for the lithosphere cracking model from several observations. The direction of extensional stress predicted by modeling of thermal and other stresses in the Pacific plate is normal to the Hawaiian chain (Fig. 3.17) (Lithgow-Bertelloni and Guynn, 2004; Stuart et al., 2007). However, there is little that is particularly anomalous about the calculated stress field in the exact neighborhood of Hawaii, so its precise location, with its curious geometric relationships to features of the distant Pacific plate boundaries remain a puzzle. It has been suggested that volcanism is aligned on a pair of *en echelon* trends separated by ~65 km (Jackson and Shaw, 1975), which would suggest shearing (Fig. 3.18). The veracity of such systematic patterns in the array of Emperor and Hawaiian volcanoes remains unconfirmed, however.

Regarding melt volume, Hawaii fits the model of lithospheric cracking qualitatively. It erupts on Cretaceous lithosphere ~50–100 km thick, and volume has tended to increase as the melt extraction locus has migrated into thicker lithosphere, the reverse of what is predicted by the Plume hypothesis. The recent surge in melt productivity onset when volcanic activity migrated from thinner to thicker lithosphere, across the Molokai fracture zone (Fig. 2.12).

Petrological observations suggest that the melt is extracted from the asthenosphere under this thick lithosphere (Sections 6.1.3 and 7.6; Presnall, 2005). The compositions of the earliest melts erupted at new volcanoes are alkalic, then changing to tholeiitic, and back to alkalic in late stages of eruption. Mineralogical phase relationships require that the interval of melt extraction spans a range of pressure conditions centered at ~5 GPa (~165 km) (Presnall et al., 2010). The early alkalic melts are thought to be extracted from high in the asthenosphere, and the tholeiites from depths down to ~110 km. Nanodiamonds are found in the late-stage alkalic lavas, indicating a depth of origin of those lavas of >180 km. Such a sequence of compositions is consistent with pre-existing melt being extracted from progressively deeper parts of the asthenosphere as each volcano matures.

How large is Hawaii compared with other intraplate oceanic volcanoes? Is it a completely unique phenomenon, or merely the largest member in its class? To answer this it must be compared with chains and seamounts that formed well away from spreading ridges. This is a minority of Pacific seamounts. Most erupted on young sea floor, i.e., at or near to the ridge where the sea floor formed (Hillier, 2007).

A full data base of the volumes of Pacific volcanoes is not yet available (Hillier, 2007). Two other edifices of comparable size are Tahiti, forming on lithosphere aged ~70 Ma, and the Savai'i/Upolu volcano pair of Samoa, forming on similarly old lithosphere. Both of these have volumes of several tens of thousands of cubic kilometers and may be of the order of half the volume of the Big Island of Hawaii. The Big Island is a composite of five volcanoes, whereas Tahiti and Savai'i/Upolu only involve three and two volcanoes, respectively. This, and the

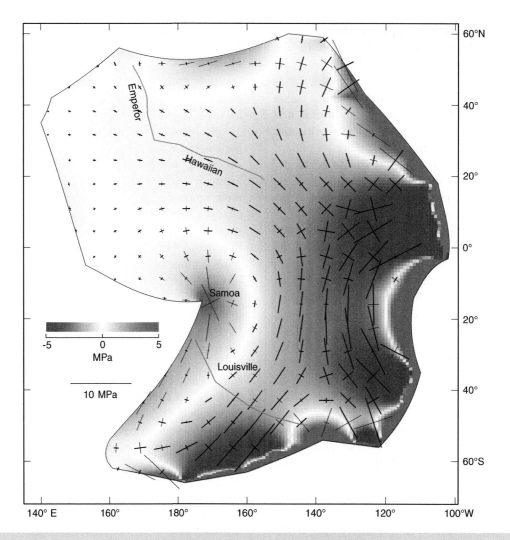

Figure 3.17 Horizontal stress field for the Pacific plate from cooling and thermal contraction. Horizontal principle stresses are shown by orthogonal lines – thick for compression and thin for tension. Shading shows tensional and compressional horizontal mean normal stress (from Stuart et al., 2007). See Plate 17

statistics of sizes of Pacific seamounts (Fig. 3.8), suggest that the Big Island is an end member in a continuum of large volcanic edifices in the Pacific, and owes its volume to a systematic process, rather than being a separate and distinct phenomenon. Its volume may be explained by scaling up of the same process that forms "petit spots" (Hirano et al., 2006) and smaller seamounts, i.e., lithospheric cracking.

3.5.3 Large-volume, brief volcanism

Only exceptionally is large-volume magmatism followed by ongoing magmatism lasting many tens of millions of years. The stereotypical model of the very rapidly emplaced flood basalt followed by a small-volume, time-progressive chain is rare and exceptional.

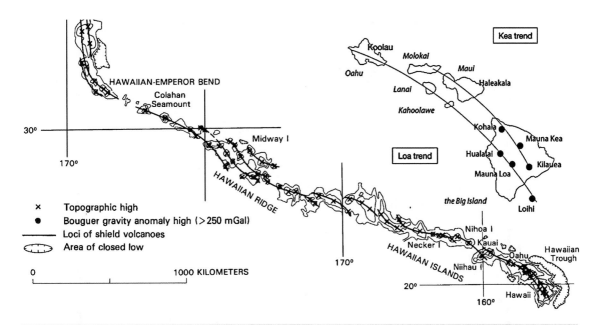

Figure 3.18 Speculative separation of Hawaiian volcanoes into two, *en echelon* trends (from Jackson et al., 1972). Detail of postulated Kea and Loa trends is from Ren et al. (2005).

Flood basalts erupt in many different tectonic contexts, including on or near to spreading ridges, in the back-arc regions behind subduction zones, and in the deep interiors of plates. The length of time over which they are emplaced varies widely, as do their total volumes and spatial patterns of magma distribution.

(a) On-ridge volcanism

The Central Atlantic Magmatic Province (CAMP)[16] erupted in association with continental break-up (McHone et al., 2005). It formed in the early Jurassic, at ~201 ± 2 Ma and may have been almost entirely emplaced during a brief period lasting only 1–2 Ma (McHone, 2000). It is spread over an area ~10 × 10⁶ km² in

size, elongated for ~6000 km along the continental margin, and extends over four continents – Europe, Africa and North and South America (Fig. 3.19).

Its volume is conservatively estimated to be ~2.4 × 10⁶ km³, making it one of the largest continental flood basalt provinces on Earth. However, much of this huge volume comprises volcanic margins up to 2000 km in length along the coast of North America and North Africa. Many of the flows in the continental interiors are thin.

The CAMP volcanic province fits the Plume hypothesis poorly. There is no evidence for precursory domal uplift around a magmatic center, and no evidence for plume tail volcanism. Although active volcanism has occurred in the ocean subsequent to flood volcanism, for example, the Canary Islands, the Cape Verde Islands, and the Cameroon line, none onset immediately after CAMP flood volcanism. The New England seamount chain has been

[16] http://www.mantleplumes.org/TopPages/CAMPTop.html

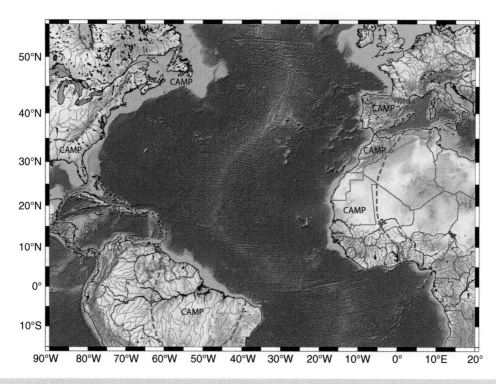

Figure 3.19 The extent of the Central Atlantic Magmatic Province (CAMP)[17]. See Plate 18

suggested as the most likely candidate, but that volcanism cannot have started earlier than ~120 Ma, at least 80 Ma after CAMP flood volcanism ceased (McHone, 1996).

The most likely explanation for CAMP volcanism is that, like volcanism in the North Atlantic, it resulted from the continental break-up process itself. Rifting was ongoing in the region for ~25 Ma before CAMP volcanism started, and the central Atlantic opened immediately afterwards. Geological observations on land call for widespread heterogeneous sources and strong eruptive control by reactivation of pre-existing lithospheric structures (Beutel, 2009; Lundin and Doré, 2005).[18] The volcanic margins formed where continental break-up

occurred along re-activated sutures. These may have provided fusible material for cycling through the melt zone, boosting the voluminous melting predicted even for normal mantle rising between rifting continents (van Wijk et al., 2001). A similar model has been proposed for the volcanism in Paraguay, including the Paraná basalts. Volcanism there was not limited to a single pulse but occurred sporadically throughout the period ~241–58 Ma. Its geochemistry cannot be explained by an asthenospheric source but is attributed to remelted continental mantle lithosphere, erupted in response to differential motions between coherent blocks in the South American plate (Section 7.6.3; (Comin-Chiaramonti et al., 2007; Gallagher and Hawkesworth, 1992; Peate et al., 1992). The Eastern Snake River Plain-Yellowstone volcanic chain is also forming at a re-activating

[17] http://www.mantleplumes.org/CAMP.html
[18] http://www.mantleplumes.org/PangaeaDikes.html

boundary between two distinct lithospheric blocks (Christiansen et al., 2002).

(b) Near-ridge volcanism

The Ontong Java Plateau is the largest flood basalt ever known to have erupted on Earth.[19] It covers an area of $\sim 2 \times 10^6 \, km^2$, some 0.4% of the surface of the Earth, and has an estimated volume of $50 \times 10^6 \, km^3$ (Fig. 2.5). Palinspastic reconstructions suggest that it may even have formed as part of a plateau as large as $100 \times 10^6 \, km^3$, that was later fragmented to form also the Hikuranga and Manihiki plateaus (Taylor, 2006). These enormous volume estimates are based on sparse seismic refraction experiments, which suggest typical crustal thicknesses of $\sim 30 \, km$, i.e., $\sim 24 \, km$ of thickness in excess of normal sea floor. As exemplified for the case of Iceland, however, volume estimates that assume the full extent of such thick layers with crustal seismic wave speeds may be suspect (Section 5.5.1).

If the Ontong Java, Hikuranga and Manihiki plateaus originally comprised a single feature, it would have been some $3000 \times 1000 \, km$ in size, extending over sea floor created in the period ~ 160–$120 \, Ma$. Such a plateau would thus have been emplaced on crust that was, at the time, ~ 0–$40 \, Ma$ in age. It is not known whether it erupted from a ridge, like Iceland, or near to one. Clearly, this information could be critical to understanding what process formed the plateau, but few data are available to help. Magnetic reversals have not been mapped, and only a few radiometric dates are available from samples retrieved from six shallow boreholes drilled up to $\sim 200 \, m$ into basaltic basement. That this makes the Ontong Java Plateau the best-sampled of all oceanic plateaus gives an idea of the paucity of data available from such features.

[19] http://www.mantleplumes.org/TopPages/OJTop.html

The radiometric dates are all close to $122 \, Ma$, with a possible spatially restricted resurgence of volcanism at $\sim 90 \, Ma$ (Fitton, et al., 2004). The few data available are thus consistent with this vast plateau having mostly erupted very quickly, and this is usually what is assumed. However, the data are absurdly few and no information regarding age is available for the deeper parts. On the Ontong Java Plateau itself, three or four regions, each the size of Iceland, are devoid of any age estimates whatsoever.

The plateau has no time-progressive volcanic chain emanating from it. It has been suggested that the Louisville chain represents plume tail volcanism following on from plume head magmatism that built the plateau itself (e.g., Table 1.4). However, this association is inconsistent with plate motion models and geochemical evidence. These factors, in addition to the absence of the predicted precursory uplift and post-emplacement subsidence (Section 2.5.5) provide a poor fit to the Plume hypothesis. That model cannot, in any case, explain the extraordinarily large magma volumes that apparently formed (Section 3.2) (Fig. 3.6).

The only model currently proposed that can explain the volume of the Ontong Java Plateau, and is consistent with the geochemistry, subdued vertical motions and lack of a time-progressive volcanic chain, is enhanced fertility near a mid-ocean ridge (Anderson et al., 1992; Korenaga, 2005; Smith and Lewis, 1999). The plateau formed at or near to a ridge, possibly a ridge-ridge-ridge triple junction, which was spreading at the super-fast rate of $\sim 15 \, cm \, a^{-1}$. Eclogite blocks from earlier-recycled subducted crust, may have been entrained in the upwelling mantle source. The volumes can be explained if $\sim 25\%$ of the upwelling material was eclogite blocks that melted almost completely. In total, an expanse of subducted ocean crust $\sim 2400 \times 2400 \, km$ in area would have to have been recycled and re-melted. Although eclogite is dense, small blocks approximately the same thickness as oceanic crust, i.e., ~ 7–$8 \, km$, can be entrained without sinking. A temperature decrease of ~100 K is predicted as a result of removal of the required latent heat of fusion (Fig. 1.13; Korenaga, 2005).

(c) Off-ridge volcanism

The continental counterpart to the Ontong Java Plateau is the Siberian Traps[20] – the largest volcanic event ever known to have occurred on land. The Siberian Traps contrast with the Ontong Java Plateau in that they did not erupt on or near to a spreading ridge that might provide a mechanism for upwelling and melt generation. They erupted in the interior of a continent, much of it Archean cratonic lithosphere that may have been up to 300 km thick (Section 2.5.5, Fig. 2.20).

The area and volume of the Traps are difficult to estimate precisely because much has eroded away or been submerged by later deposits in the 225–265 Ma that they have lain on the surface of the Earth. Some probably also lie beneath the Kara Sea in the north. They presently cover a land area of ~7 × 10^6 km^2 that is up to ~3500 km wide. They are thought to have originally had a volume of ~4 × 10^6 km^3 and to have been up to 6.5 km thick. This clearly contrasts with the CAMP province, for example, where intracontinental flows may have been only a few hundred meters thick.

The Siberian Traps comprise two large lobes with the thickest volcanics in the eastern lobe. They are not associated with any time-progressive volcanic chain. It is commonly assumed that they erupted very rapidly, perhaps in no more than ~1 Ma at the Permian/Triassic boundary. While the largest pulse of volcanism may have been brief, a wide range of radiometric dates, palaeomagnetic data, and geochemical similarities point to several pulses extending throughout a period of up to ~40 Ma. As for the Ontong Java Plateau, our knowledge is still incomplete. Huge regions have not been dated precisely and current knowledge is based on just a few dates from a relatively restricted area (Ivanov, 2007).

The precursory subsidence beneath the thickest part of the Traps (Section 2.5.5) and the absence of an associated time-progressive volcanic chain is strong evidence against a plume origin. In addition, a plume model cannot explain the melt volumes. The most voluminous melts erupted through Archaean cratonic lithosphere so thick that a plume arriving at its base would have produced no melt, regardless of its composition (Fig. 3.6) (Cordery et al., 1997; Davies, 1999), and numerical modeling of thermal lithospheric erosion has shown that plumes cannot erode the cold lower lithosphere, even with exceptionally high temperatures (e.g., Farnetani and Richards, 1994).

A model involving catastrophic lithospheric loss fits the observations better. This process can account for the precursory subsidence (Fig. 2.4), the absence of a time-progressive volcanic chain and, critically, the melt volumes (Elkins-Tanton, 2005; 2006; 2007; Elkins-Tanton and Hager, 2000).[21] In this model, the lower lithosphere transforms to dense eclogite and sinks. As it descends, asthenosphere rises to take its place, decompresses and melts (Fig. 1.21). A pulse of magmatism of the order of a million cubic kilometers or more in volume can be generated in a time as short as ~1–2 Ma. A second pulse of magmatism is predicted to occur as the sinking lithospheric instability dehydrates and fluxes the surrounding mantle with volatiles, lowering its solidus and enhancing melt production. The total duration of melt production is several million years. Hydration of the mantle may also contribute to initiating the instability. Such hydration might have resulted from the dewatering of slabs subducted in the ocean trenches that are thought to have surrounded the region prior to Trap eruption (Ivanov, 2007; Ivanov et al., 2008).

[20] http://www.mantleplumes.org/TopPages/
SiberiaTop.html

[21] http://www.mantleplumes.org/LithGravInstab.html

(d) Cratonic flood basalts

Both the thermal Plume hypothesis and the catastrophic lithospheric loss models assume that the melt erupted in flood basalts is formed on the same time-scale as it erupts. In order for this to occur, the lithosphere must be thinned to allow asthenosphere to rise to sufficiently shallow depths for the necessary decompression to occur. However, there are some continental flood basalts where lithospheric thinning apparently did not occur.

In southern Africa, four large flood basalts erupted onto cratonic lithosphere in the period 2.71–1.88 Ga, the Ventersdorp, Great Dike, Bushveld and Soutpansberg (Barton and Pretorius, 1997; Cawthorn et al., 1991; Marsh et al., 1992; Nelson et al., 1992). Each erupted synchronously with a major orogen, apparently in association with formation of a collisional rift. Each is elongated to exploit the lithospheric anisotropy produced by the related orogeny, i.e., there was strong lithospheric control of magma emplacement.

All these flood basalts apparently erupted without lithospheric thinning, either by extension or lithosphere removal. Diamonds, erupted through kimberlite pipes in the flood basalts, and dated using their silicate and sulfide inclusions, have ages that pre-date the oldest flood basalts. Diamonds form in the lower lithosphere, and thus they show that it was still intact after the flood basalts had erupted.

The timing of the flood-basalt emplacements, all synchronous with orogens, renders a plume explanation unlikely. In addition, the absence of major lithosphere thinning renders the plume, delamination and gravitational instability models implausible. Under such circumstances it seems that rupturing of the entire lithosphere and draining of pre-formed sublithospheric melt reservoirs is required (Fig. 1.22); (Silver et al., 2006).

For such a mechanism, long-term supersolidus conditions are required. A relatively high geotherm might exist as a result of thermal blanketing by overlying continents, long-term convective upwelling, or fluxing of the upper mantle by volatiles released from subducted slabs. The latter has been suggested to produce mantle melting leading to volcanism in east Asia (Ivanov, 2007; Zorin et al., 2006). Supporting evidence for this has been provided by seismic tomography, which shows structure beneath Changbai volcano in east Asia to be restricted to shallow depth (Zhao et al., 2009).[22] Similar models to have been proposed for the several flood basalts associated with the break-up of Gondwana (Sears, 2007).[23] Rapid drainage events were then facilitated by orogenic rupturing of the entire lithosphere providing a pathway for the melt. Such a model implies the existence of large, long-lived reservoirs of magma beneath thick continental lithosphere, features which should then exist at the present day and be detectable seismically.

3.5.4 Small-volume, sustained volcanism

If the largest-volume melting anomalies can be explained without appealing to anomalously high temperatures, there should be no problem doing the same for smaller volume anomalies. It is clear that small volumes can leak up through lithospheric fractures (Hirano et al., 2006). The only thing in need of explanation is, then, why fractures form where they do.

(a) On- and near-ridge volcanism

The Easter seamount chain has been built by small-volume, sustained magmatism from an expanse of oceanic sea floor stretching from the Easter microplate to the island of Sala y Gomez, approximately 700 km further east (Fig. 2.10). Sea floor spreading is almost the fastest anywhere

[22] http://www.mantleplumes.org/Changbai.html
[23] http://www.mantleplumes.org/EarthTess.html

on Earth along this part of the East Pacific Rise, which has a full spreading rate of ~15 cm a^{-1}.

Excess volcanism has built an elongate ridge ~3000 km long running easterly across the Nazca plate. The excess volcanic rate over the last ~10 Ma has been ~0.006 km^3 a^{-1} (Rappaport et al., 1997), equivalent to a thickening of the oceanic crust beneath the chain by ~4%, or 250 m (Fig. 3.2). Morphologically, the ridge comprises several *en echelon* chains ~200–500 km long and up to ~80 km wide, trending at 5–15° oblique to the overall ridge trend (Kruse et al., 1997; Rappaport et al., 1997). It is mirrored on the Pacific plate to the west of the East Pacific Rise by the little-studied Crough Island chain.

The Easter seamount chain is relatively well dated. It is age-progressive, ages to the east, and generally formed on the east flank of the East Pacific Rise by eruptions on 0–10 Ma crust. Very few dates are available for the Crough Island chain, so it is not known whether that chain is also age-progressive or not. A plume model has been proposed for the Easter seamount chain, both before and after it was found to be age progressive. Nevertheless, it lacks an initial flood basalt and its close spatial relationship with major structures associated with the rotating Easter microplate is accorded the uncomfortable status of coincidence within that model.

The chain is well explained by excess volcanism at a tectonically complex part of the ridge where the spreading axis is displaced laterally by ~300–400 km. This part of the ridge forms the southern edge of the rotating Easter microplate (Schilling et al., 1985b; Searle et al., 1995). It performs a similar function to a transform fault, though it is considerably more complex and includes overlapping ridge segments. Because it formed in this region, the entire seamount chain now lies along a fracture zone in the Nazca plate. Other models have been proposed, including diffuse extension, possibly as a result of the concave embayment in the Chile trench into which the chain is subducting (e.g., Sandwell et al., 1995; Winterer and Sandwell, 1987). Age dating evidence for diffuse volcanism is required

to support those models, but this has not yet been found.

The Easter seamount chain is one of the larger of many seamount chains that extend at right-angles out from the East Pacific Rise. In addition, numerous other linear features exist in the Pacific, ranging from substantial island chains to subtle corrugations of the sea floor on the flanks of the East Pacific Rise. Several of the larger island chains have been shown by age dating to be not time progressive but to have erupted along their entire lengths for prolonged periods, for example, the Cook-Austral and the Line islands chains (McNutt et al., 1997). They bear virtually no resemblance to what is predicted by the Plume hypothesis. The corrugations that flank the East Pacific Rise are small-scale, subtle features that are only revealed by high-precision satellite bathymetry. They can only plausibly be explained by thermal or stress-related mechanisms associated with the adjacent ridge. Small-scale sublithospheric convection has been modeled numerically and can successfully reproduce the irregular patterns of eruption and melt volumes of these volcanic systems (e.g., Ballmer et al., 2007; 2009).

(b) Off-ridge volcanism

Relatively small-volume, persistent melting anomalies forming distant from spreading ridges are common. Some of the most notable examples include Samoa, the Cameroon line, the Canary islands, the Cape Verde islands, the European Cenozoic Volcanic Province, Hoggar, Reunion, Tristan da Cunha and Yellowstone.

European Cenozoic volcanism is sparse, small-volume, and distributed throughout a broad region extending from the Massif Central in France, through the Rhenish Massif in central Germany, the Bohemian Massif in the Czech Republic and Lower Silesia in south-west Poland (Fig. 3.20) (Lustrino and Carminati, 2007). Volcanism has been ongoing for ~77 Ma (Abratis et al., 2007), and now covers a total area of ~20,000 km^2. It typically comprises small,

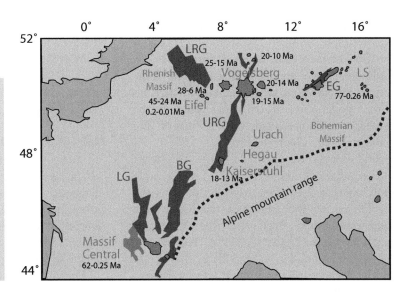

Figure 3.20 Map of the Cenozoic volcanic rocks of central Europe and rift-related sedimentary basins in the Alpine foreland: LRG – Lower Rhine Graben; URG – Upper Rhine Graben; EG – Eger Graben, BG – Bresse Graben, LG – Limagne Graben; LS – Lower Silesia. Numbers indicate eruption age data from Abratis et al. (2007) and Lustrino and Wilson (2007).[24]

monogenetic centers (e.g., Eifel), scattered necks, plugs and dikes, and a few central volcano complexes. The Vogelsberg is the largest of these, but even this covers no more than ~2500 km².

The region has commonly been explained using a mantle plume model, despite the fact that it fulfills none of the predictions at all (Granet et al., 1995; Ritter et al., 2002). Precursory domal uplift and flood basalt volcanism did not occur, there is no evidence for high magmatic temperatures, and no seismic evidence for a plume-tail structure extending to the core-mantle boundary. Volcanism has always been scattered throughout the region, with no regular time progression (Abratis et al., 2007). Plume variants include dividing the province into sub-regions, each proposed to be underlain by a small plume, for example, beneath the Eifel area. That model fits no better. The Eifel volcanic field covers an area of only 600 km² – <1% of what is expected for plume-head-related flood volcanism, and activity has persisted at the same place for 45 Ma, with no time progression and a long period of quiescence. Seismic observations do

not reveal underlying structure consistent with either high temperatures or a conduit-like feature extending down into the lower mantle (Section 5.5.1).

European Cenozoic volcanism is clearly influenced by Alpine convergence further south.[24] Deformation of the European hinterland clearly shows that the stress field associated with Alpine collision is felt at distances up to ~2000 km from the collision front (Dezes et al., 2004; Chalot-Prat and Girbacea, 2000). Volcanic activity in the sub-regions has been associated with active rift systems and sedimentary basins, which are themselves spatially and temporally correlated with events in the Alpine orogeny (Ziegler and Dèzes, 2007). Some of the volcanic fields lie within the rift systems, and others are on uplifted basement massifs adjacent to the rifts, as predicted by stress-field modeling (Ellis and King, 1991).

3.5.5 Small-volume, brief volcanism

(a) On-ridge volcanism

In some cases relatively brief volcanic episodes, producing trivial volumes of excess magma, have

[24] http://www.mantleplumes.org/Europe.html; http://www.mantleplumes.org/Carpathians.html

been proposed to require a deep mantle plume. An example close to a spreading ridge is the Bouvet melting anomaly (Morgan, 1971). Although Bouvet appears on few modern plume lists and fulfills none of the primary predictions of that model, it is still commonly assumed to be caused by a plume, largely on the basis of geochemistry (e.g., Georgen, 2008).

The Bouvet region lies on the southwest Indian ridge, ~300 km east of the Bouvet triple junction (Fig. 2.7). This triple junction is the confluence of the Mid-Atlantic Ridge, the southwest Indian ridge and the American-Antarctic ridge (Ligi et al., 1999). Tectonically the region is complex and rapidly evolving, and includes propagating and dying ridge segments, pseudofaults, a sheared microplate with internal bookshelf tectonics, *en echelon* overlapping ridge segments and possibly has a history of ridge jumps.

The Bouvet melting anomaly itself comprises a transform-fault-bounded section of the southwest Indian ridge ~100–200 km long that is in general ~1 km shallower than neighboring ridge segments. It peaks at an elevation of ~800 m above sea level at Bouvet Island, an uninhabited, volcanically active island ~50 km^2 in area. The crustal thickness, and thus the quantity of excess melt production, is not known in detail but the relatively minor bathymetric anomaly suggests that it cannot be large. No time-progressive volcanic trails emerge from the ridge, and excess melt production has been short lived. The section of shallow ridge is mirrored by a similar region on the American-Antarctic ridge, the western branch of the triple junction.

The Bouvet melting anomaly and its counterpart on the American-Antarctic ridge are well-explained by lateral flow of asthenosphere toward the triple junction, which by itself can easily explain the observations. The proximity of ridge-transform intersections, along with enhanced extension induced by the complex tectonics of the triple junction, may encourage magmatism still further. Geochemistry has shown that Precambrian garnet-bearing rocks,

possibly from recycled lower crust, contribute to the lavas, and may enhance the fusibility of the source (Kamenetsky et al., 2001).

(b) Off-ridge volcanism

Away from spreading ridges, Bermuda provides an intriguing example of relatively brief, small-volume volcanism.[25] While it could simply be dismissed as permissive volcanism through a lithospheric crack, it is associated with several enigmatic features for which no comfortable explanation seems to be currently available (Section 2.5.6; Fig. 2.23) (Vogt and Jung, 2007).

The volcanic construct at Bermuda comprises four major volcanoes aligned in a northeast-trending array. They originally rose to heights of perhaps 3 km above sea level, but are now eroded down and truncated a little below sea level. Their total extrusive volume, estimated from their bathymetric morphology, is roughly 20,000 km^3. They are sampled only by shallow boreholes and thus their entire lifetime of eruption is not well known, but if they formed over ~5 Ma, the average eruption rate would have been ~0.004 km^3 a^{-1}.

The volcanoes form an isolated cluster in the middle of a broad, regional, bathymetric swell. Their nearest neighbor is the Muir seamount, some 250–300 km to the northeast. Muir seamount may be as old as the Cretaceous – much older than the ~30–35 Ma Bermuda volcanoes. Still further north are the New England seamounts. With ages of ~100–70 Ma, they are also too old to be associated with Bermuda volcanism (Duncan, 1984).

Bermuda cannot be explained by a plume model. Fixed hotspot reference frames predict a time-progressive volcanic chain aligned at right-angles to the Bermuda volcanic ridge and extending westward onto North America. Selected volcanic features there, dating back as far as 115 Ma, have been proposed as compo-

[25] http://www.mantleplumes.org/Bermuda.html

nents of such a chain, but such proposals lack plausibility (Vogt and Jung, 2007). If the swell on which Bermuda rests results from plume-head-related uplift, then the absence of large-volume magmatism, i.e., an oceanic plateau, is unexplained, along with the absence of time-progressive volcanism to the east during the 47–40 Ma subsequent to swell formation.

There is no obvious explanation for the collective Bermuda phenomenon that does not appeal to *ad hoc* events. It seems inescapable that the volcanism must be related to the bathymetric swell in some way, but why was there a time delay of some 10 Ma between formation of the swell and volcanism? Although the volumes erupted are small compared, for example, with Hawaii, nevertheless a substantial volcanic archipelago, with the largest island ~50 km long and possibly several kilometers high, must once have existed. In the absence of any obvious related tectonic event, one is left concluding that the volcanism must be attributed to lithospheric cracking releasing pre-existing melt from the asthenosphere. However, why such cracking should occur at this particular time and place is unclear. Shallow-mantle convection processes such as EDGE and "eclogite engine" have been suggested, but independent tests of these for Bermuda have yet to be made (Anderson, 2007a; King and Anderson, 1995; 1998).

3.6 Plume variants

Mismatches between observations and the predictions of the Plume hypothesis, along with clear associations of volcanism with shallow structures, have inspired variants of the basic model.[26] Unexpected eruption rates have been attributed to plume pulsing, for example, at Hawaii and Iceland, where late-stage or multiple flood basalt eruptions occurred (O'Connor

[26] http://www.mantleplumes.org/ DefinitionOfAPlume.html

et al., 2000). The chevron ridges about the Reykjanes ridge have been attributed to temperature or chemical pulses in a plume beneath Iceland (Fitton and Hardarson, 2008; Jones et al., 2002; White et al., 1995). It is not explained, however, why these are not accompanied by bursts of volcanism close to the postulated plume center, nor why such pulsing is not common at other proposed plume localities.

Unexpected distributions of volcanism, for example, in the region of scattered volcanism southeast of the Azores, have been attributed to complex plume structures such as multi-headed-, or numerous closely spaced plumes. It has been suggested that many "plumelets" arise from the heads of the African and Pacific "superplumes" (Fig. 1.2; Courtillot et al., 2003). European volcanic sites such as Eifel have been attributed to "baby plumes" rising from a large plume head stalled beneath the transition zone under Europe. In this respect, plumes are undergoing the same evolution as large igneous provinces, and geosynclines before them, where increased knowledge is revealing the inadequacy of a simple model and a complex classification scheme involving many sub-categories is replacing it.

Lateral flow from distant plumes has also been suggested to account for unexpected spatial distributions of eruptives. The Iceland-Faeroe ridge is postulated to have formed via ~700 km of lateral flow from an Icelandic plume centered beneath Greenland (Vink, 1984). Volcanism between the island of Sala y Gomez and the southern edge of the Easter microplate has been attributed to lateral flow from a plume beneath Easter island along a "leaky channel" (Kruse et al., 1997). Continued volcanism on the Cocos ridge has been explained by lateral flow from a plume beneath the Galapagos, and the eastern lobe of the Siberian Traps has been attributed to a flow of up to 2000 km from a plume beneath the west Siberian basin (Saunders et al., 2005). In the latter two cases, melt is required to flow from thinner to thicker lithosphere, violating the "upside down drainage" concept (Section 2.5.5; Sleep, 1996) and, in the case of Siberia, to force its way through Archean craton ~200 km thick.

Clear associations of magmatism with near-surface structures such as mid-ocean ridges and sutures have been attributed to the interaction of plumes with these features, either by "capture" or by coincidence. "Plume-ridge interaction" is invoked to explain why some parts of the spreading plate boundary produce more melt than others. The arrival of a plume cannot break continents up, so passive volcanic margins have been attributed to lithospheric rifting occurring coincidentally over pre-existing, abnormally hot asthenosphere from plumes (White and McKenzie, 1989; 1995). Since volcanic margins are the rule rather than the exception, such a view would require that most of the asthenosphere is plume-derived (Yamamoto et al., 2007a, b), an ironic development given that plume material was originally invoked to account for anomalies that were viewed as being relatively rare.

3.7 Discussion

There is extreme diversity of magma volume, spatial distribution and time history of eruption among melting anomalies. Some are apparently rapidly erupted, and others long lived. Some are spatially restricted, whereas others are distributed over areas thousands of kilometers wide. Any combination of these characteristics may be displayed by melting anomalies both vast and small, and there appears to be a smooth gradation in all these features. Volume ranges from the largest flood basalts down to the smallest isolated seamounts and volcanic regions. For the latter, a thermal plume model makes no sense because it fails all reasonable tests.

Where should the line be drawn to separate those anomalies where it is reasonable to test a plume model further, from those where it can be rejected out of hand? If some melting anomalies result from mantle plumes and others from different mechanisms, a clear division into more than one population based on some criterion is expected. There is no evidence that this criterion can be either volume, spatial distribution, or time-history of eruption.

How can models for the very largest eruptive volumes be tested? The key may lie in the very high rates at which some large pulses of volcanism occur. A sustained high volume rate of ~1 km^3 a^{-1} for a million years or more cannot be maintained by a rising diapir with realistic temperatures, stalled at the base of thick lithosphere. More melt is produced in plumes if they are hotter or rise to shallower depths, but such scenarios are rarely tenable. There is no unequivocal evidence for exceptionally high temperatures at any melting anomaly (Chapter 6), and plumes cannot effectively thin the lithosphere by thermal erosion.

Some plate-related processes can thin the lithosphere and allow asthenosphere to rise to shallow depths. These processes include continental break-up, continental intraplate extension, and catastrophic lithosphere loss. Where asthenosphere that was formerly 100–200 km deep is allowed to rise to shallow depths, large volumes of melt will be produced without anomalously high mantle temperatures. Such processes cannot, however, explain large, rapid eruptions where there is no evidence for lithospheric thinning.

Large volume, rapid eruptions through thick lithosphere almost certainly require the release of pre-existing melt that accumulated in the mantle over a period much longer than the eruption time. It is not traditionally expected that large quantities of melt can be retained in the mantle. Molten rock has a lower density than the host rock, and if it forms an interconnected network it will tend to drain upward. Partial melt of degrees higher than ~1% are not traditionally expected to be retainable (Faul et al., 1994; McKenzie, 1989).

Nevertheless, there is substantial evidence from independent sources that large quantities of retained melt do exist in the mantle. Partial melt is required to explain the seismic "low velocity zone" in the upper mantle above the transition zone (Section 5.5.1 and 6.1.3). Melting anomalies exist that cannot reasonably be explained except by the extraction of pre-existing melt, such as the "petit spots" near the Japan trench (Hirano et al., 2006) and the Proterozoic flood basalts of southern Africa (Silver et al., 2006). The question then becomes how large a volume of melt can accumulate, where and in what form it resides, and what the release mechanism is. A chemical discontinuity such as the base of the lithosphere may comprise a barrier to upward-migrating melt where the solidus of the subcontinental mantle lithosphere is higher than the temperature of the melt.

The largest single eruptions known to have occurred on the Earth's surface are a few thousand cubic kilometers in volume. The Rosa member of the Columbia River Basalts, for example, is known to be monogenetic and to have a volume of ~1300 km^3. If such a lava represented 5% of the volume of a spherical magma chamber, that chamber would have a diameter of ~35 km. A ~10^6 km^3 flood basalt could be produced by draining ~1% of partial melt from a region ~1000 km in diameter and 100 km thick, by continually refilling and repeatedly emptying such a chamber.

These numbers are similar to those often quoted for plume heads. The question then becomes whether the observations support most strongly a model of the sudden arrival of a hot body that produces melt at the same rate at which it is erupted at the surface, or one whereby the melt accumulated over tens or even hundreds of millions of years, to then be released on a much shorter time-scale by rupturing of the lithosphere as a consequence of tectonic processes.

Large volume magmatism is also represented by "giant" dikes. Dikes considered worthy of the designation "giant" are typically 300 km long and 30 m wide, though they may be as much as 1000 km long and 100 m wide (McHone et al., 2005). Such dikes may represent single magmatic events and have volumes of thousands of cubic kilometers (Ernst et al., 1995). The largest and most famous examples comprise the ~1267 Ma Mackenzie dike swarm in Canada.

It has been suggested that giant radial dike swarms[27] represent the feeder systems of the flood basalt volcanism interpreted as melt from plume heads. The dikes are postulated to emanate from localized centers that can be pinpointed to an accuracy of 50–100 km (Ernst and Buchan, 1997).[28] Such a model is inconsistent with the Plume hypothesis, however, which predicts that plume head volcanism is erupted from the asthenosphere throughout a region ~2000 km in diameter (Section 3.2). Furthermore, there are no coeval dike swarms where the members are broadly distributed throughout a full 360°. The closest thing to such a pattern that exists are broad arcs, which are seen in ~15% of giant dike swarms.[29] This feature must be a response to regional variations in the orientation of stress in the crust when the dikes were emplaced. Dikes intrude normal to the direction of least principle stress, and this varies in different tectonic provinces, which a single giant dike may traverse. An example is in the south and central eastern margin of North America, where dikes ranging in strike from northwest through north to northeast occur. Dikes with these different strikes are, however, mixed together and were also intruded in a regular order and controlled by variations in stress direction associated with the opening of the Central Atlantic ocean (Beutel, 2009).[30]

In some cases, the members of postulated radiating dike swarms differ widely in age. For

[27] http://www.mantleplumes.org/TopPages/GiantDikesTop.html

[28] http://www.mantleplumes.org/GiantRadDikeSwarms.html

[29] http://www.mantleplumes.org/GiantDikePatterns.html

[30] http://www.mantleplumes.org/PangaeaDikes.html; http://www.mantleplumes.org/CAMP.html

example, some of the "radiating dikes" that define the Karoo triple junction, formerly assumed to be coeval, were found instead to be separated in age by billions of years when actual dating was done (Jourdan et al., 2004; 2005).[31] In addition, the geochemistry of parts of the Karoo volcanic province that differ in age by up to ~900 Ma, have essentially the same geochemistry, consistent with a source in the continental lithospheric mantle (Section 7.6.3). Magmatism with a constant composition, erupted in the same region for such an extended period, is inconsistent with a relatively fixed mantle plume source (Jourdan et al., 2009).

No plume center postulated on the basis of giant dike orientations has a time-progressive volcano chain progressing from it. Giant dikes are fundamentally simply large intrusions whose shapes are controlled by the crustal stress field. They do not provide independent evidence for plumes.

Plume heads are predicted to have diameters of ~2000 km at the time flood basalt eruption begins. It has been claimed that flood basalts have similar dimensions (though this is not always true – the Columbia River Basalts, for example, are only ~500 km in diameter; Fig. 2.9), and this has been cited as evidence in support of a plume origin. However, much of the lateral extent of flood basalts may result from gravity flow over low-relief topography and not reflect the distribution of subsurface feeder conduits. Lavas of the Columbia River Basalts flowed for hundreds of kilometers from the feeder dikes from which they issued. The longest known lava flow may have emanated from the Deccan Traps and flowed for 1000 km to form the Rajahmundry Traps on the opposite side of the Indian continent (Fig. 2.6; Self et al., 2008).

The common, clear associations of magmatism with tectonic events or structures are an impetus for plate-related explanations. Many melting anomalies are associated with features such as continental break-up, rifts, microplates,

propagating ridges, triple junctions and back-arc areas (Sheth, 1999a, b). Many variants of the Plume hypothesis are available, and any surprising association can be explained in these terms, or by appealing to coincidence. But is the hypothesis then predictive, useful or needed? Many melting anomalies lie near some of the most remarkable and elegant tectonic features on Earth, for example, the Galapagos islands and Easter Island (Fig. 3.16; Fig. 2.10). There and elsewhere, numerical modeling is needed to explore whether the melt extraction loci, and the volumes produced, can be explained by extensional stress.

3.8 Exercises for the student

1 What is "normal" and what is "anomalous" volcanism?

2 Is the concept of a "large igneous province" helpful?

3 Can a plume model be ruled out for a melting anomaly lacking an initial flood basalt?

4 Are the areas, volumes and eruption rates of melting anomalies distributed fractally, or do they comprise more than one statistically distinct population, that might reflect distinct genesis processes?

5 Are different processes required to explain volcanic fields of different sizes?

6 Would modern models for explaining melting anomalies have fit the observations well at various times in the geological past?

7 Does the "upside-down drainage" model for lateral flow of plume material improve consistency of observations for the global melting anomaly constellation as a whole?

8 How much of a topographic anomaly is Iceland, when the global-scale bathymetric swell that fills much of the North Atlantic is removed? How much crustal thickening is required to support this topographic anomaly isostatically?

9 How much "excess" melt is there at Iceland?

[31] http://www.mantleplumes.org/Karoo.html

10 Can progressive, radial collapse of the Iceland lava dome explain the massive volcanoes in its interior?

11 Map the bathymetry of the igneous basement in the Iceland region.

12 Is there any continental crust beneath Iceland?

13 Where is the submerged, older crust that lies beneath the surface Icelandic lavas, and what is its composition?

14 How can the low elevation of Iceland be explained in view of the seismic crustal thickness of up to ~40 km?

15 Model the stress field around the Cocos-Nazca spreading axis.

16 Construct a data base of information on Pacific volcanoes.

17 Model numerically formation of the Ontong Java Plateau by the upwelling of eclogite beneath a spreading ridge.

18 Are there space-time correlations between magmatism in the European Cenozoic Volcanic Province and events in the Alpine Orogeny?

19 Why is excess volcanism produced at the south margin of the Easter microplate?

20 Why is there excess volcanism at the Azores, Afar and Bouvet triple junctions but not at the Galapagos and Rodrigues triple junctions?

21 What is the nature of continental basaltic volcanism behind subduction zones?

4

Time progressions and relative fixity of melting anomalies

Nullus quippe credit aliquid nisi prius cogitaverit esse credendum.[1]

– Augustine of Hippo (354–430)

4.1 Introduction

The term "hot spot" originally arose to express the concept that the excess volcanism that created the Hawaiian chain resulted from the Pacific lithosphere moving over a hot region in the mantle. However, it was the perception that there are numerous such "hot spots" in the Earth and that they are fixed relative to one another, which underpinned the more elaborate Plume hypothesis. The basis for the relative fixity proposal was the perception that melting anomalies have volcanic tracks leading from them, that these tracks lie parallel to the directions of motion of the plates on which they lie, and that the volcanoes age with distance from the youngest one (Figs 1.6 and 4.1). It was predicted that, once the motions of the plates had been corrected for, the melting anomalies would be found to have maintained fixed positions relative to one another since their inceptions.

In order for long-lived melting anomalies to remain fixed relative to one another, it was thought that they had to arise from a region

[1] No-one believes anything unless he decided to believe it beforehand.

deeper than the shallow mantle, which was presumed to be convecting vigorously in association with plate motion. This was the rationale for why plumes were proposed to be rooted in the deep mantle. The view that they must rise from the core-mantle boundary followed because it is the only known thermal boundary layer in the deep mantle.

Since the entire surface of the Earth is thought to be mobile, as a consequence of plate drift, a fixed frame of reference to which all motions could be compared would be a convenience. The possibility that the global constellation of melting anomalies could provide this was thus an attractive prospect. Such a reference frame would offer a significant advantage over the palaeomagnetic pole, the natural alternative, because the latter offers only latitudinal control, whereas a fixed-melting-anomaly reference frame would also provide longitudinal control. Additional advantages would be higher precisions and reduced need for the collection of rock samples and detailed laboratory geophysical analysis.

There are three possible scenarios for the collective migration behavior of melting anomalies.

1 They are fixed relative to one another, and to the geomagnetic pole. Only the plates are moving.

2 Melting anomalies are fixed relative to one another, but the entire constellation moves as a whole rela-

Plates vs. Plumes: A Geological Controversy, 1st edition. By Gillian R. Foulger.
Published 2010 by Blackwell Publishing Ltd.

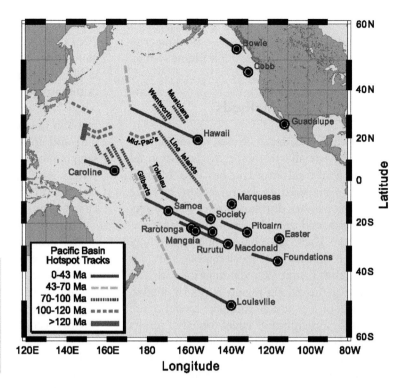

Figure 4.1 An example of volcanic chains used to test the fixed-plume hypothesis. Circles indicate locations of currently active melting anomalies at one end (adapted from Koppers et al., 2003).

tive to the geomagnetic pole ("true polar wander" – Section 4.2.3). The separations between contemporaneous volcanoes in different tracks thus remain constant, but the latitudes and longitudes at which the volcanoes on a single track form may change. This implies that the whole mantle, along with the plumes postulated to be embedded in it, and the drifting overhead plates, experience that component of motion.

3 Melting anomalies move relative to one another and do not provide a stable frame of reference to which the motions of other elements may usefully be referred.

What are the results of research done to date? Testing relative fixity and seeking explanations for the results is the subject of a large literature within which many conflicting conclusions can be found. It is difficult to rationalize these diverse conclusions for a number of reasons. Many studies use very different subsets of data, including different groups of melting anomalies, age dates and plate motion models, and which data were used, and how error budgets were assessed,

is often not stated. Very short-lived volcano chains may be included, and melting anomalies that fulfill virtually no other predictions of the Plume hypothesis on which the concept of relative fixity is so dependent, for example, Samoa. Large numbers of free parameters may be included in data inversions so that apparently good data fits are achieved but it is unclear if the results are statistically significant.

Despite these factors and the wide variety of different approaches that may be taken, it is clear that the relative fixity hypothesis fails and this cannot be accounted for by true polar wander. The failure is most extreme for the primary Iceland and Emperor-Hawaii melting anomalies, and much current work focuses on explaining these misfits. Ironically, "ridge capture" by the Mid-Atlantic Ridge is favored as an explanation for the misfit of the Iceland melting anomaly (Vink, 1984), whereas "ridge escape" has been proposed to explain the misfit of the Emperor melting anomaly (Tarduno et al., 2009). The failure of the latter melting anomaly to comply with the fixed-"hot spot" hypothesis is particularly noteworthy, because the Emperor-Hawaii system is the type example plume system.

4.2 Methods

Several kinds of data can be used to test for relative fixity, but all are subject to uncertainties and the final total error can be large. Melting anomalies proposed to be underlain by deep mantle plumes are distributed over several tectonic plates, and thus plate motion must be corrected for. There are considerable differences between candidate plate motion models. In particular, the history of deformation in the Antarctica-Australasia region is poorly known and the choice of which candidate model to use can have a significant effect on the results (Raymond et al., 2000; Steinberger et al., 2004).

The ages and locations of the volcanoes that make up the chains are also required. The locations of volcanic features on oceanic plates are well known from sea-floor bathymetry. Good bathymetric maps are now available from satellite gravimetry (Sandwell and Smith, 1997; Smith and Sandwell, 1997). Ages of the required accuracy need radiometric dating, however, which is more problematic.

4.2.1 Radiometric dating

The most widely used techniques are the potassium-argon (K-Ar) and Ar-Ar methods.[2] Whole-rock K-Ar dates are the least reliable, and single-crystal Ar-Ar dates the best.

The K-Ar technique is applied to either whole rocks or to single crystals extracted by hand from crushed samples. Such dates may be as accurate as 1–2%, but there are many potential problems that can diminish this accuracy. All the components of the rock sample used must be primary and unaltered. This is rarely the case, especially

[2] http://www.mantleplumes.org/ArAr.html

for basalts dredged from the sea floor (Baksi, 2007a, b). No Ar should have been lost from or gained by the sample when it erupted. Ar is a noble gas, mobile, and can easily be lost. Alternatively, atmospheric Ar can be trapped when lava crystallizes. Contamination of melt by older material will lead to dates that are too old, and incomplete extraction of Ar during sample processing will produce ages that are too young. Finally, the apparent age of a rock may be reset if it is reheated *in situ*, for example, by later volcanic activity.

The single-crystal method is free from some of the problems that can degrade whole-rock dates. For example, single crystals do not trap atmospheric Ar when they crystallize, and they are not thought to be susceptible to contamination by older material. However, the method is reliant on crystals that are high in K, for example, biotite or anorthoclase.

The highest quality Ar-Ar dating involves irradiating a single crystal with neutrons to cause some ^{39}K (present as a known fraction of the total K in the rock) to decay to ^{39}Ar. The ratio $^{40}Ar/^{39}Ar$ (as a proxy for ^{40}K) is then measured. The most accurate way of doing this is by heating the sample in increments (step heating), which releases argon from different regions of the crystal. $^{40}Ar/^{39}Ar$ changes with each step, and when 80% or more steps are within acceptable error limits, the age of the crystal can be calculated. A more primitive approach is "total fusion", where $^{40}Ar/^{39}Ar$ is measured from Ar released in a single step. In addition to its greater geological robustness, the Ar-Ar system also has major analytical advantages over K-Ar dating. It is not necessary to measure the ^{40}K and ^{40}Ar separately, nor is complete extraction of the Ar necessary, as only the ratio $^{40}Ar/^{39}Ar$ is needed and not the total amount of gas present.

In practice, many existing radiometric dates from volcanic ridges and chains were obtained using older methods, for example, K-Ar, and are not accurate enough to be useful for testing the relative fixity hypothesis. The single-crystal Ar-Ar method is fairly new, and many samples are currently being redated. In some cases this has resulted in the rate of progression of volcanism along ridges and chains deduced on the basis of earlier dates being significantly revised (e.g., Koppers et al., 2004; Pringle et al., 2008).[3]

To these technical problems may be added the issue of the relatively long lifetime of activity of many volcanoes. Hawaiian volcanoes, for example, are known to have been volcanically active for as at least 5 Ma, including the post-erosional rejuvenated stage (Section 7.6.2). A wide range of ages is thus expected for samples from any seamount or volcano, reflecting the time it took to build the edifice. This adds to the difficulties of assessing age progressions, especially where stratigraphic relationships are poorly known, as is generally the case, for example, for submarine edifices.

4.2.2 Earth's palaeomagnetic and spin axes

Palaeomagnetic measurements can reveal the latitude at which lavas erupted and cooled. The inclination and declination of the permanent remnant magnetization acquired by lavas when they cooled through their Curie temperatures is measured. The method requires knowledge of the orientations of the samples drilled from the flows.

It is usually assumed that the Earth's magnetic axis has been, on average, coincident with its spin axis throughout geological time. This is known to be true for the most recent few tens of thousands of years from measurements of magnetic directions in Quaternary and Neogene formations whose geographical latitudes are known accurately. There is also evidence that it has been true for most of geological time, for example, from corals and the equatorial sediment bulge in the Pacific. However, there are fewer precise data, and thus more uncertainty, further back in time. The magnitude of the non-

[3] http://www.mantleplumes.org/PacificAges.html

dipolar component may be as large as ~10% of the total field, and this can potentially introduce significant errors into studies of the relative fixity of melting anomalies on different sides of the planet. These errors add to others, such as analytical uncertainties and errors in the measured orientations of the drilled cores.

Palaeomagnetic measurements have the drawback that they cannot measure longitude, so east-west motions cannot be constrained. Plate motion models may also be based partly on palaeomagnetic data. Thus there is no true independence between palaeomagnetic measurements of the latitudes of volcanoes when they formed and calculations from plate models of the latitude of the sea floor on which they erupted. This problem is greater for motions older than the oldest present-day sea floor, which is ~200 Ma. Palaeomagnetism nevertheless provides an independent method of testing the relative fixity hypothesis and does not need the radiometric dates that are so expensive and difficult to measure to the required degree of accuracy. If melting anomalies are fixed relative to one another, and to the palaeomagnetic pole, then the latitudes at which all the volcanoes on a single ridge or chain formed are expected to be the same.

4.2.3 True polar wander

It was known as early as 1980 that the volcanoes of the Emperor chain did not all form at the same latitude. Younger volcanoes were known to have formed at progressively more southerly latitudes from differences in the climactic provenance of biota dredged from them and from palaeomagnetic measurements (Section 4.5.4) (Karpoff, 1980; Kono, 1980). But did other melting anomalies move in unison? Did Earth's palaeomagnetic pole drift with respect to a fixed-"hot spot" reference frame? The concept of "true polar wander" permits an additional degree of freedom to be solved for in data inversions, that may reduce observed misfits (Goldreich and Toomre, 1969). True polar wander should not to be confused with "apparent polar wander", which is the motion of the geomagnetic pole relative to a single plate. Apparent polar wander occurs primarily as a result of plate motion, and is different for each plate.

4.3 Predictions of the Plume hypothesis

The Plume hypothesis predicts that melting anomalies remain fixed relative to one another during their lifetimes. As a consequence they provide a fixed reference frame to which other motions, for example, of tectonic plates, may be referred. The constellation of melting anomalies is not necessarily fixed with respect to the geomagnetic pole, as true polar wander may occur (Section 4.2.3).

4.4 Predictions of the Plate hypothesis

The predictions of the Plate hypothesis depend on the mechanism of melting anomaly formation. The question does not arise in the case of short-lived volcanism, for example, flood basalt formation resulting from continental break-up,[4] catastrophic lithospheric thinning,[5] and sublithospheric melt ponding and draining (Section 1.9) (Silver et al., 2006). It does not apply either in broad volcanic provinces such as the European Cenozoic Volcanic Province, which result from diffuse intraplate extension.[6] It is only relevant to processes that build long, narrow, regularly time-progressive volcanic constructs terminated by localized eruptive centers.

Small-scale sublithospheric convection builds linear volcanic ridges beneath oceanic crust.[7] This volcanism is not time-progressive but

predicted to occur along the length of the ridge throughout the active period (Ballmer et al., 2007; 2009). Processes that build time-progressive ridges and volcano chains emanating from plate boundaries are fertility at mid-oceanic ridges and enhanced volcanism at plate boundary junctions.[8] Those melting anomalies are not predicted to be relatively fixed, however, because mid-ocean ridges migrate relative to one another. Some plates are deficient in subduction zones and growing, for example, the African plate, whereas the Pacific plate is almost entirely surrounded by subduction zones and as a consequence is shrinking (Fig. 1.5). Melting anomalies tied to ridges, for example, Iceland and Easter, are thus expected to form time-progressive constructs but they are not expected to be fixed relative to one another or to melting anomalies in the interiors of plates (e.g., Doglioni et al., 2005).[9]

Melting anomalies that result from intraplate extension in oceanic and continental crust may under some circumstances be quasi-stable relative to one another. This is expected to be rare, because it requires the special circumstance of persistent extensional stress in individual, localized regions. Persistent melting anomalies in the interiors of plates occur in response to plate-wide stress fields that are governed primarily by plate driving forces and thermal contraction as the lithosphere cools (Lithgow-Bertelloni and Richards, 1995; Stuart et al., 2007). They occur where extension results in lithospheric cracking. They are predicted to retain their relative positions as long as the pattern of stress stays the same. Their loci are thus stable relative to the plate boundary as a whole, and will remain so as long as the plate boundary configuration remains stable, and the orientations of the volcanic tracks produced will be parallel to the plate motion direction.

If the plate boundary configuration changes, for example, if ridges are annihilated, subduction

[4] http://www.mantleplumes.org/
VM_DecompressMelt.html
[5] http://www.mantleplumes.org/LithGravInstab.html
[6] http://www.mantleplumes.org/Europe.html
[7] http://www.mantleplumes.org/SSC.html

[8] http://www.mantleplumes.org/RTIntersections.html
[9] http://www.mantleplumes.org/Hotspots.html

zones created or transform faults evolve into different types of boundary, the intraplate stress field will change in both orientation and magnitude. Areas that were previously in compression may become extensional and vice versa, and both the speed and direction of motion of the plates may change. At these times, magmatism at some localities may cease, onset, or change in rate, or the orientations of the volcanic tracks produced may change (Smith, 2007). Melting anomalies in the plate interiors are then expected to migrate with respect to one another. A radical reorganization of the plates occurred in the period ~55–40 Ma (Rona and Richardson, 1978),[10] but plate motions have remained relatively stable subsequently. The stress conditions in plates are thus expected to have changed significantly during that reorganization but to have been fairly stable from that time up to the present.

The Plate hypothesis predicts that, of the several tens of melting anomalies on global lists, only a few may be quasi-fixed relative to one another and these will tend to lie on the same plate. Major changes in relative motions between melting anomalies will accompany major tectonic reorganizations. Where long-lived intraplate melting anomalies lie on different plates, their relative positions are expected to change in concert with the relative motions of the plates on which they lie and with the stress fields that permit them to exist. They are not expected to be fixed relative to one another, but to typically move at rates of the order of the migration rates of plate boundaries.

4.5 Observations

4.5.1 Melting anomalies without tracks

The Plume hypothesis predicts that the arrival of a plume is accompanied by flood basalt erup-

tion, followed by continuation of "plume tail" magmatism to form a time-progressive volcanic track. Such a sequence is a rare exception rather than the rule (Anderson, 2005b). Of the 49 "hot spots" listed in Table 1.4, only 17 are cataloged as definitely having an associated volcanic ridge or chain, and 19 are listed as questionably having one. Of this total of 36, 22 are listed as having no associated flood basalt. Remarkably, 11 melting anomalies from the catalog of 49 are listed as having neither volcanic track nor flood basalt, leaving little to remain. Volcanic tracks did not form following eruption of the Earth's largest flood basalts, including the Siberian Traps,[11] the Central Atlantic Magmatic Province,[12] and the Ontong Java Plateau (Fig. 2.5).[13] Coherent tracks are also lacking in broad regions of diffuse volcanism such as the West Pacific ocean,[14] the European Cenozoic Volcanic Province, East Asia[15] and Hoggar in North Africa. At Hoggar, a ~1000-km-long track is predicted by the Plume hypothesis, but no such track exists (Liégeois et al., 2005).[16]

Is Iceland associated with a time-progressive volcanic track? The region has had a complicated magmatic and tectonic history that includes multiple episodes of volcanism, lateral jumps of the spreading axes, migration of enhanced volcanism along the plate boundary, and spreading along parallel pairs of ridges. The presently active Iceland region is a diffuse spreading plate boundary. The history of volcanism at Iceland cannot be realistically described without recognition of these facts, and it cannot be explained plausibly in terms of a simple model of a migrating point source.

[10] http://www.mantleplumes.org/Rona_letter.html

[11] http://www.mantleplumes.org/Siberia.html
[12] http://www.mantleplumes.org/TopPages/CAMPTop.html
[13] http://www.mantleplumes.org/TopPages/OJTop.html
[14] http://www.mantleplumes.org/SWPacific.html
[15] http://www.mantleplumes.org/Mongolia.html
[16] http://www.mantleplumes.org/TopPages/HoggarTop.html

The numerous suggestions that have been made regarding what volcanics correspond to the postulated Icelandic plume head are an indication of the complexity of the region and the poor fit of the model. These suggestions include the Siberian Traps, which erupted at ~225–265 Ma (Lawver et al., 2002), the Alpha Ridge in the Arctic Ocean, which formed at ~80 Ma (Forsyth et al., 1986), lavas in the Davis Strait emplaced at ~61 Ma (McKenzie and Bickle, 1988), and the ~2600-km-long volcanic margins of the present-day North Atlantic Ocean, which formed at ~54 Ma (Morgan, 1981). Times of inception of ~90 Ma (Morgan, 1983) and ~130 Ma (Lawver and Muller, 1994) have also been proposed (Fig. 3.10).

Following the large-volume volcanism at ~54 Ma, magmatic rates dwindled everywhere except at the latitude of present-day Iceland (Section 2.5.4). During the following ~6 Ma, localized enhanced volcanism at the spreading ridge built the Iceland-Faeroe ridge with its crustal thickness of ~30 km (Fig. 3.12) (Bott and Gunnarsson, 1980). At ~44 Ma a new northeast-orientated spreading center formed in Greenland, north of the Iceland-Faeroe ridge and ~100 km from the coast. Spreading about this new center caused a sliver of continental crust to migrate into the Atlantic Ocean, where it subsided below sea level to form the Jan Mayen microcontinent. Sea-floor spreading continued along spreading centers on both sides of this microcontinent until ~26 Ma, when the easternmost Aegir ridge became extinct. At about this time, to the south, enhanced volcanism expanded throughout ~600 km of the Mid-Atlantic Ridge and the Iceland platform began to form (Foulger and Anderson, 2005; Foulger et al., 2005a).

Spreading along a parallel pair of ridges probably occurred from the beginning of development of the Iceland platform. The existence of two ridges since ~17 Ma is documented with certainty from geological mapping in Iceland (Jóhannesson and Saemundsson, 1998; Saemundsson, 1979). It follows naturally from palinspastic reconstructions that a microplate of captured crust – oceanic, continental or a com-

bination of both – must exist beneath Iceland today (Fig. 3.12) (Foulger, 2006). In the absence of lateral jumps, the spreading centers on either side of this microplate would progressively migrate to the east and west, away from the single Mid-Atlantic Ridge axis in the ocean. Instead, the spreading centers have repeatedly jumped inward toward central Iceland, thereby minimizing their lateral offset from the oceanic Kolbeinsey and Reykjanes ridges. Two ridge jumps are known for each of the western and eastern spreading axes (Fig. 3.12, lower right panel). In addition to this, eastward migrations of the Reykjanes ridge are thought to have occurred at ~44 Ma and ~7 Ma (Bott, 1985; Foulger et al., 2005a; Saemundsson, 1979), and may still be occurring in small steps today (Hey et al., 2007; 2008). Such migrations of the ridge may be the cause of the chevron ridges about the Reykjanes ridge (Fig. 3.14; Section 3.5.2).

The location of the volcanic track predicted to be associated with an hypothetical Icelandic plume has been calculated (Fig. 3.10) (Lawver and Muller, 1994). The calculations assume a plume is presently located beneath south-central Iceland and fixed relative to postulated plumes on the African, Indian, North and South American and Australian plates. The results predict that a time-progressive volcanic track extends from Iceland, west along the Greenland-Iceland ridge, across Greenland and into Baffin Bay. An extension back to 130 Ma has been calculated, which would place such a plume beneath Ellesmere Island at that time. Such old onset ages are inconsistent with the concept that the postulated Icelandic plume head corresponds to the volcanism that built the volcanic margins at ~54 Ma.

There is virtually no observational evidence for such a track. Thulean line volcanism in west and east Greenland and Britain started simultaneously at ~62 Ma, when the postulated plume is predicted to have underlain central Greenland. Volcanism was widespread in the North Atlantic during much of the period ~54–44 Ma, and does not fit a model of a point source. The observations are, instead, consistent with flood volcanism associated with continental break-up. Volcanism in east Greenland revived at ~44 Ma when the Jan Mayen microcontinent split off Greenland. The volcanism that built the Greenland-Iceland and Iceland-Faeroe ridges occurred at the Mid-Atlantic Ridge itself (Lundin and Doré, 2004). These ridges are anomalous only by virtue of their crustal thickness, not by the chronology of the eruptives that built them. They are time-progressive only in the sense that oceanic crust everywhere is time-progressive. There is no observational evidence whatsoever to support the suggestion that the Siberian Traps comprise the head of a plume currently beneath Iceland (Bailey and Rasmussen, 1997).

Within Iceland itself, the picture is one of persistent widespread volcanism throughout a diffuse spreading plate boundary, with little evidence for any preferred direction of migration or of volcanism associated with a migrating point source. Volcanism has typically occurred along multiple ridges. The claim that the spreading ridge repeatedly jumped east leaving a wake of extinct ridges in western Iceland is often reiterated (e.g., Hardarson et al., 1997). However, this is a gratuitous oversimplification of the observations that ignores the facts that there have usually been two parallel ridges active in Iceland at any one time, and that the eastern ridge repeatedly jumped to the west (Jóhannesson and Saemundsson, 1998). The overall pattern of volcanism is one of a broad province that experienced multiple migrations of volcanism from both east and west, consistent with maintaining colinearity with the Kolbeinsey and Reykjanes ridges.[17]

4.5.2 Short-lived melting anomalies

The Pacific plate is rich in short-lived volcanic ridges and chains, many of which trend quasi-parallel to one another and comprise just a few,

or even only two, seamounts (Fig. 4.1). The Samoa, Marquesas and Pitcairn melting anomalies, for example, have been volcanically active for no more than ~10 Ma.[18] On the fast-moving Pacific plate, such volcanism may form a track up to ~1000 km long. For many systems of this kind, a linear volcano chain is the only prediction of the Plume hypothesis that is fulfilled – an associated flood basalt and unequivocal evidence for high magmatic temperatures are absent and it is not presently logistically practical to seek a "plume tail" using seismology. Nevertheless they have commonly been included in studies to test the hypothesis that deep mantle plumes are relatively fixed (Fig. 4.1) (e.g., Koppers et al., 2003).

The time progressions associated with short-lived Pacific volcano chains range from regular, through erratic to unknown (Clouard and Bonneville, 2001; 2005). The ~200-km-broad, ~400-km-long Marquesas archipelago is well-dated, for example, and mostly time-progressive with only a few exceptional dates (Fig. 12 in Clouard and Bonneville, 2005). The 500-km-long Society archipelago essentially comprises two clusters of volcanoes, an older one to the northwest and a younger one to the southeast (Fig. 17 in Clouard and Bonneville, 2005). Both have rocks dated at 2–3 Ma, although the Plume hypothesis predicts the eruptives should be separated in age by 3–4 Ma. Such short archipelagos cannot reasonably be used to test the Plume hypothesis. The ~900-km-long Samoa chain illustrates irregular time-progression, with dates of 23 Ma, 10 Ma and 1.6 Ma closely clustered at its northwestern end, which is predicted to be the oldest by the fixed-Plume hypothesis (Fig. 4.2).

4.5.3 Melting anomalies with long chains that are not time-progressive

Testing the relative fixity hypothesis is most heavily dependent on the longer volcanic

[17] http://www.mantleplumes.org/TopPages/ IcelandTop.html

[18] http://www.mantleplumes.org/TopPages/ SamoaTop.html

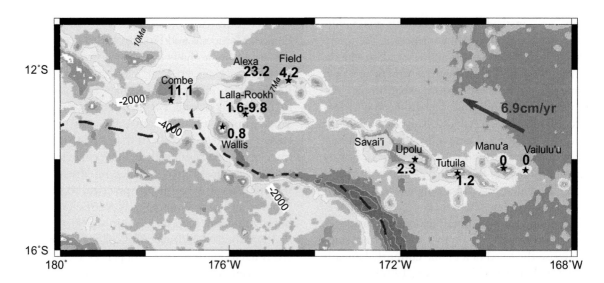

Figure 4.2 Ages for basalts from the Samoan chain. Dashed line represents the plate boundary. Arrow represents present-day Pacific plate motion direction in the region (DeMets et al., 1994) (from Clouard and Bonneville, 2005).

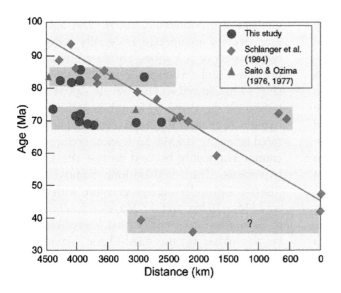

Figure 4.3 Radiometric ages vs. distance from the Line-Tuamato volcano chain. Diagonal line represents predicted volcanic propagation rate for a fixed mantle plume (from Davis et al., 2002).

chains. Of these, some have approximately the right azimuth but do not have regular time-progressions along them. Notable examples are the Line Islands and the Austral-Cook islands in the Pacific ocean (Fig. 4.3) Two major episodes of volcanism occurred at the Line Islands, each

lasting ~5 Ma. During these, volcanism occurred synchronously over distances of 1200 and 4000 km along the main chain, and also at nearby seamounts that do not form part of the chain (Davis et al., 2002). The ~2000-km-long Austral-Cook island chain comprises three

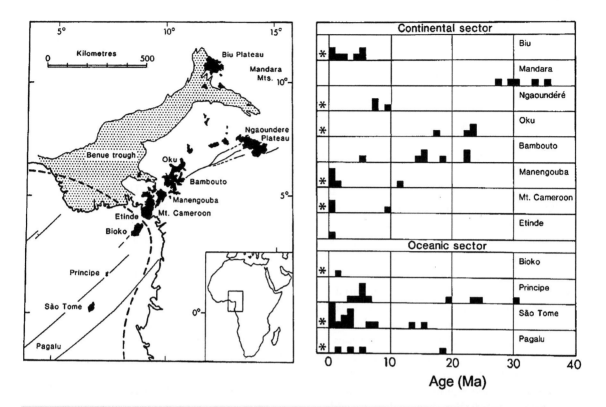

Figure 4.4 Left: Outcrop of the Cameroon volcanic rocks and sediments of the adjacent Benue trough (stippled). Right: K-Ar age data for Cameroon line volcanic rocks (from Fitton, 1987).

distinct, parallel, time-progressive sub-chains with dates inconsistent with one another (McNutt et al., 1997). The ~1600-km-long Foundation chain is an example of a long chain that exhibits partial time progression. That chain is well-dated, and mostly, but not entirely, regularly time-progressive.

In the Atlantic Ocean the ~1600-km-long Cameroon chain straddles the continental margin of West Africa (Figs 2.11 and 4.4). It began to form at ~35 Ma and has subsequently been chronically volcanically active along much of its length. No time-progression is detectable (Fitton, 1987). The Cameroon chain continues oceanward as the St Helena seamounts, which fan out and become increasingly scattered to the southwest. Sparse data are consistent with broad time-progression along them (O'Connor et al.,

1999). The Cameroon chain and the St Helena seamounts are traditionally assumed to be distinct and separate entities, though there is no fundamental reason why they should be so viewed. Indeed, there is considerable morphological similarity between them and the southwesternmost part of the Tristan-Walvis Ridge system further south (Fig. 2.7).

4.5.4 Melting anomalies with long, time-progressive tracks

The linear volcanic constructs on which the hypothesis of relative fixity is most heavily dependent are the exceptional few that span large age ranges and are thought to have regular time-progressions along them. This group comprises essentially the Walvis Ridge on the African plate, the Lakshadweep-Chagos-Réunion ridges on the Indian plate, and the Louisville and Emperor-Hawaiian volcano chains on the Pacific plate.

(a) Tristan-Walvis Ridge

The currently active melting anomaly associated with this system is considered to be the Tristan island group and Gough island, which lie ~500 km east of the spreading ridge in the South Atlantic (Fig. 2.7). The associated volcanic track is considered to be the Walvis Ridge, which extends for ~3400 km northwest to the Etendeka volcanic province on the coast of southwest Africa. The system is broad, variable in morphology along its length, and departs strongly from a narrow, linear volcano chain. Its currently active southwestern end comprises a ~500-km-wide region of scattered islands and seamounts and active volcanism is distributed over a 400-km-wide region. The Walvis Ridge varies from ~100–400 km in breadth. In addition, there is a scatter of seamounts further south, some of which are younger than the sea floor on which they sit and thus formed in the plate interior and not on the spreading ridge. In the context of the Plume hypothesis, there is considerable ambiguity where the postulated linear volcanic track is observed to lie.

Approximately 20 whole-rock $^{40}Ar/^{39}Ar$ ages have been reported from the region, mostly for basalts dredged from the sea floor (O'Connor et al., 1999). The majority of these ages are insufficiently reliable to test for time-progression. Some of the results have ill-defined plateau ages and/or excess scatter in their isochron diagrams, suggesting large amounts of contamination by atmospheric argon and/or large amounts of ^{39}Ar recoil and alteration in the samples (Baksi, 1999; 2005). Only a single date from the Walvis Ridge is reliable, and none from the Rio Grand Rise, a complementary ridge on the South American plate, west of the Mid-Atlantic Ridge (Fig. 2.7).

Despite the inadequacy of the data, the results from the Walvis Ridge have nevertheless been interpreted as showing a unidirectional time progression. The apparent rate is variable, with discrepancies of up to ~20 Ma between the measured ages and those predicted by linear time progression. An overall average rate of ~2–3 cm/a has been deduced and used in studies of relative fixity. However, the Tristan-Walvis Ridge system is, in truth, of little use for testing for relative fixity. The lack of reliable dates, coupled with the ambiguity regarding where the proposed plume is currently located and the wide range of possible locations for an older volcanic track, mean that this observation set is too weak to rule out many candidate models.

Several lines of evidence suggest an explanation for Walvis Ridge and Rio Grande Rise volcanism in plate-scale diffuse deformation. On the basis of high-resolution satellite gravity data, morphological features on the Walvis Ridge and the Rio Grande Rise have been interpreted as wrench and shear faults, comprising reactivated, older faults (Fairhead and Wilson, 2005). Fault movements on the Walvis Ridge correlate with breaks in stratigraphy in sedimentary basins in Africa, which indicate tectonic deformation events. Tectonism is also accompanied by alkaline volcanism, which has been sporadic and widespread throughout Africa for at least 2 Ga, and igneous episodes in widely separated parts of Africa are commonly synchronous. Volcanism re-activates old, pre-existing structures in the plate interior, in particular rifts and rift intersections, some of which have a record of repeated activity for the full 2 Ga. The African plate has moved long distances over the deeper mantle over this time, and yet time-progressive volcanic tracks are not observed. The faults on the Walvis Ridge and Rio Grande Rise, along with the volcanism that they permit, likely comprise elements of this vast deformation system, dissipating stress changes within the African and South American plates (Fig. 4.5) (Bailey, 1992;

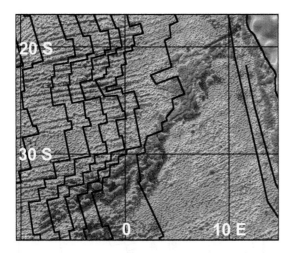

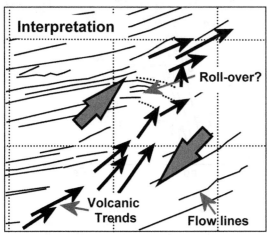

Figure 4.5 Top: Free air gravity map of the Walvis Ridge. Bottom: Interpretation in terms of stress release at discrete times, resulting in dextral shear movement (broad, gray arrows) and the formation of roll-over structures. Black arrows represent volcanic lineaments that propagate away from the Mid-Atlantic Ridge at the time of magmatism (from Fairhead and Wilson, 2005).

Bailey and Woolley, 1995; 1999; Moore et al., 2008).[19]

[19] http://www.mantleplumes.org/Africa.html;
http://www.mantleplumes.org/SouthernAfrica.html;
http://www.mantleplumes.org/SAtlantic.html;
http://www.mantleplumes.org/Argentina.html;
http://www.mantleplumes.org/Brazil.html

(b) Lakshadweep-Chagos-Réunion system

The Deccan Traps is postulated to represent flood volcanism associated with an arriving plume head, and the ~2500-km-long Lakshadweep-Chagos-Réunion Ridge to represent subsequent plume tail volcanism (Fig. 2.6).[20] At its southern end lie the Mascarene islands – Réunion, Mauritius and Rodrigues. This system comprises one of only three apparently time-progressive volcanic tracks that arguably emanate from flood basalts (Table 1.4), the others being the Tristan-Walvis Ridge and Yellowstone systems.

The Deccan Traps mostly formed in the eight-year period ~67–60 Ma. The chronology of volcanism along the Lakshadweep-Chagos-Réunion Ridge between the Deccan Traps and currently active Réunion Island is mostly constrained by single-crystal Ar-Ar dating of samples from shallow holes drilled in the sea floor at 5–10 sites. As for the Walvis Ridge, several of these samples are altered, or contain large amounts of atmospheric argon, rendering the ages unreliable, or insufficiently accurate to be useful for testing for time progression (Baksi, 1999; 2005). Only one reliable, accurate age of 31.5 Ma is available from the Mascarene plateau, and two less-reliable ages of 50 ± 2 Ma and ~53–56 ± ~8 Ma from the Lakshadweep-Chagos ridge. As for the Walvis Ridge, new, high-quality radiometric dates are needed for this system to be used for testing the relative fixity hypothesis.

Again, despite their unreliability, the results have been interpreted as indicating progression in age from zero at Réunion Island to ~65 Ma at the Deccan Traps. The rates of progression deduced are $3.7\,\text{cm}\,\text{a}^{-1}$ for the period 0–45 Ma, and $13.5\,\text{cm}\,\text{a}^{-1}$ for the period 45–65 Ma (Duncan and Hargraves, 1990).

The system is of special interest because the currently active melting anomaly, Réunion, is on a different tectonic plate from the early part

[20] http://www.mantleplumes.org/TopPages/DeccanTop.html

of the presumed volcanic track (Fig. 2.6). It is thus one of a subset of three melting anomalies suggested to be "ridge-crossing" (Sleep, 2002). The counterpart to the Réunion system east of India, is the Ninetyeast Ridge-Kerguelen system. Five high-precision $^{40}Ar/^{39}Ar$ from that system show linear age progression at a rate of $11.8 \pm 0.5\,cm\,a^{-1}$ for the period 77–43 Ma, revising earlier estimates of $0.86 \pm 1.4\,cm\,a^{-1}$ based on dates obtained using older techniques (Pringle et al., 2008). The Ninetyeast Ridge, which lies approximately normal to the Southeast India Ridge, formed along a large and complex transform offset between the Wharton Ridge and the Indo-Antarctic Ridge during a period of repeated southward jumps of the latter (Fig. 4.6) (Krishna et al., 1995). It is as yet unclear how the dating results fit with this complex tectonic history. It has also been proposed that the New England seamounts form a "ridge-crossing" chain (Morgan, 1983).

"Ridge-crossing" melting anomalies have been cited as evidence for melt supply systems that are relatively unaffected by the arrival of a migrating ridge and thus rooted deeper than mid-ocean ridge volcanism. There are several features that such an hypothesis must account for at the Lakshadweep-Chagos-Réunion system, however. The northern part of the volcanic ridge, the Lakshadweep Ridge, comprises stretched, thinned continental crust overlain by volcanics, and is thought to have formed as a volcanic rifted margin when India rifted from Madagascar in the Cretaceous and later from the Seychelles in the early Tertiary (Chaubey et al., 2008).

Like the Ninetyeast Ridge, the Lakshadweep-Chagos Ridge formed along a fracture zone created by a long transform fault that offset the spreading ridge that built the floor of the Indian Ocean. The Lakshadweep-Chagos-Réunion and Ninetyeast volcanic ridges are thus not in random places. There is no evidence for "capture" of the postulated ridge-crossing Réunion plume by the approaching Central Indian spreading ridge, despite the fact that such a process has

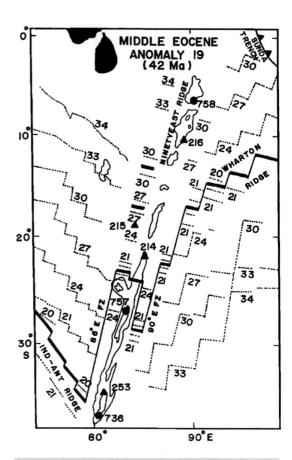

Figure 4.6 Schematic reconstruction of the spreading ridge and magnetic anomalies in the region of the Ninetyeast Ridge at ~42 Ma (from Krishna et al., 1995).

been widely invoked elsewhere, for example, at Iceland, the Galapagos and the Azores. Lastly, basalts from the Mascarene Plateau and the Chagos Ridge differ so radically in composition that it has even been suggested that they are formed by two different plumes (Burke, 1996). An alternative model for such volcano geometries has been suggested that is linked to migrating ridge-transform offsets (Beutel, 2005).[21]

[21] http://www.mantleplumes.org/JumpingHotspots.html

(c) Louisville

The Louisville seamount trail is a ~4300-km-long line of ~70 seamounts in the south Pacific, extending from the Tonga-Kermadec trench in the northwest toward the Pacific-Antarctic ridge in the southeast (Fig. 4.7). The northwestern half is clearly defined by a chain of large seamounts. In contrast, the younger half is defined by far fewer, sparser, smaller seamounts and for several long intervals the chain does not exist. The youngest volcanism considered to be part of this chain is a small, isolated seamount cluster ~500 km northwest of the Pacific-Antarctic ridge.

The chain is broadly arcuate in shape and lacks the sharp bend of the Emperor-Hawaiian pair of chains. It changes direction where it crosses two major seafloor features, the Wishbone Scarp and the Tharp Fracture Zone. Its younger part runs close to the Eltanin Fracture Zone, across which the age of the sea floor changes by some 10 Ma. Opportunities to study the Louisville chain are limited because research cruises are infrequent to this remote location, which is often plagued by hostile weather.

The Louisville chain provides a test of the repeatability of age dating results. Total fusion $^{40}Ar/^{39}Ar$ dates were initially obtained for samples from ~10 localities (Watts et al., 1988), and subsequently re-dated using the more accurate incremental-step-heating method (Koppers et al., 2004). The two methods agree to within a few million years, except for the oldest sample which was found to be 10–12 Ma older than originally measured. In general, re-dating of older samples using the latest techniques typically results in ages up to ~15% older (Wessel and Kroenke, 2009).

The older dates suggested a highly linear age progression along the chain. The newer, more accurate dates revised this conclusion to reveal significant variations (Koppers et al., 2004). In the period ~77–60 Ma, age progression was fast, subsequently slowing and becoming irregular, and even reversing in the period ~45–30 Ma. Only two dates are available for the period ~30–0 Ma. These are consistent with a linear

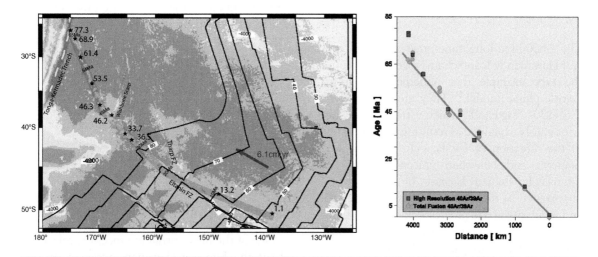

Figure 4.7 Left: The Louisville seamount chain, sea-floor ages from magnetic anomalies, bathymetry and radiometric ages of seamounts measured using the high-resolution incremental heating $^{40}Ar/^{39}Ar$ method (from Clouard and Bonneville, 2005). Right: Comparison of re-dated, high-resolution, incremental heating $^{40}Ar/^{39}Ar$ dates with earlier, less accurate total fusion ages (from Koppers et al., 2004).

time progression but do not provide strong constraint.

The Louisville chain is age-progressive to a first-order, with the caveats that only few data are available and there is considerable second-order variation in the apparent migration rate. More age dates are needed. The Louisville chain resembles the Emperor-Hawaiian chains in both age and narrowness, but it differs from them in that magmatism has greatly dwindled with time, whereas for the former it has surged.

The Louisville and Emperor-Hawaii chains are well suited to test for relative fixity, and they are the only pair that can be used for the period prior to ~55 Ma. Relative fixity is within ~200 km for the period 0–55 Ma, but large deviations occur for older times, increasing with age to ~550 km for 80 Ma (Wessel and Kroenke, 2009). Notwithstanding this, the ability of the Emperor-Hawaiian and Louisville chains to discriminate between the Plate and Plume hypotheses is limited because they both lie on the same plate, and both hypotheses allow relative fixity under that circumstance.

(d) Emperor-Hawaii system

The ~6000-km-long time-progressive Emperor and Hawaiian volcano chains together comprise the type example that underpins the fixed-mantle-plume model.[22] They are separated by the ~130° "big bend", which is traditionally but erroneously explained as resulting from a change in the direction of Pacific plate motion. The Emperor and Hawaiian volcano chains are intensively studied because of their pivotal importance to the Plume hypothesis. The rate and regularity of age progression, and the stability of the geographical location of the melt extraction locus, are particularly relevant.

[22] http://www.mantleplumes.org/TopPages/HawaiiTop.html

One of the most remarkable features of the Emperor and Hawaiian chains is the unilateral time progression of the onset of volcanism along them. An extensive body of radiometric dates obtained from the 1960s through the 1980s suggested that this progression was fairly regular. As for the Louisville chain, however, many ages were obtained using inaccurate techniques such as the K-Ar and total fusion $^{40}Ar/^{39}Ar$ methods. Later re-dating using the incremental-step-heating $^{40}Ar/^{39}Ar$ method showed that the rate of migration of the melt extraction locus across the Pacific plate has been less regular than previously assumed (Fig. 4.8) (Sharp and Clague 2006). After the Emperor chain began to form at ~82 Ma, its migration rate varied from 17 ± 5.4–$4.8 \pm 0.4\,cm\,a^{-1}$ – more than a factor of three. The rate was more constant during formation of the Hawaiian chain, but still varied from 6.5 ± 1.1–$4.6 \pm 1.0\,cm\,a^{-1}$.

Although large-volume, tholeiitic volcanism has generally been confined to a segment of the chain no more than ~100 km long, post-erosional alkalic volcanism may persist for several million years (Section 7.6). As a result, active volcanism may extend for distances of up to ~550 km along the chain. The onset of tholeiitic volcanism progresses relatively regularly along the chain, but the later post-erosional alkalic volcanism may migrate back and forth along it (Clague and Dalrymple, 1987). Volcanism also occurs on the crest of the surrounding flexural arch, at distances of ~200 km from the chain, making the true breadth of the melting anomaly ~400 km (Bianco et al., 2005). This compares with ranges of volcanism of ~350 km in Iceland, with ~200 km highly active (equivalent to 35 Ma of plate motion) and 1000 km in the Galapagos region, also with ~200 km highly active.

One of the most intriguing recent revelations about the Emperor-Hawaiian system concerns the geographical wandering of the melt extraction locus. The original Plume hypothesis predicts that it remained at approximately the same latitude while the chains formed. Palaeomagnetic

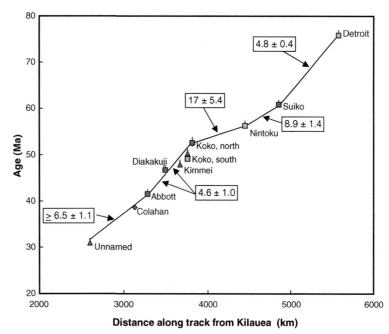

Figure 4.8 ^{40}Ar/^{39}Ar ages of Hawaiian-Emperor volcanoes vs. distance along the chain from Kilauea volcano. Boxed values are volcanic migration rates in cm/year (from Sharp and Clague, 2006). "*50-Ma initiation of Hawaiian-Emperor bend records major change in Pacific plate motion*", Sharp, W.D. & Clague, D.A. **313**: 1281–1284, 2006. Reprinted with permission from Science.

and palaeontological evidence showed as early as 1980 that the locus migrated south as the Emperor chain formed (Karpoff, 1980; Kono, 1980), and this was confirmed to a high degree of certainty by later studies (Tarduno et al., 2007).

A progressive southerly migration of the locus of eruption of ~13° (1500 km) was calculated from measurements of palaeomagnetic direction in cores drilled from several Emperor seamounts (Tarduno et al., 2007). A major southerly migration is also supported by independent calculation of the apparent polar wander path of the Pacific plate. That result requires no northward motion for the Pacific plate at all for the period 82–50 Ma, suggesting that the Emperor melt extraction locus migrated south across a latitude-stationary Pacific plate (Fig. 4.9) (Beaman et al., 2007; Sager, 2007). Even allowing for uncertainties of 300–400 km in that method, southerly migration of the volcanic locus of a minimum of ~2000 km, or 85% of the length of the Emperor chain, is still required. In contrast, volcanoes of the Hawaiian

chain have all erupted at roughly the same latitude, i.e., ~19°N.

Simply put, these results imply that, between ~82 Ma and 50 Ma, the volcanic locus migrated south at rates of up to 17 cm a^{-1} across a Pacific plate that was latitude-stationary with respect to the North Pole. Then, at ~50 Ma, the south component of its motion ceased abruptly at the same time as the Pacific plate began to move north at approximately the same rate as the volcanic locus had previously migrated south, i.e., a few centimeters per year.

What are the implications for the famous bend? Palaeomagnetic measurements alone cannot constrain palaeolongitudes. However, global plate motion models, which include other data, for example, transform fault orientations, do have the power to constrain palaeolongitude.

It has been commonly assumed, and widely taught for many years, that the bend is due to a change in plate motion, possibly resulting from the collision of India and Asia. This is a frequently perpetuated myth, as is unambiguously

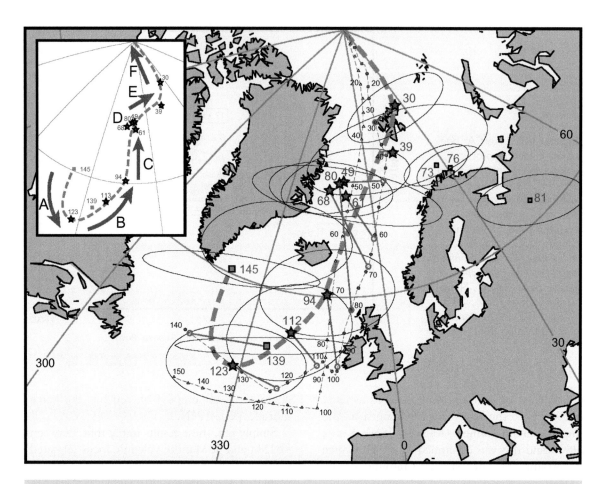

Figure 4.9 Pacific apparent polar wander path. Stars, with 95% confidence ellipses and ages in Ma, denote pole positions defining the most likely apparent polar wander path. Squares show poles determined from magnetic lineation skewness. Thin dashed lines show predicted polar wander path from plate/hotspot motion models (Duncan and Clague, 1985; Wessel et al., 2006). Lines show offset between palaeomagnetic and plume model predicted poles. Inset shows interpreted phases of polar wander, with pole ages in Ma (from Sager, 2007).

shown by global plate motion models that fail to permit such a change in motion (Fig. 4.10). There are uncertainties in plate motions in the Australia-Antarctic region, and different models can reduce the misfit. However, no model can convincingly reproduce the bend. This, coupled with the southward migration of the Emperor melt extraction locus, ironically make the Emperor-Hawaii volcanic chain system arguably the one that fits the relatively-fixed Plume

hypothesis the worst in the entire global constellation of melting anomalies.

4.6 Hotspot reference frames

How do melting anomalies behave as a whole? Are they fixed relative to one another within the observational errors, do they migrate in unison with respect to Earth's palaeomagnetic pole, or

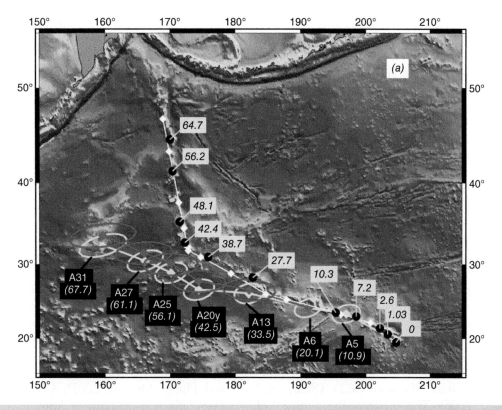

Figure 4.10 Locations of seamounts and volcanoes of the Emperor and Hawaiian chains, with ages. Lines with 95% confidence error ellipses: positions for seamounts and volcanoes predicted assuming relative fixity with Indo-Atlantic melting anomalies. Black labels A# denote magnetic anomalies with ages in millions of years given in parentheses. The ages of islands and seamounts are given in white boxes (from Raymond et al., 2000).

do they wander relative to one another? Two geometric aspects need to be considered – the locations of linear volcanic systems as a whole, and the age progressions along them.

The situation is clearest for times older than ~50 Ma. There are few volcanic chains of this age, the Emperor and Louisville seamount chains, and the Tristan-Walvis Ridge and Lakshadweep-Chagos-Réunion ridges being the most important ones. In this period, the Emperor melt extraction locus migrated south with respect to the Earth's magnetic field by ~2000 km. This could be explained by true polar wander if the other melting anomalies migrated in concord. However, the position of the Emperor

chain as a whole is predicted wrongly if relative fixity with Indo-Atlantic melting anomalies is assumed, which rules out both relative fixity between melting anomalies and true polar wander (Fig. 4.10).

Comparisons of the movements of melting anomalies on different plates require a plate motion model. Calculated relative motions can be minimized by choosing the most compliant data sub-set and incorporating a large number of free parameters such as the location of the assumed plume root and its age of inception. However, a large misfit cannot be eliminated (Steinberger et al., 2004). The relative fixity hypothesis can be rejected for the period prior to ~50 Ma. This conclusion, drawn only from the trajectories of volcanic ridges and chains as a whole, is reinforced if the rates of age progression along them are also taken into account. The progression rates along the Emperor and Louisville chains, for example, do not match. These two volcanic loci moved independently of one another (Koppers et al., 2004). Too few high-quality radiometric dates are available from the Walvis Ridge and the Lakshadweep-Chagos-Réunion ridges for similar comparisons to be made.

For times younger than ~50 Ma, more linear volcanic constructs exist and more data are available. The coherence of the results is obscured by several basic problems, however. First, many different kinds of data are involved, each associated with its own errors, and these may sum to large total uncertainties. For example, volcanoes are built over a finite period of time – several million years in the case of Hawaiian volcanoes, including the "post-erosional" stage (Section 7.6.2) (Sharp and Clague, 2006). This increases the difficulty of assigning an exact age to any point on a ridge or chain. Where volcanism is diffuse and widespread there will be ambiguity in any proposed current melting anomaly position. Uncertainties exist in palaeomagnetic directions, radiometric dates, and plate circuit models used. The larger the total error, the less able the data set is to test relative fixity.

Second, getting from question to answer may require multiple, subjective logical and computational choices to be made. These may include, for example, what subset of melting anomalies to use of the many hypothesized to be formed by plumes ("hot spot culling"; Wessel and Kroenke, 2009). As few as four, and more than 50 ridges and chains have been used in recent studies (Morgan and Phipps Morgan, 2007; Steinberger et al., 2004). The choice of which plate motion model to use is subjective and decisions regarding which radiometric dates to use are not necessarily guided by objective consideration of their statistical reliabilities.

Many additional free parameters may be introduced, for example, dividing ridges and chains into segments and assuming that plate motion was different for each (Tarduno et al., 2009). The more free parameters solved for in data inversions, the better the model fit will be, but the improvement may not be statistically significant.

Many different tests of the relative fixity hypothesis have been made, but the results are problematic to compare and assess. The assumptions made and data used in individual studies are often not described in detail. Studies may be motivated by the *a priori* assumption that melting anomalies are indeed fixed relative to one another, and describe only the subset of possible results most consistent with this. The use of questionable approaches compounds the difficulty of comparing the results of different studies. Some tests use as data only the azimuths of ridges and chains, assuming many to be time-progressive without evidence (Morgan and Phipps Morgan, 2007). Chains that are essentially non-existent, or known to be not time-progressive, for example, Hoggar and the Cameroon line, may also be used (Morgan and Phipps Morgan, 2007). Very short-lived chains, for which there is essentially no evidence for an underlying plume, for example, the Marquesas and Pitcairn chains, may also be included. Multiple plumes may be assumed where time progressions are complex, for example, along the Cook-Austral chain,[23] or plume clusters in diffuse volcano provinces, with chains defined on a pick-and-choose basis. Age data may be simply discarded on the grounds that they are discordant with dominant trends.

In view of these complexities, it unsurprising that a great variety of conclusions has been reached. Some studies report discrepancies of several hundred kilometers in the positions predicted for Indo-Atlantic melting anomalies, suggesting that melting anomalies are not relatively

[23] http://www.mantleplumes.org/Cook-Austral.html

fixed for times younger than 50 Ma (Torsvik et al., 2002). Others contend that, within the errors, true polar wander can account for the motions of Pacific melting anomalies and that all hotspots on the Pacific and African plates have remained relatively fixed within the errors since 50 Ma (e.g., Gordon et al., 2005a, b). An intriguing finding is that, although many Pacific volcano chains have bends comparable to the Emperor-Hawaiian bend, they formed at different times and, when corrected for distance from the plate pole of rotation, the bend angles are different (Koppers and Staudigel, 2005).

For the period 0–50 Ma, the clearest and most significant misfit is Iceland. That melting anomaly has remained centered on the Mid-Atlantic Ridge ever since it began to develop at ~54 Ma. No calculations are thus needed to show that it cannot have remained fixed relative to other on- or near-ridge melting anomalies such as Easter and the Galapagos. Calculations that show how a melting anomaly would have moved, had volcanism resulted from a relatively fixed plume, require a ~1500-km-long time-progressive volcanic track extending across Greenland for which there is no evidence (Lawver and Muller, 1994). No model can argue plausibly that the Iceland melting anomaly was fixed relative to others and because of this it is usually simply left out of reconstructions.

If Iceland is eliminated from the post-50-Ma data set, relative mobility between the remaining primary melting anomalies is less extreme than for times prior to 50 Ma. The largest known discrepancy concerns the Réunion system. The Deccan Traps erupted at ~65 Ma at a latitude of ~30°S, whereas the island of Réunion today lies at 21°S. This requires a migration of the melt extraction locus of ~1000 km to the north, motion that is not shown by the Kerguelen melting anomaly and which cannot be explained by true polar wander (Torsvik et al., 2002).

In conclusion, despite a great diversity of analysis approaches, and variety in the results found, it is clear that the hypothesis of relative fixity does not fit many primary observations.

Eliminating the worst-fitting data and increasing the number of free parameters enables a better model fit to be obtained and an even closer approximation can be achieved by reducing the set of melting anomalies to only those on a single plate. In addition to being required by the Plume hypothesis, this result is, however, also permitted by the Plate hypotheses. Thus, it is consistent with both models and cannot rule out either.

4.7 Plume variants

The original Plume hypothesis predicts that melting anomalies are fixed relative to one another as a result of their deep-mantle roots being relatively fixed (Morgan, 1971). Thus, explanations are needed for why this prediction is not borne out by observation. In particular, explanations are needed for the most extreme misfits – Iceland and the Emperor chain.

4.7.1 "Mantle wind"

It was pointed out early on that the surface expressions of deep mantle plumes are not expected to be fixed relative to one another, even if their roots are, because the mantle through which they rise is convecting (McKenzie and Weiss, 1975). Attempts have thus been made to model quantitatively the distortion of plume conduits by large-scale mantle convection currents.

Convection models have been erected from whole-mantle seismic wave-speed tomography structures (Section 5.6.1) (Steinberger et al., 2004). In order to do this, it is assumed that shear-wave speed in the mantle is directly related to density. The largest upwellings are then the so-called "superplumes" – two vast low-wave-speed bodies, one beneath the south Pacific and the other beneath the South Atlantic and Africa (Section 5.6.1). In the Pacific, calculations made on this basis show a large convection circuit with upwelling associated with the south Pacific "superplume", balanced by sinking beneath the subduction zones that ring the Pacific (Fig. 4.11).

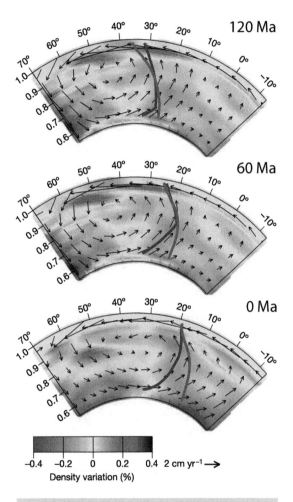

−0.4   −0.2   0   0.2   0.4   2 cm yr⁻¹ →

Density variation (%)

Figure 4.11 North-south cross-section through the Pacific mantle, and showing calculated deflections by "mantle wind" of two hypothetical Emperor-Hawaii plume conduits. Mantle density anomalies are computed assuming a direct relationship with seismic wave speed. Calculated mantle flow is shown as arrows. An initiation age of 170 Ma is assumed for the plume, since this yields the best fit to observations of melting anomaly migration (from Steinberger et al., 2004).

The convection models so derived can explain some of the relative motion between melting anomalies (Fig. 4.12). In particular, approximately half of the southerly migration of the Emperor melting anomaly and part of its positional mismatch relative to Indo-Atlantic melting anomalies can be explained. However, there are two major problems with the basic approach. The most serious is that seismic wave-speed cannot be assumed to indicate density, despite the widespread, erroneous assumption to the contrary. Seismic wave speed is dependent on three effects – phase (both mineralogical and state, i.e., whether the material is partially molten or not), composition and temperature (Section 5.1.2). The latter is the weakest effect. Critically, the low seismic wave-speeds of the "superplumes" are not caused by high temperature and those bodies are not buoyant. On the contrary, seismic normal-mode analysis, which can constrain density, shows that the low wave-speeds of the "superplumes" are due to composition and not high temperature. The "superplumes" are dense, not hot and rising. The term "superplume" is a misleading misnomer and it is unfortunate that it has fallen into widespread use (Section 5.6.1) (Brodholt et al., 2007; Trampert et al., 2004).

The second difficulty with modeling mantle convection is ambiguity. Many free parameters can be adjusted and a large suite of different results can be obtained. This reduces the power of the method to provide an objective test. Adjustable parameters include:

- The plate motion model used. There is significant uncertainty in the motion of the African plate (and thus Indo-Atlantic melting anomalies) relative to the Pacific plate (and thus Pacific melting anomalies), because of uncertainties in deformation in the Antarctic and Australasian regions.

- The seismic tomography model used.

- The dates of onset of hypothesized plumes. Old assumed onset ages for which there is no evidence, for example, 170 Ma for the presumed Emperor plume, may influence the results significantly.

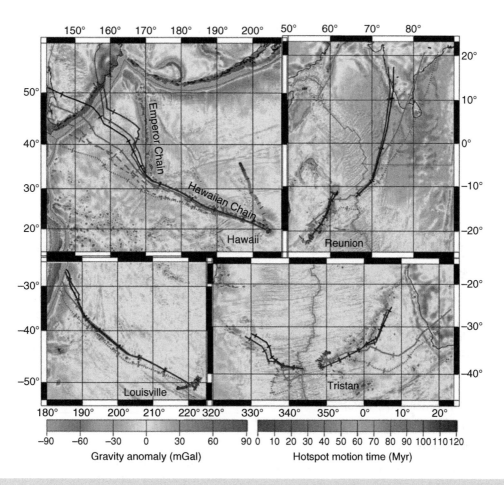

Figure 4.12 Modeled positions of volcano ridges and chains superimposed on Free-Air gravity maps of the Emperor-Hawaii, Louisville, Lakshadweep-Chagos-Réunion and Tristan-Walvis Ridge regions. Suites of lines show measured motions of volcanic loci with respect to the geomagnetic pole. Dotted lines assume the melting anomalies are relatively fixed. Short-dashed line assumes only relative motion in the African hemisphere. Long-dashed line, only motion in the Pacific hemisphere. Continuous lines, motions of all melting anomalies allowed (from Steinberger et al., 2004).

- The number of stages into which a volcanic ridge or chain is divided. A different plate pole of rotation may be assumed for each stage.

- Which melting anomalies are used.

- The mantle viscosity structure used. This controls the horizontal speed of convection.

Formal uncertainties cannot be calculated for the results because some of the choices are sub-

jective. The wider the range of possible results, the less useful mantle convection modeling becomes as a test of relative fixity. This is because the exercise becomes one of assuming fixity and then selecting the model that fits best. However, even given the wide range of possible results and the additional option of assuming an arbitrary amount of true polar wander, mantle convection still cannot explain the locations of

Iceland and much of the Emperor volcanic chain (Fig. 4.12).

4.7.2 Other variants

A large suite of other adaptations of the Plume hypothesis have been suggested to explain non-fixity or broad distributions of volcanism. These include:

- Lateral flow of plume material, in particular "capture" by spreading ridges.

- "Plume escape" – the rapid return to a vertical orientation of a plume as it breaks away from ridge "capture" (Tarduno et al., 2009).

- The deflection of plume conduits by distortions of the mantle flow field caused by mid-ocean ridges, down to depths as great as 1500 km.

- Migration of the plume base, for example, as a result of flow in the core-mantle boundary layer (Hager, 1984; Hansen et al., 1993).

- Jumps of volcanism from ridges to nearby intraplate localities, and vice versa.

- Interactions between plumes and lithospheric structures.

- "hot lines", to explain simultaneous eruption throughout long chains.

- The existence of several "constellations" of plumes, each with independent motion, for example, separate Pacific and Indo-Atlantic constellations, with Iceland forming a third constellation with one member (Molnar and Atwater, 1973).

- Later, secondary volcanism resulting from non-plume processes that obscure plume-related time-progressions (Koppers et al., 2008; Natland, 1980).

- Non-dipolar components of the geomagnetic field.

- A broad surface distribution of plume tail volcanism.

- Multiple plumes, or "plumelets" rising from a broad plume head, for example, for the St Helena, Tristan and Galapagos melting anomalies.

- Headless, thermochemical plumes, which are deflected by mantle convection faster than plumes with heads.

- Diffuse intraplate deformation not included in plate motion models.

Beyond the issues of mantle convection, errors and adaptations lies the final question of tolerance of misfit. How exact a fit of positions, ages and bends in linear volcanic constructs is required to be considered reasonably consistent with the hypothesis? Relative motions of melting anomalies of up to a few centimeters per year may be written off as "geological noise", and considered sufficiently good to be consistent with relative fixity. However, such rates are also comparable to the relative motions between plate boundaries. Thus, they are also consistent with the expectations of the Plate hypothesis, and such a result is incapable of testing one hypothesis against the other.

4.8 Discussion

The study of relative fixity of melting anomalies is beset by logical difficulties. On the one hand, studies that conclude relative fixity within the errors are considered to have confirmed the predictions of the Plume hypothesis. On the other hand, studies that find that melting anomalies move relative to one another are not considered to be inconsistent with the hypothesis because it can be claimed that relative fixity is not expected, for example, because of deflection by mantle convection. A large suite of potential rationalizations for the observed non-fixity has been developed.

Many studies assume *a priori* that the roots of melting anomalies are relatively fixed, and that the task at hand is to explain the observed wanderings of their surface expressions. Studies of relative fixity rarely seek to test the Plume hypothesis. The shedding of data that do not comply with the initial assumptions, such as discordant radiometric ages or entire melting anomalies such as Iceland and Cameroon, is not appropriate, unless it is acknowledged that the Plume hypothesis is rejected for them.

The issue of postulated ridge-crossing plumes is particularly interesting. Both examples, Réunion and Kerguelen, lie in the Indian Ocean. Unfortunately there are few high-quality data from them and there is thus little with which to examine the postulate that the volcanism on both sides of the ridges was produced by persistent, coherent magmatic systems. Alternative models include evolution of ridge-transform systems causing excess magma production on one ridge flank to change sides (Beutel and Anderson, 2007) and complex ridge migrations (Krishna et al., 1995).

Depending on the process, the Plate hypothesis permits, but does not require, melting anomalies on a single plate to be approximately fixed relative to one another. Many melting anomalies are expected to be directly linked to processes at plate boundaries. Thus their relative mobility is expected to be governed by the relative motions of plate boundaries. The rates of relative motion observed for melting anomalies and plate boundaries are, indeed, of the same order.

Diffuse volcanism, for example in the younger parts of the St Helena and Tristan-Walvis Ridge systems, is unsurprising in the context of the Plate hypothesis, and is viewed as merely reflecting where the lithosphere is in extension. Narrow, linear, regularly time-progressive volcanic ridges and chains are expected to be the exception rather than the rule, and this is observed to be the case. Ideal volcanic tracks are, contrary to popular perception, rare. There are at present only two reliably dated, fully reported examples of this type, the Emperor-Hawaii and Louisville chains. On the basis of present knowledge, narrow volcanic chains that are not time progressive are more common than those that are. The Hawaiian system is not a "typical hot spot" as is often claimed, but a melting anomaly that is unique on Earth.

The Emperor-Hawaiian system is arguably the most enigmatic of all. The melt extraction locus migrated rapidly across the plate while the latter was essentially latitude-geo-stationary, but as soon as the plate began rapid northward motion at ~50 Ma it halted. The only thing that did not change radically was the relationship between the melt extraction locus and the plate itself. This behavior has aptly been described as "unhotspot-like".[24] It seems that the only thing the Emperor-Hawaiian melt extraction locus senses is the plate itself. What process can account for this, except one that is in essence rooted in the plate itself – a persistent region of extension related to the stress field within the plate itself?

More modeling is needed of the predictions of the Plate hypothesis concerning the migrations of melting anomalies. Tectonic events are predicted to induce magmatism, but at the same time magmatism is an indicator of extensional stress and might thus conceivably be used to constrain tectonism. Both forward and reverse modeling approaches might thus be possible. The concept of a surface melting anomaly as a single, persistent, coherent entity in its own right is not helpful. It makes no sense to consider volcanoes that erupt hundreds of kilometers away from one another, or be separated in time by millions of years, to be in some sense a single object. Volcanism is a consequence of lithospheric processes, regardless of the deeper origins of the melt.

4.9 Exercises for the student

1 How spatially restricted are melt extraction loci?

2 Are postulated plumes randomly distributed on the surface of the Earth?

3 Why are there so few long, time-progressive volcano chains on Earth?

[24] http://www.mantleplumes.org/HawaiiBend.html

4 What caused the bend in the Emperor-Hawaiian volcano chain?

5 How could a lithospheric fracture beneath the Hawaiian chain be sought?

6 What is the global relationship between reactivated sutures and fault zones, and melting anomalies?

7 It has been hypothesized that Pacific seamount volcanism changed from diffuse to focused in linear chains, because the stress field in the plate evolved to require this. How may this be tested?

8 Is the Louisville volcano chain linked to the Eltanin fracture zone?

5

Seismology

知之为知之，不知为不知，是知也。[1]

– Confucius (551–479 BC)

5.1 Introduction

Seismology is essentially the only tool available for looking into the Earth's interior with any degree of focus. Other geophysical data such as gravity, geoid height and magnetics are ambiguous and of limited usefulness for revealing the sizes and shapes of geological bodies beneath melting anomalies. Geochemistry is even weaker, and has almost no power at all to measure the depth of origin of lavas erupted at the Earth's surface. Were it not for seismology, most of the interior of the Earth would be essentially unknown to us.

The seismic parameters most widely used to study the origin of melting anomalies are the compressional and shear seismic wave-speeds – V_P and V_S. Other parameters include anisotropy – the directional variation in wave-speed, and anelastic attenuation – the absorption of energy and its conversion to heat as seismic waves travel through the Earth. Anisotropy is typically interpreted in terms of crystal orientation, from which the direction of viscous flow in the Earth is surmised (Montagner and Guillot, 2000),

[1] Wise people are honest about what they know and what they don't know.

whereas attenuation may give information on the presence of partial melt. These two parameters are more difficult to measure than wave-speeds, and less often used, though they may be critical for reducing the ambiguity in interpreting wave-speeds alone.

How can seismology contribute to studying the origins of melting anomalies? Let us first briefly review the average, large-scale structure of the Earth's interior (Fig. 5.1). The main seismological divisions are the crust, mantle and core. The crust extends from the surface down to the Mohorovičić discontinuity, or "Moho", which is an abrupt jump in V_P from ~7.2–8.2 km s^{-1} (Mohorovičić, 1909). The Moho typically underlies continents at depths of 35–40 km. Beneath the oceans it commonly lies at a depth of 5–10 km (Fig. 3.1) and is usually interpreted as the bottom of the basaltic layer formed at mid-ocean ridges. Oceanic crustal thickness is thus usually assumed to indicate the amount of melt extracted from the mantle beneath, though this has not yet been confirmed by drilling in the oceans.

Beneath the Moho lies the mantle. This is thought to consist largely of peridotite, but its composition varies considerably as a result of the many dehomogenizing processes associated with plate tectonics. The crust and underlying mantle should not be confused with the lithosphere and underlying asthenosphere. The crust and mantle are defined on the basis of seismic wave speeds.

Plates vs. Plumes: A Geological Controversy, 1st edition. By Gillian R. Foulger.
Published 2010 by Blackwell Publishing Ltd.

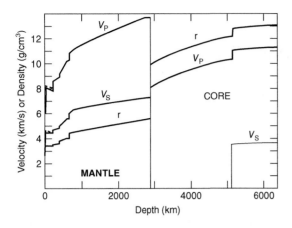

Figure 5.1 The Preliminary Reference Earth Model (PREM) (from Dziewonski and Anderson, 1981).

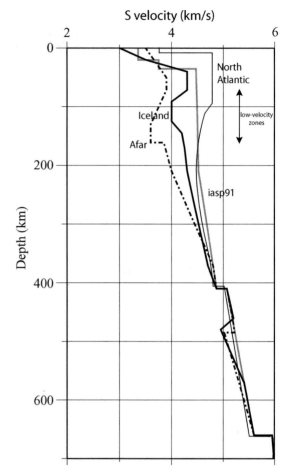

Figure 5.2 V_S for Iceland and Afar, derived using receiver functions, and from the North Atlantic, compared with the IASP91 average global model (from Vinnik et al., 2005).

The lithosphere and asthenosphere are rheological subdivisions originally proposed on the basis of gravity observations. They represent the cold, outer layer of the Earth that deforms elastically or by brittle fracture in earthquakes, and a warmer, underlying layer that flows viscously over long time periods. In recent years the original definitions have become confused with petrological concepts and seismic observations. The term "lithosphere" has been used for the high-wave-speed "LID" that overlies the seismic low-velocity layer in the shallow mantle, and for several other concepts (Anderson, 2007b), and the top of the peridotite section in ophiolites has been referred to as the "petrological Moho". Using the same word to mean several different things introduces confusion in the same way that using the word "plume" to refer to numerous, radically different things renders impossible any meaningful debate regarding whether or not they exist.

Seismic wave-speeds tend to increase with depth, primarily because of progressive increase in pressure. An exception to this general rule is the "low-velocity zone", a global layer in the shallow mantle several tens of kilometers thick (Fig. 5.2). It has reduced wave speeds and high attenuation and anisotropy and has sometimes been equated with the asthenosphere. The low-velocity zone is a likely repository for ubiquitous partial melt in the shallow mantle.

Wave-speeds increase particularly rapidly at depths of ~410 and 650 km, where V_P increases by 8 and ~6%, respectively. These high wave-speed gradients are caused by pressure-induced changes in the crystal structures of olivine and garnet (Anderson, 1967). The discontinuity at ~410 km depth is attributed to the phase transformation from olivine to wadsleite, and that at ~650 km depth to the olivine phase changes from ringwoodite to perovskite and magnesio-wüstite. An additional phase change in garnet at ~650 km depth, the transformation to the perovskite crystal structure, may also contribute. The region between ~410 and ~650 km is known as the mantle transition zone. This zone should not be confused with the "transition region", which is Bullen's region C and extends down to ~1000 km, the original depth at which the top of the lower mantle was placed (Bullen, 1963).

The pressures at which the phase changes occur, and thus the depths of the seismic discontinuities associated with them, are dependent on temperature and composition and can thus potentially give information about candidate source regions for the volcanism at melting anomalies (Bina and Helffrich, 1994).[2] The Clausius–Clapeyron relation defines the line that separates two phases on a pressure-temperature diagram (Fig. 5.3). The sign of the Clapeyron slope determines whether the depth of the phase change (and thus the seismic discontinuity) is elevated or depressed by temperature anomalies. It also indicates whether the phase change will encourage or discourage the passage of approaching cold sinkers (e.g., slabs) and hot risers (e.g., thermal diapirs).

The phase change in olivine at ~410 km depth has a positive Clapeyron slope. It is expected to be elevated by ~8 km/100 K for low-temperature anomalies and depressed by the same amount by high-temperature anomalies. This encourages the passage of cold sinkers and hot risers because as such bodies approach the discontinuity they transform to the new phase early, and the resulting changes in density encourage their onward motion.

In contrast, the phase change in olivine at ~650 km has a negative Clapeyron slope of ~5 km/100 K. It thus has the opposite effect, and will impede the passage of cold sinkers and hot

[2] http://www.mantleplumes.org/TransitionZone.html

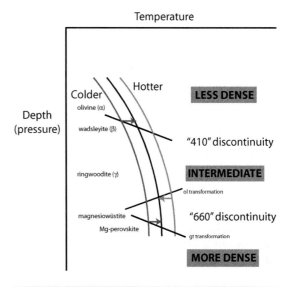

Figure 5.3 Schematic illustrating the Clausius-Clapeyron relation, which describes the transition from one mineralogical phase to another in pressure-temperature space.

risers with similar compositions. However, the nearby phase change from garnet to perovskite has a positive Clapeyron slope, which may partly or wholly cancel out the effect of the phase change in olivine (Vacher et al., 1998). As a result it is presently unclear what the total effect on mantle convection of both phase changes together may be (Deuss, 2007).

The depth of the phase change at ~410 km is also dependent on composition (Fig. 5.4). It deepens by several kilometers for each mole percent of additional Mg (Katsura et al., 2004; Schmerr and Garnero, 2007). Because the composition of olivine in the mantle is likely to vary by several percent around the $Mg_{0.9}Fe_{0.1}SiO_4$ generally assumed to be the average, compositional variations may cause tens of kilometers of topography on the seismic discontinuities – comparable to the effect expected for temperature anomalies of a few hundred degrees. Variation in the water content of the transition zone also has effects of similar magnitude at ~410 km, shallowing the phase transformation

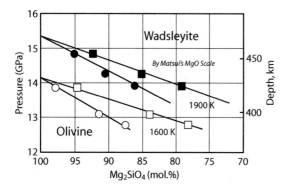

Figure 5.4 Variation in the pressure of the olivine-wadsleyite phase transformation vs. composition at 1600 and 1900°C (from Katsura et al., 2004).

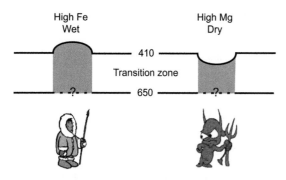

Figure 5.5 The effect on the transition zone discontinuities of composition, water content and temperature.

there by ~4km/100 ppm increase in water (Smyth and Frost, 2002; Wood, 1995).

In summary, a shallow 410-km discontinuity is expected for a fertile, wet, cold mantle and a deep one for a barren, dry, hot mantle (Fig. 5.5). The depth may vary by up to a few tens of kilometers. The picture is less clear for the 650-km discontinuity, which results from several phase changes, the net effect of which is poorly understood. That discontinuity is now usually taken to mark the base of the upper mantle, though this boundary was originally placed at ~1000km (Anderson, 2007b), yet another

semantic source of confusion in particular in debates regarding the depth of subducting-slab penetration.

The upper mantle is rich in structural variation. In the top few hundred kilometers, wave-speeds vary laterally by up to ~ ±10%. High wave-speeds clearly reveal subducting slabs beneath trenches and low wave-speeds characterize the volatile-fluxed mantle wedges above the slabs and the mantle beneath major volcanoes. A major change in seismic structure occurs across the 650-km discontinuity. Around this depth many subducting slabs flatten and become horizontal (Fukao et al., 2009; Goes et al., 2000; Hamilton, 2007b; Li et al., 2000a). Underneath, lateral wave-speed variations in the lower mantle are weak, and bear little resemblance to those in the upper mantle above. In other words, few seismically coherent structures cross the 650-km discontinuity (Dziewonski, 2005; Gu and Dziewonski, 2001; Gu et al., 2001).

At the very bottom of the mantle is a ~200-km-thick layer known as D" ("D double primed").[3] It has highly variable seismic wave speeds (Julian and Sengupta, 1973), which may require the presence of partial melt. A phase change in the pyroxene mineral enstatite from the perovskite crystal structure to the denser, post-perovskite phase occurs at these depths, which may account for some of the seismic observations (Iitaka et al., 2004; Murakami et al., 2004; Oganov and Ono, 2004). Beneath D" lies the outer core. V_P decreases from ~14 to ~8kms^{-1} on passing from the mantle to the core. Shear waves are not transmitted at all, indicating that the outer core is liquid. Temperature there is thought to be ~1000°C higher than in the mantle above. This makes the core-mantle boundary the second-strongest thermal boundary layer in Earth, the strongest being the Earth's surface, where the temperature rises by ~1300K across a region 50–100km thick. The core-mantle boundary region is where plumes are postulated to be rooted.

[3] http://www.mantleplumes.org/DDub.html

5.1.2 Seismology is not a thermometer

It is a critically important point, and one frequently ignored, that seismology can, with few exceptions, only reveal seismological structure, for example, wave-speeds, reflecting horizons and anisotropy. It does not measure petrology, composition, mineralogical phase, degree of partial melt, flow direction or temperature.[4] Variations in all these factors combine to affect seismic wave-speeds. For plausible geological situations, the presence of low-degree partial melt potentially affects V_P and V_S the most. Composition has the second strongest effect – a 1% reduction in the Mg/(Mg+Fe) content of olivine reduces seismic wave-speed by as much as a temperature reduction of 70 °C (Fig. 5.6). Expected variations in upper mantle composition are sufficient to explain essentially all the global variation in seismic wave-speeds observed (Chen et al., 1996).[5] Lithologies such as eclogite may have low wave-speed anomalies equivalent to peridotite with temperatures anomalies of 200 °C. Temperature has the weakest effect on seismic wave-speeds (Table 5.1). Anisotropy may be caused by cracks, fissures, petrological fabric, alignment of magma-filled lenses caused by stretching (Kohlstedt and Holtzman, 2009), or alignment of olivine in the lithosphere caused by shearing related to plate movements over the underlying mantle.

To reduce the ambiguity in interpreting seismological images such as tomography cross-sections, independent data are required. Unfortunately, such data are rarely available. As a result, heavy reliance is commonly placed on simplifying assumptions, for example, that

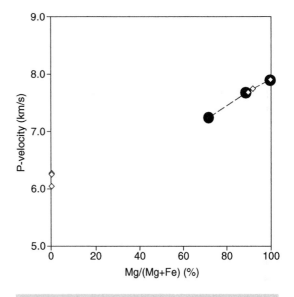

Figure 5.6 Compressional-wave speed along the olivine *b*-axis vs. Mg/(Mg+Fe) (adapted from Chen et al., 1996).

Table 5.1 Typical reductions in V_P and V_S for plausible variations in composition, degree of partial melt and temperature in the mantle.

Phase	Partial melt (per 1% increase in melt content)	Composition (per 10% reduction in Mg/(Mg+Fe) in olivine)	Temperature (per 100 K increase)
V_P	2–3%	7%	1%
V_S	3–10%	12%	1.5%

[4] http://www.mantleplumes.org/TomographyProblems.html
[5] http://www.mantleplumes.org/InterpretingSeismicVelocity.html

seismic wave-speed variations are proportional to temperature variations and that nothing else varies. Such assumptions are indefensible and the inevitable outcome is unreliable interpretations. Tomographic cross-sections are not geological cross-sections and "redium" and "blueium" are not lithologies. Seismology can contribute to investigations of the origins of melting anomalies, but it must be applied to problems that it can potentially solve, and not to ones which it virtually never can, such as mapping temperature anomalies.

5.2 Seismological techniques

Seismology offers a diverse suite of methods for imaging Earth structure. For most of the 20th century, seismology relied on analog instruments, often recording on paper or magnetic tape from which measurements were made by hand. Many local or regional seismometer networks involved simple instruments with only a single, vertical-component sensor. These are no longer widespread problems. Almost all modern seismology now uses digital data and involves complex computer data processing. The use of three-component seismometers is now standard in earthquake seismology, and the instrumentation of choice is broadband seismometers, which record signals over a wide range of frequencies. These signals can be filtered to extract different kinds of data suitable for different analyses.

Despite major advances in instrumentation, methodology and data-processing, all seismological experiments still face inherent problems, limitations and ambiguities. These difficulties tend to be de-emphasized in reports of seismological results, which commonly present inadequately statistical errors, degree of disagreement with other experiments, and ambiguity in interpretations in terms of physical and geological structure. An example of this is maps of seismic tomography results where well-resolved regions are not distinguished from poorly resolved regions that are devoid of crossing rays and

where wave-speeds have not been changed from the initial starting values. Incomplete presentation of all possible interpretations of the data leads to inconsistencies between the results of different experiments and enigmas that are often left unresolved. In order to combat these problems, this section includes descriptions of the limitations of seismic methods, along with their strengths.

5.2.1 Explosion seismology

Seismic energy from explosions is reflected and refracted at geological interfaces. It is recorded by sensors deployed at the surface on land or on the sea floor, or by strings of sensors towed behind survey ships. The method has the advantages over earthquake seismology that the locations and times of the explosions are known, and it can be used in regions where earthquakes do not occur.

There are two main types of explosion seismology – refraction and reflection surveying. Refracted energy travels along layer interfaces as "head waves" in the lower layer. It is refracted up everywhere and recorded by strings of surface sensors generally laid out in a straight line. The speed at which the head wave propagates along the sensor string equals the wave-speed in the deeper layer. Reflection surveying relies on waves reflected upward from interfaces below the sensor array. These waves do not penetrate far into the deeper layer. They yield information on the depths to reflecting horizons, but reveal little about the wave-speeds in the lower layer.

Both methods suffer from several limitations:

- Usually only information on V_P is obtained, so it may be difficult to compare results with those of other seismic methods, which yield only V_S.
- Detecting low-velocity layers, i.e., layers overlain by material with higher seismic wave speed, is more difficult than detecting layers where velocities increase progressively with depth. Amplitude data are required to detect low-velocity layers, which are

more difficult to measure accurately than phase arrival-time data.

■ In general only very shallow depths can be probed, even in the most ambitious experiment. A profile of sensors hundreds of kilometers long, with powerful explosions at either end, only yields information about the upper few tens of kilometers. Rare exceptions to this were profiles shot using nuclear explosions. The structures beneath Siberia, and to a lesser degree North America, were studied using this technique in the 1960s and 1970s. The signals were strong enough to be recorded throughout profiles ~2500 km long, yielding information on structure as deep as ~800 km (Egorkin, 2001; Julian and Anderson, 1968).

Where basaltic crust is thick, and its lower part sufficiently deep that it is neither porous nor affected by surface alteration, explosion seismology can potentially give information about the temperature of the source from which it melted. The higher the source temperature, the thicker the crust will be. Also, the degree of melting, MgO, olivine content and V_P in the lower crust will be elevated. Thus, the average V_P of the lower crust will correlate positively with total crustal thickness. For a source with normal temperatures but enhanced fertility, lower MgO and olivine content will result in a negative correlation between V_P and total crustal thickness (Fig. 5.7) (Korenaga et al., 2002).

5.2.2 Tomography

Seismic wave-speed tomography involves using a network of sensors in the area of interest to determine two- or three-dimensional underlying Earth structure from the arrival times of compressional (P) and shear (S) waves (Iyer and Hirahara, 1993). Earthquakes or explosions may be used as energy sources. A starting structure is assumed, which may be the best one-dimensional model available for the region. This structure is then perturbed, usually several times, to minimize the misfit between measured seismic-wave arrival times and those calculated for the starting model.

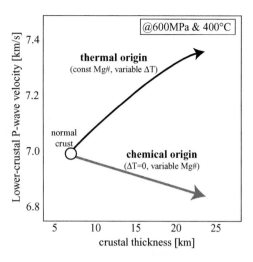

Figure 5.7 Predicted relationship between crustal thickness and lower-crustal V_P at a temperature and pressure of 400 °C and 600 MPa. When thick crust is produced by anomalously hot mantle, a positive correlation results because a higher degree of melting yields magma with a higher MgO content. Thick crust can also be created by melting of fertile (low MgO) mantle, resulting in a negative correlation (from Korenaga et al., 2002).

Tomography can be used on many spatial scales from local regions a few kilometers across to the entire mantle. The quality of the resolution depends critically on the data set. Seismic waves traveling in many different directions through the region of interest are required, providing rays that cross within the volume of interest. Where rays propagate in one direction only, volumes with anomalous seismic wave-speeds cannot be resolved well, and anomalies are typically elongated along the ray bundle. Where there are no penetrating rays, no information is available and computer programs may perturb the initial, assumed starting value in an arbitrary manner. Tomography experiments on two scales are most relevant to melting anomalies – a scale of a few hundred kilometers, and the entire mantle.

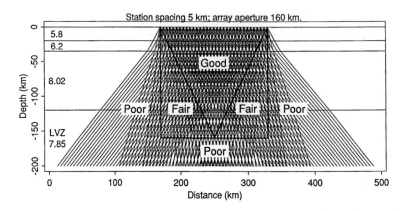

Figure 5.8 Schematic cross-section illustrating a teleseismic tomography experiment and potential ray sets that ideally approach the seismic network from below. Regions where rays cross may yield well-resolved structures. Good resolution is possible in the volume directly beneath the network, and resolution degrades as depth and lateral distance from the center of the network increases (from Evans and Achauer, 1993).

5.2.3 Teleseismic tomography

Teleseismic tomography involves a network of seismic stations deployed throughout a region of the order of hundreds of kilometers wide, and recording large earthquakes that occurred thousands of kilometers away ("teleseisms").[6] Seismic waves from such earthquakes pass through the deep interior of the Earth and approach the network traveling steeply upward (Fig. 5.8).

Three-dimensional images of both V_P and V_S beneath the network, extending down to a depth roughly the same as the network diameter, can be obtained. This depth limitation can be increased to ~1.5 times the network diameter by using techniques that go beyond simple ray theory and take into account the finite wavelengths of teleseismic waves. Such techniques are known as "finite-frequency" or "banana-doughnut" tomography (Dahlen et al., 2000; Hung et al., 2000). The structural resolution that can be achieved is typically ~100 km.

The question of the continuity or lack thereof of structures through the transition zone and into the mantle immediately beneath is impor-

tant to distinguishing between the Plate and Plume hypotheses. In order to study the transition zone, images extending down to 700–800 km are needed, with a resolution of ~50 km. To achieve this, dense arrays of ~200 seismic stations are required. Such experiments can only be done as part of national instrument deployments, and have not yet been staged to study single melting anomalies.

Teleseismic tomography suffers from several limitations:

- Depth is difficult to resolve because rays from teleseisms approach the network traveling steeply upward. This geometric situation results in a strong tendency to overestimate the vertical extents of structures. This problem is particularly unfortunate in the case of melting anomalies, because the dimension of most interest – the depth extent of structures – is the one with the largest error.

- Resolution is relatively low and anomalies significantly smaller than ~100 km are likely to be missed.

- Structure far from the network is assumed to be homogenous, although this is clearly not true. Because of this simplifying assumption, the effects of distant structure can corrupt the image obtained for the volume of interest. This effect can be partially eliminated, but at the expense of any significant

---

[6] http://www.mantleplumes.org/Seismology.html

linkage with absolute Earth structure. Thus, teleseismic tomography tends to only yield the variations in structure within an imaged block. It cannot be assumed, for example, that the average wave-speed in the surrounding region is the same as that at the periphery of the imaged block, or the average within it.

5.2.4 Whole-mantle tomography

This technique yields absolute wave-speeds for the entire mantle, rather than relative wave-speeds for some local region. It uses recordings of large earthquakes on permanent global seismometer networks. There are many computational approaches, but most rely on the measured arrival times of compressional and shear waves at seismic stations, surface wave data, and perhaps entire waveforms.

Like teleseismic tomography, whole-mantle tomography also has limitations, including:

■ *The non-uniform global distribution of earthquakes and seismic stations:* Most large earthquakes occur around the rim of the Pacific Ocean and in the Alpine-Himalaya belt, and are associated with subduction and continental collision zones. Earthquakes occur as deep as ~700 km in some places. Smaller, shallower earthquakes occur at mid-ocean ridges and transform zones. However, few large earthquakes occur in the interiors of the plates that make up the majority of the Earth's surface. The distribution of seismic stations that make up the global networks is also non-uniform. Most are deployed on land, and the two-thirds of the Earth's surface that is ocean is essentially uninstrumented except at a few oceanic islands, for example, Hawaii. As a result, much of the Earth's mantle is poorly sampled or even entirely devoid of through-passing seismic waves.

■ *Low resolution:* Stations of the global networks may be separated by hundreds of kilometers, and the wavelengths of seismic waves used may be as long as ~100 km. As a result, resolution may be little better than ~1000 km. Such resolution is inadequate to detect structures on the scale of a few tens, or even a few hundreds of kilometers, that have been postulated, for example, plume tails.

■ *Low-wave-speed anomalies* are more difficult to image than high-wave-speed anomalies. This is because seismic waves passing through low-wave-speed anomalies arrive at seismic stations later than waves that travel around them. Thus, through-going waves do not give rise to the earliest arrival and tend not be measured by seismologists.

5.2.5 Presenting tomography results

Display and interpretation of the results are particularly critical issues.[7] Very different visual impressions may be given, depending on the color-scale chosen, and the "zero anomaly" value assumed (i.e., the wave speed at which the results change from being mapped in "cooler" colors to "warmer" colors). The "zero anomaly" is typically chosen to be the average value at each depth. In the case of teleseismic tomography, where absolute wave speeds are not available, this choice typically results in images that show both low- and high-wave-speed anomalies, regardless of their true relationship to the global average. Sometimes, different color scales are used for different parts of a study volume, for example, for the upper and lower mantles. Different choices of the depths, orientations and extents of horizontal and vertical cross-sections can also make major differences to the appearance of the results (Section 5.5.1).

In most experiments, some regions are poorly sampled, or not sampled at all by seismic waves. This is particularly true of the peripheral parts of regions studied using teleseismic tomography, and for much of the Pacific Ocean in whole-mantle tomography (Julian, 2005). Ideally such regions should be blanked out in maps and cross-sections, but they are often shown, instead, with their unperturbed starting wave-speed values. Where the dearth of data is extreme, cross-sections that apparently show interesting variations in seismic wave-speed may, in reality, only be showing where data exist and where they do not.

Incomplete descriptions of results may also mislead. Often only the highest and lowest

[7] http://www.mantleplumes.org/Seismology.html

wave-speeds detected are quoted, giving the impression that much stronger anomalies are widespread than is, in truth, the case. It is more reasonable to quote the average value, and the volume that is significantly different from the global background, rather than simply the peak strength. Tomography tends to smear anomalous regions, making them appear broader and weaker than they actually are.

5.2.6 Receiver functions

The receiver-function method is sensitive to zones of rapid variation in seismic wave-speed. When teleseismic waves propagate steeply up through the crust and mantle they are reflected and converted from P- to S- and from S- to P-waves at discontinuities producing complex reverberations (Fig. 5.9). As a result a series of

multiple arrivals spread out in time is recorded at each seismic station. The receiver-function method analyzes these multiples to calculate the layered structure beneath the station (Langston, 1979). Digital data recorded on three-component seismometers are required.

The method can be applied to both the crust and the mantle. Where the Moho is sharp and well-developed it will be a particularly strong reflector. Receiver functions are thus particularly well-suited to measuring crustal thickness. They can also measure the depths to the 410- and 650-km discontinuities and topography on them.

Limitations of receiver functions include:

- The requirement for a good *a priori* crustal model from an independent source, for example, surface waves. Structures from explosion seismology are not suitable because they generally specify V_P whereas receiver functions are dominated by V_S structure. The V_P/V_S ratio in the Earth is variable and unknown,

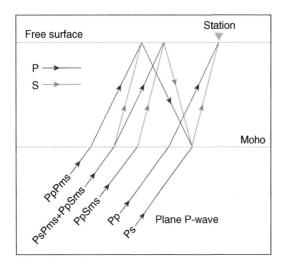

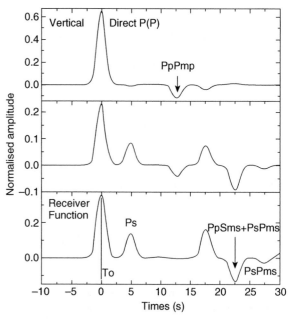

Figure 5.9 Simplified schematic illustrating the principle of receiver functions. Left: Ray diagram showing the major *P*-to-*S* converted phases that contribute to the waveform measured at a seismic station. Right: Idealized vertical (upper) and radial (middle) seismograms predicted, and the corresponding receiver function calculated (lower).

- so a V_P structure cannot be accurately converted to V_S.
- Receiver functions can only perturb a starting model. Furthermore, they are not sensitive to absolute wave-speeds, but only to rapid wave-speed variations.
- The signal-to-noise ratio may be very small. For example, as many as 100 seismograms must be stacked in order to reduce the errors in the depth to the 410-km discontinuity to a level of ~10 km.

5.2.7 Shear-wave splitting

Virtually all minerals and rocks are anisotropic, i.e., their properties vary with direction (Montagner and Guillot, 2000). Anisotropy complicates seismic-wave propagation in many ways, for example, by causing wave-speeds to vary with propagation direction. One of the easiest ways to measure the effects of anisotropy is to assess the "splitting" of a shear wave that has transformed into two independent waves, with different propagation speeds and polarization directions. By analyzing the motion of the ground produced by a shear wave, it is possible to determine this orientation and to measure the total time difference between the two waves, which depends upon the integrated strength of anisotropy along the path of the wave.

The mantle above ~250 km is particularly strongly anisotropic, an effect that is usually assumed to be caused by dislocation creep aligning olivine crystals so that their crystallographic *a*-axes (100-axes), corresponding to the polarization direction of the faster shear wave, are parallel to the flow direction. Because of this situation, measurements of shear-wave splitting are often interpreted in terms of the directions of flow in the upper mantle.

5.3 Predictions of the Plume hypothesis

The arrival of a "plume head" at the Earth's surface is predicted to cause the eruption of a flood basalt (Campbell, 2006). Postulated examples include the Columbia River Basalts, the Paraná basalts, the Siberian Traps, the Ontong Java Plateau and volcanic margins such as those bordering the North Atlantic Ocean. These flood basalts commonly comprise basaltic layers up to several tens of kilometers thick. They are predicted to have melted in a hot source, which would increase their olivine and MgO content relative to FeO. This compositional difference would tend to elevate seismic wave-speeds and lead to a positive correlation between lower crustal V_P and total crustal thickness (Fig. 5.7). Both this correlation and the total thickness of the basaltic layer are potentially measurable using seismology.

High melt production rates are expected only in plumes where they rise to relatively shallow depths. For lithosphere ~100 km thick, melt production rates – even in large plume heads – are predicted to be no more than ~0.07 km^3 a^{-1} (Fig. 3.5) (Cordery et al., 1997). Thus, melt production rates much higher than this imply a lithosphere no more than a few tens of kilometers thick. This can potentially be measured using seismology.

The Plume hypothesis predicts that melting anomalies are fed via conduits that extend from the surface down to the core-mantle boundary – "plume tails". These conduits are zones of hot, upwelling material. They rise from the thermal boundary layer at the base of the mantle where the temperature increases by ~1000 °C. The regions of D" where they are expected to be rooted are predicted to have low wave-speeds. Estimates of the temperature anomalies in plume conduits are variable, but a few hundred degrees is commonly expected. Elevated temperatures will reduce seismic wave-speeds by an amount that varies with depth (Karato, 1993). In the shallow mantle, a temperature anomaly of 300 °C is expected to lower V_S by ~4.5%, decreasing to ~1.5% at ~1000 km depth (Table 5.1; Fig. 5.10).

Where plume conduits approach or pass through the transition zone they are predicted to interact with the 410- and 650-km discontinuities (Section 5.1). It is unclear what

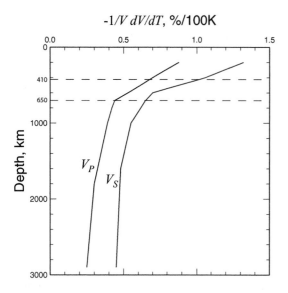

Figure 5.10 Sensitivity of seismic wave speeds to temperature at different depths in the Earth's mantle (from Julian, 2005).

the net effect at the 650-km discontinuity will be on a rising plume, as the phase changes in olivine and garnet are thought to have opposite Clapeyron slopes, and it is not known which dominates. If the phase change in olivine dominates, a rising plume of the same composition as the transition zone will pancake beneath the 650-km discontinuity, and if some of the flow penetrates, it will deflect the discontinuity upward. A net Clapeyron slope of −2 MPa/K will allow a plume to pass through little impeded, whereas a net Clapeyron slope of −3 MPa/K will block it completely from rising further (Davies, 1999). The 410-km discontinuity is predicted to be warped downward by ~24 km for a plume with a 300 K temperature anomaly.

5.4 Predictions of the Plate hypothesis

The Plate hypothesis attributes melting anomalies to processes of relatively shallow provenance

that are driven by heat loss from the Earth's top thermal boundary. The sources of melts need not have particularly elevated temperatures compared with mantle material at similar depths elsewhere. The Earth's surface thermal boundary layer is ~100–150 km thick, however, and material drawn from deep within it is expected to have higher temperatures than material drawn from shallow levels. MgO and olivine are not expected to be elevated in the basaltic crust and V_P in the lower crust is expected to correlate negatively with total crustal thickness (Fig. 5.7).

Many plate-based mechanisms for melting anomalies involve rifting or stretching of the lithosphere, which allows asthenosphere to rise passively and to partially melt on decompression. Continental break-up, enhanced fertility at mid-ocean ridges, enhanced volcanism at plate-boundary junctions and continental intra-plate extension are mechanisms of this kind. In these cases, underlying mantle structures that might be detectable using seismology are expected to scale up with the lithospheric process at work. Where rifting is extreme and the lithosphere thick, underlying structures in the mantle are expected to be wide and to extend deeper than where the lithosphere is thin and the stretching minor.

Shallow convection in the mantle includes small-scale sublithospheric convection and EDGE convection (Anderson, 1998a). Such convection is expected to be seismically detectable (King and Ritsema, 2000). Catastrophic lithospheric thinning would require that a large portion of lithosphere, which was perhaps originally 100–200 km thick, was removed by sinking and detachment or delamination due to gravitational instability, possibly taking with it part of the lower crust. This process causes the lithosphere to be unusually thin at the time of flood basalt formation.

Cracking of the entire lithosphere as a result of far-field tectonic stresses may enable melt, pre-existing beneath the lithosphere, to rise to the surface. This process may account for intra-plate melting anomalies such as the "petit spots" near the Japan Trench (Hirano et al., 2006), oceanic islands such as Hawaii, Tahiti and Samoa, and flood basalts that erupted through intact, thick lithosphere, for example, Karoo. Small-volume intraplate melting anomalies such as seamounts are widespread throughout the interiors of the plates, suggesting that extra-ctable melt is also widespread throughout the mantle. Where melting occurs, lowered seismic wave speeds are expected, with V_S depressed more than V_P. Where voluminous melting occurs, for example, at Hawaii, stronger and more widespread lowered seismic wave-speeds are expected.

The supply of melt large enough to form flood basalts on time-scales as short as a few million years is attributed largely to the release of melt previously accumulated and ponded beneath the lithosphere. If such accumulations existed in the past, there seems no reason why they should not exist today, and be detectable using seismology. Melt ponds might comprise layers containing a high degree of partial melt. Depending on how large the degree of melting is, the top surfaces might be strong reflectors of seismic energy. The layers themselves would be expected to have low seismic wave-speeds, high V_P/V_S ratios, and high attenuation, particularly of shear waves.

When partial melt forms in a peridotite mantle source, garnet and clinopyroxene are the first minerals to melt. The source is thus depleted in these minerals when melt is extracted. These minerals have relatively high densities but low seismic wave speeds. The residuum left behind is thus expected to be relatively buoyant but to have higher seismic wave-speeds than the original fertile peridotite. Melting anomalies that are fed by extensive melting in the mantle beneath are thus expected to be underlain by buoyant material that not only provides eleva-tion support but also has relatively high seismic wave-speeds.

The Plate hypothesis does not require melting anomalies to be underlain by structures that are continuous throughout the mantle. Related structures, and associated seismic anomalies, are predicted to be confined to the shallow mantle and to be associated with formation and extr-action of melt and its emplacement at and near the surface. In this context, "shallow" means the top ~1000 km of the mantle. The extension of seismic anomalies to that depth is expected only in exceptional cases, however. In most cases seismic anomalies are expected to be confined to the mantle above ~410 km, occasionally extend-ing down into the transition zone itself. Related effects, for example, on discontinuity topogra-phy, may occasionally exist at 410 km, but these are not expected to be mirrored by similar effects on the 650-km discontinuity.

5.5 Observations

A surprising thing about the hunt for plumes is how few localities have been thoroughly studied using seismology.[6] In order to discriminate between the structures predicted by the Plate and Plume hypotheses, the mantle should ideally be imaged at high resolution through the transi-tion zone and into the mantle beneath. A dense, high-quality seismic network several hundred kilometers wide deployed in the area of interest is needed for this purpose.

Until recently, such networks could be deployed only on continents. At sea, ambi-tious and expensive ocean-bottom-seismometer deployments are needed. Because very few cur-rently active melting anomalies are surrounded by extensive landmasses, many key melt-ing anomalies such as Réunion, Tristan and Louisville have hardly been studied at all using seismology. In the following sections, melting anomalies will be categorized and discussed according to the detail in which they have been studied.

5.5.1 Well-studied melting anomalies on extensive land masses

Only three melting anomalies have been studied in detail using seismology – Iceland and the North Atlantic Igneous Province, Eifel in Germany, and Yellowstone in the USA.

(a) The North Atlantic Igneous Province

Iceland and the North Atlantic Igneous Province have been well studied seismically as the result of a unique combination of circumstances.[8] The passive margins of the North Atlantic have been heavily researched for hydrocarbon exploration. Iceland itself is the largest oceanic volcanic island in the world, with a landmass ~450 × 300 km in area. Furthermore, it is surrounded at distances of no more than ~1000 km by Greenland, Scandinavia and the British Isles.

Explosion seismology has been used since the 1960s to probe the crust beneath the Greenland-Iceland-Faeroes ridge. Early studies in Iceland comprised simple, mostly short, refraction profiles (Flovenz, 1980; Palmason, 1980). No clear Moho was detected, and it was concluded that the base of the crust lay at ~10 km depth where the wave-speed gradient suddenly decreased. The material beneath had a V_P of ~7.0 km s^{-1} and was interpreted as mantle with unusually low wave-speeds as a result of being hot and partially molten (Björnsson et al., 2005). This – the "thin/hot" crustal model – was considered to be consistent with the Plume hypothesis on account of the hot mantle it envisaged.

In the 1980s and 1990s, explosion profiles hundreds of kilometers long were shot on the submarine Greenland-Iceland and Iceland-Faeroe ridges, and in Iceland itself (Bott and Gunnarsson, 1980; Darbyshire et al., 1998; Holbrook et al., 2001). The crust beneath

[8] http://www.mantleplumes.org/TopPages/IcelandTop.html

the submarine Greenland-Iceland and Iceland-Faeroe ridges was found to be ~30 km thick. Refracted head waves revealed a clear Moho beneath the latter ridge.

In Iceland, a reflective horizon originally detected in 1960 at ~20–40 km depth and interpreted as the Moho (Bath, 1960), was re-discovered (Bjarnason et al., 1993; Björnsson et al., 2005). As a result, the material previously thought to be hot, partially molten mantle was re-interpreted as sub-solidus, gabbroic lower crust. The "thin/hot" model was thus superceded by the "thick/cold" crustal model. That model is supported by low seismic attenuation and normal V_P/V_S ratios in the lower crust, which indicate that the upper ~40 km beneath Iceland are sub-solidus and cool (Section 6.5.2) (Menke and Levin, 1994; Menke et al., 1995).

However, far from radically changing views of the fundamental origin of Iceland, this complete reversal of perception of the crust made essentially no difference to mainstream opinion. The "thick/cold" model was considered to be consistent with the Plume hypothesis on the grounds that a thick crust represented an anomalously large volume of melt. This example illustrates well how observation-independent the Plume hypothesis is as a practical matter.

Assuming that the thick, low-V_P layer does indeed correspond to solidified melt, its origin in either an anomalously hot or an anomalously fertile source was investigated by correlating seismic wave speed with crustal thickness for the Greenland-Iceland ridge (Korenaga et al., 2002). A negative correlation was found, consistent with a cool, fertile source (Fig. 5.11).

In the 2000s, dense, broadband seismometer networks covering all Iceland were deployed for several years (Foulger et al., 2000; 2001; Wolfe et al., 1997). What is probably the best earthquake data set ever collected at any melting anomaly on Earth was acquired and used for numerous analyses. Crustal thickness over all Iceland was mapped using receiver functions (Fig. 3.11), a major improvement over the explosion profiles used earlier, which had only

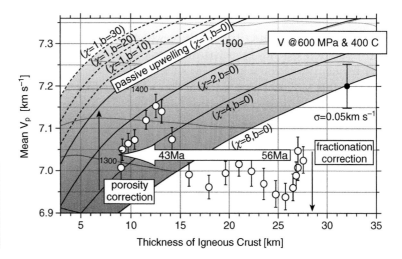

Figure 5.11 Mean velocity of the lower crust as a function of whole crustal thickness (open circles with error bars) (from Korenaga et al., 2002).

yielded deep information from beneath the central parts of a few long profiles.

Two major discoveries emerged from this work. First, no true Moho exists beneath most of Iceland (Foulger et al., 2003). The rise from crust-like wave-speeds with $V_S = 4.1 \, km \, s^{-1}$ ($V_P \sim 7.1 \, km \, s^{-1}$) to mantle-like wave speeds with $V_S = 4.5 \, km \, s^{-1}$ ($V_P \sim 7.8 \, km \, s^{-1}$) is not abrupt but occurs gradually across a layer up to ~8 km thick. This is not typical of the true Moho. This finding is consistent with the explosion seismology data. The deep interface detected by that work is nearly everywhere a supercritical reflector. Almost no head waves refracted along it were observed, suggesting that it is not a widespread sharp increase in seismic wave speed.

A gradational crust-mantle boundary accords with the conclusions drawn from the low elevation of Iceland (Section 3.5.2). If the crust is assumed to be thick, the density contrast between it and the mantle beneath can only be <90 kg m³ and not the <300 kg m³ typical for oceanic crust and mantle. No plausible petrology has yet been found to account for this small density contrast (Gudmundsson, 2003) – this enigma is one of many that still remain about Iceland, despite many decades of study using the most ambitious and technologically advanced experiments performed anywhere on Earth.

The second major finding was a detailed map of the thickness of the layer with crust-like wave-speeds. This layer is ~30 km thick beneath eastern Iceland, ~20 km beneath western Iceland, and about ~40 km thick in a block ~10,000 km² in area beneath central Iceland (Fig. 3.11). This pattern is not what is expected for a mantle plume. If a plume either migrated from the west or was continually centered beneath the spreading ridge, the thickest crust would form a belt traversing Iceland from west to east. The results are consistent with those from seismic anisotropy, which suggests flow generally north-northeasterly, parallel to the spreading plate boundary (Li and Detrick, 2003).

The ~40-km-thick block beneath central Iceland is a feature unique along the entire Greenland-Iceland-Faeroe ridge. Nowhere else is such thick crust detected. It has a unique structure, with its lower part being a thick low-velocity layer (Du and Foulger, 2001). This fact suggests that it has an origin different from the rest of the Greenland-Iceland-Faeroe ridge. It may represent the submerged, trapped microplate that is required by palinspastic reconstructions of Iceland (Fig. 3.12; Section 4.5.1) (Foulger, 2006).

Several teleseismic tomography images of the mantle have been calculated (Foulger et al.,

2000; 2001; Hung et al., 2004). The most sophisticated of these maximizes the depth to which good resolution extends by accounting for the finite frequencies of the seismic waves (Section 5.2.3). In this way, data from the ~450-km-wide network yielded images with good resolution down to ~650 km (Fig. 5.12).

Iceland is underlain by a low-wave-speed anomaly that extends from the surface downward through the upper mantle, where it ends abruptly near the base of the transition zone. The sharpness of the cut-off is remarkable, and resolution tests show that if the anomaly extended beneath the 650-km discontinuity it

(a) P-velocity model

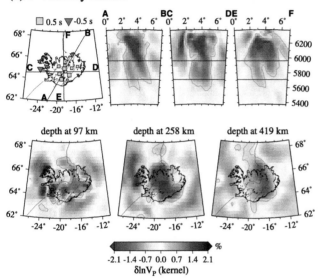

(b) S-velocity model

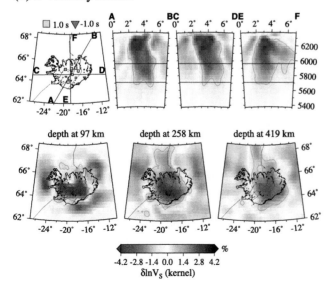

Figure 5.12 Teleseismic tomography images of compressional- and shear-wave-speed perturbations beneath Iceland (from Hung et al., 2004). See Plate 19

would be detected. The anomaly is strongest beneath Iceland but continues north along the Kolbeinsey ridge. Throughout most of its volume it has a strength of ~2% in V_P and 4% in V_S relative to regions to the east and west (i.e., it has a strength of ~1% in V_P and 2% in V_S relative to the average for the volume studied).

The North Atlantic has also been imaged using global tomography (e.g., Megnin and Romanowicz, 2000; Ritsema et al., 1999). Essentially all studies show that the region is underlain by a low-wave-speed anomaly that terminates at the base of the transition zone (Fig. 5.13). The strength of the anomaly is typically ~1.5% in V_S. Anomalies detected in the lower mantle are much weaker and their locations and shapes are different in different studies, i.e., they are not reliable.

Published images of the mantle beneath the Iceland region provide striking examples of how similar results can be presented in ways that give radically different impressions. It has recently been claimed that whole-mantle tomography shows a low-V_P anomaly extending from the surface down to the core-mantle boundary, which has been interpreted as a mantle plume (Fig. 5.14) (Bijwaard and Spakman, 1999). Surprisingly, the structure shown in Fig. 5.14 is essentially the same as that shown in Fig. 5.13. The appearance of a mantle-traversing low-V_P anomaly in Fig. 5.14 was achieved by saturating the color scale at the low value of ± 0.5%.

If the data shown in Fig. 5.14 are re-plotted using a color scale with greater range, and

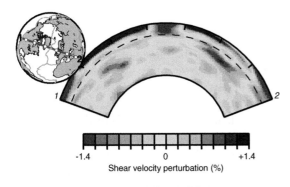

Figure 5.13 Whole mantle tomography cross-section through Iceland, calculated using earthquakes that occurred 1980–1998 and were recorded on several global networks. Over a million surface-wave phase velocity measurements and 50,000 hand-picked compressional- and shear-wave arrival time measurements for a large variety of seismic phases were used. This image is currently the best available for the global transition zone. The results show a low-wave-speed anomaly filling the whole North Atlantic, but extending no deeper than the transition zone (from Ritsema et al., 1999). See Plate 20

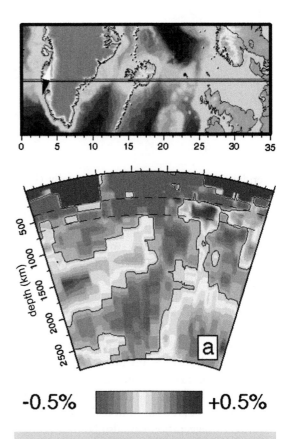

Figure 5.14 Whole-mantle tomography model purporting to show a mantle plume extending from the surface down to the core-mantle boundary under the North Atlantic (from Bijwaard and Spakman, 1999). See Plate 21

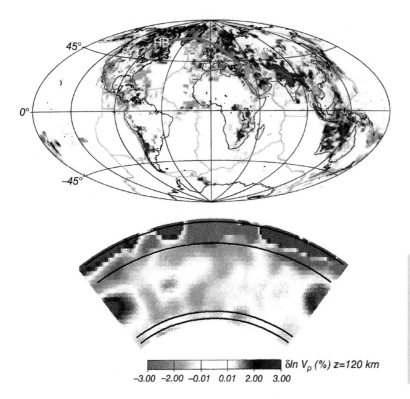

Figure 5.15 Model shown in Fig. 5.14 replotted with the color scale saturated at an anomaly strength of ±3%, and the line of section extended to underlie Canada and Scandinavia. HB – Hudson Bay. See Plate 22

$\delta ln\ V_p$ (%) z=120 km

-3.00 -2.00 -0.01 0.01 2.00 3.00

extending the line of section horizontally (Fig. 5.15), several things immediately become clear. First, the anomaly in the lower mantle is much weaker than that in the upper mantle, and the illusion of a continuous structure vanishes. Second, the form and shape of the anomalies in the lower mantle bear little resemblance to those of Fig. 5.13 – there is poor agreement between different studies. Third, a lower-mantle anomaly of similar strength and arguably more plume-like in shape is found beneath Hudson Bay in Canada, where no plume is expected. This feature is not seen in Fig. 5.14 because in that plot the line of section is truncated at distances of ± ~2,500 km from Iceland.

Vertical exaggeration can also be misleading. Fig. 5.16 shows the structure beneath the North Atlantic from Fig. 5.13, plotted with a vertical exaggeration of 5 (Ritsema, 2005). This imparts a tall, plume-like appearance to the low-wave-speed anomaly, obscuring the simple fact that it

is actually wider than it is tall. This is not to say that its shape is certain – global tomography cannot resolve anomalies narrower than ~1000 km, and so smaller features appear artificially widened. The true width of this anomaly is unknown.

Many critical questions remain unanswered. What is the real strength of the anomaly? What are its true lateral dimensions? Although its depth extent is confidently known, its width is not. Teleseismic tomography can only detect variations in wave-speeds within the study volume, not absolute wave-speeds. Images from teleseismic tomography are typically designed such that the arithmetic mean of the anomalies at each depth is zero, and the color scale is fully utilized. The method cannot determine, nor the images illustrate, the extent of anomalies relative to an average global model, nor can it determine how far anomalies of interest extend outside the study volume.

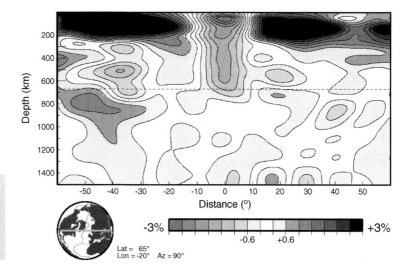

Figure 5.16 Whole-mantle tomography cross-section through Iceland, with a vertical exaggeration of ~5 (from Ritsema, 2005).

Whole-mantle tomography cannot answer these questions either, because of its limited resolution. The upper-mantle anomaly beneath the North Atlantic shown in Figs 5.13–5.16 is ~2000 km wide, but resolution tests permit it to be as little as 300 km wide if it has a strength of 6% throughout its entire volume (Ritsema, 2005). Such a strong anomaly is not detected by the teleseismic tomography, although a receiver function study suggests that anomalies as much as 10% lower than the global average in restricted areas cannot be ruled out (Du et al., 2006; Vinnik et al., 2005). However, current data permit the anomalous region in the North Atlantic to be considerably wider than Iceland itself.

What happens at 650 km depth beneath Iceland? Does the seismic anomaly extend through the transition zone and continue downward with a diameter too narrow to resolve by tomography, or is it truly confined to the mantle above 650 km? This problem has been addressed by studying the transition zone using receiver functions.

Large-scale studies detect a 410-km discontinuity in the Iceland region that is elevated or at normal depth (Deuss, 2007; Gu et al., 2009). Local studies show that both the 410- and the 650-km discontinuities lie at normal depths beneath most of Iceland. An early local study reported that the transition zone thins beneath southeast Iceland by ~20 km (Shen et al., 1998) and interpreted this thinning as a result of a temperature anomaly of ~150 °C depressing the 410-km discontinuity and raising the 650-km discontinuity (Shen et al., 2002). Later work, taking into account wave-speed variations and analyzing the two discontinuities separately, showed that the local thinning results from depression of the 410-km discontinuity only, and that the 650-km discontinuity is flat (Du et al., 2004). Thus, whatever causes the thinning of the transition zone – a compositional, water or temperature anomaly – does not perturb the 650-km discontinuity. Unfortunately, the 650-km discontinuity may be caused by multiple mineralogical phase changes and the effects of variations in composition, melt content and temperature are not yet well known (Section 5.1). The absence of topography on the 650-km discontinuity may thus have little ability to constrain the depth extent of bodies.

This history of study of the transition zone beneath Iceland provides another interesting example of how seismic results are interpreted. This history involves conflicting results, assumption-driven interpretations, a radical about-turn in opinion regarding the basic observations, and finally questioning of whether the investigative approach has, in truth, any significant power to distinguish between candidate models for the Iceland melting anomaly.

(b) The European Cenozoic Volcanic Province

Small-volume, isolated volcanic fields occur throughout much of Western Europe (Fig. 3.20).[9] These have been variously attributed to one or more mantle plumes, a partially molten layer representing a "fossil plume head" from which small "baby" or "finger" plumes rise, or to permissive volcanism within or beneath the flanks of extensional rifts forming in response to Alpine orogenic events.

The Eifel volcanic field is the most intensively studied seismologically (Ritter, 2007). It is small – no more than ~70 km across and ~20 km³ in volume – and comprises a larger western and a smaller eastern part. It is interesting to note that this tiny field, and the insular shelf of Iceland, which is ~1000 km in diameter and ~10^7 km³ in volume, have both been attributed to the same phenomenon – "upper mantle plumes". Eifel was studied in a multi-faceted, multi-national experiment involving ~250 seismic stations deployed throughout an area ~500 km wide. Several techniques were applied to the vast data set gathered, including teleseismic tomography, receiver functions, shear-wave splitting and attenuation.

Teleseismic tomography reveals both low-V_P and low-V_S anomalies beneath Eifel between the near surface and the transition zone (Fig. 5.17). The low-V_P anomaly is centered beneath west Eifel. Relative to the average for the region studied, it has a strength of ~1% throughout much of its volume, peaking at ~2% locally (Ritter et al., 2001). It is elongated in an east-

west direction, and widens with depth from ~50 km near the surface to ~300 km at the top of the transition zone.

In contrast, the low-V_S anomaly is centered ~50 km further to the east and has a very different shape. Its strongest part is a quasi-spherical anomaly that extends from the surface down to ~150 km. Relative to the average of the study region, it has a strength of ~1.5 ± 1% throughout much of its volume, peaking locally at ~3% (Keyser et al., 2002). It is vertically discontinuous. No low-V_S anomaly occupies the depth range ~150–250 km. Below 250 km, and further east, weak, low-V_S anomalies are imaged.

The attenuation structure is again very different. Isolated volumes of high attenuation underlie east Eifel in the depth range ~0–50 km, and west Eifel in the depth range ~80–120 km (Ritter, 2007). The shear-wave splitting study detected variable orientations for the highest shear wave-speed, in the range ~0–90 °E.

Despite the large, 500-km aperture of the seismometer network, teleseismic tomography still has not revealed whether the low-wave-speed anomalies continue through the transition zone and into the lower mantle. Nor does topography on the transition zone resolve this question. The 410-km discontinuity is depressed beneath Eifel by ~20 km but the 650-km discontinuity lies at its normal depth (Weber et al., 2007).

What can be concluded from these seismic results? The blade-like low-V_P anomaly that extends from the near surface down into the transition zone could be caused by a compositional variation, hydration of minerals, partial melt, high temperatures or, at shallow depths, gas-filled pores. Free gas cannot exist below ~60 km depth because it will dissolve in minerals at such high pressures. Information about V_S and anelastic attenuation can potentially reduce this ambiguity. Hydration, partial melt and high temperature are all expected to reduce V_S by roughly twice as much as V_P and to increase attenuation.

The main difficulty of interpretation arises from the poor correlation between the different

[9] http://www.mantleplumes.org/Europe.html

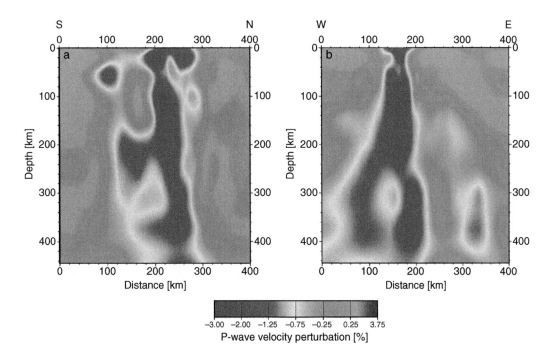

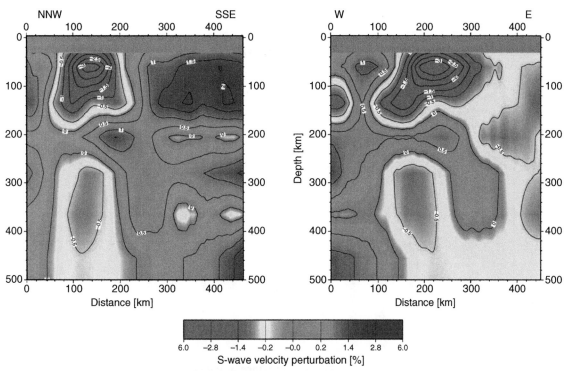

Figure 5.17 Cross-sections through the V_P (top) and V_S (bottom) teleseismic tomography models from Eifel (from Keyser et al., 2002; Ritter et al., 2001). See Plate 23

types of results. The V_P anomaly is strong and broad in the depth range 150–250 km, where there is no V_S anomaly. The strongest V_S anomaly lies in the top ~150 km beneath East Eifel, but most does not coincide with a corresponding low-V_P anomaly. Low V_S anomalies unaccompanied by V_P anomalies also occur to the east of the volcanic field. Regions of high anelastic attenuation are uncorrelated with the wave-speed anomalies and of no help in reducing the ambiguity. The down-warping of the 410-km discontinuity suggests an anomaly in either composition, hydration or temperature there, but it is not clear which.

Although the surface volcanic field bears virtually no resemblance to that of Iceland, the structure imaged beneath is strikingly similar. Both regions are underlain by low-V_P anomalies that broaden with depth to a width of several hundred kilometers at the top of the transition zone. In both cases, the 410-km discontinuity is

warped down by ~20 km but no topography is detected on the 650-km discontinuity.

The tiny Eifel volcanic field shows essentially none of the features predicted for a mantle plume. There is no flood basalt. Co-ordinated uplift and volcanism did not occur. Volcanism occurred in two phases at 45–24 and 0.2–0.01 Ma. There is no time-progressive volcanic trail, no heat-flow anomaly and no petrological evidence for high temperatures. There is, however, a persistent and voluminous outpouring of CO_2, sufficient to support a thriving industry in carbonated water production. It is estimated that some 70,000 tonnes/day degas from the surface (Fig. 5.18). The volcanism and the seismic anomalies might thus result from CO_2 (and perhaps H_2O) being drawn out of the transition zone in response to permissive escape at the surface. These volatiles could affect the seismic wave-speeds through mineral hydration, low-degree partial melting – carbonatitic due to

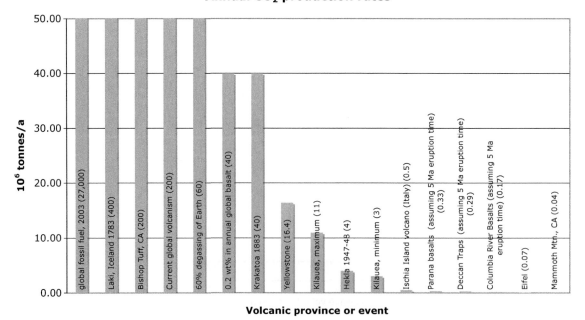

Figure 5.18 Annual CO_2 degassing rates for a variety of processes and regions.

fluxing the upper mantle and lowering its solidus–and the presence of CO_2 in pores at shallow depths.

Other volcanic fields in Europe that have been investigated seismologically include the Eger graben in the Bohemian Massif and the French Massif Central. Like Eifel, these fields have also been volcanically active for several tens of millions of years, show no time-progressive migration and are also minor in volume. However, because only V_P teleseismic tomography has been conducted, physical interpretations remain ambiguous.

The 300-km-long Eger rift has produced young eruptions and it also degasses large volumes of CO_2. No tall, narrow low-V_P anomaly similar to that beneath Eifel exists in the upper 250 km (Plomerová et al., 2007). Instead, there is a broad, low-V_P anomaly thought to indicate shallowing of the asthenosphere in response to rift-related lithosphere thinning. Beneath the Central Massif, a low-V_P anomaly extends from the surface down to 270 km depth (Granet et al., 1995). This anomaly is stronger than those beneath Eifel and Iceland, having an average value of about $1.5 \pm 1\%$, and reaching 4.7% at the shallowest levels.

A variable picture thus emerges for the European Cenozoic Volcanic Province. In only one case (Eifel) have V_P and V_S, along with anelastic attenuation and anisotropy been determined, and in that case the results correlate so poorly that they do little to reduce the ambiguity in interpretation. Beneath Eifel, wave-speed anomalies extend into the transition zone but there is no evidence that they continue on through it. These observations are not consistent with a major thermal diapir.

The small European volcanic fields are probably associated with intrusions into the grabens associated with the European rifts, and their flanks. These grabens are themselves extensional features induced by far-field stresses originating in Alpine subduction and orogenesis. Stress field modeling predicts that the maximum extensional stress occurs beneath the footwall of

normal faults, which is consistent with the common occurrence of volcanism outside of the rifts (King and Ellis, 1990).[10] An analogous model, involving magmatic activity in rift zones reactivated by changes in distant plate boundary tectonics, has been suggested for the volcanism in the West Antarctic Rift System (Rocchi et al., 2002; 2003; 2005)[11].

(c) Yellowstone

Yellowstone is the most comprehensively studied continental melting anomaly[12]. Unlike Eifel, it is the site of one of the deep mantle plumes originally proposed by Morgan (1971), largely on the basis of the ~12–0 Ma time-progressive chain of silicic calderas arrayed along the eastern Snake River Plain (Fig. 2.9). The youngest of these is the currently active Yellowstone volcano.

The intra-continental setting of Yellowstone allows large-aperture seismometer arrays to be deployed that are potentially able to resolve mantle structure well into the transition zone using teleseismic tomography. Several such experiments aimed at detecting the proposed plume have been undertaken since the 1970s. In addition, receiver functions computed using data from these experiments have been used to map topography on the transition zone discontinuities.

In the 1970s, a 57-station network of vertical short-period seismometers collected data over an area 430×250 km in size (Iyer et al., 1981). Teleseismic tomography detected a low-V_P anomaly of 15–20% in the upper crust, decreasing in strength to 5–10% in the lower crust and upper mantle (Fig. 5.19). This anomaly extended no deeper than ~200 km. Resolution analysis showed additional weak, discontinuous anomalies tilted at ~20–30° to the vertical and

[10] http://www.mantleplumes.org/Europe.html

[11] http://www.mantleplumes.org/Antarctica.html

[12] http://www.mantleplumes.org/TopPages/YellowstoneTop.html

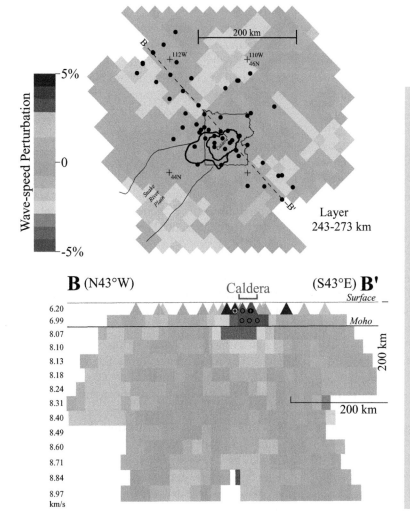

Figure 5.19 Teleseismic tomographic V_P structure beneath Yellowstone. Top: dots show the seismic stations used, the boundary of Yellowstone National Park, the calderas of the Yellowstone Plateau volcanic field, the edges of the eastern Snake River Plain, and line of cross-section BB' shown in lower panel. Grayscale in top panel indicate wave-speed variations in the depth interval 243–273 km, where a deep plume-like structure would be imaged if one exists. Lower panel: cross-section through the model at the northeast edge of the caldera (from Christiansen et al., 2002). "Upper mantle origin of the Yellowstone hotspot. *Bulletin of the Geological Society of America*", Christiansen, R.L., Foulger, G.R. & Evans, J.R. **114**: 1245–1256, 2002. Reprinted with permission from GSA. See Plate 24

extending to 400 km depth, to be artifacts caused by "smearing" along the ray bundles, which have the same orientation. The strong, shallow, low-V_P anomaly is contained within the sublithospheric low-velocity zone, which extends down to ~200 km depth. The anomaly extends west-southwestward along the eastern Snake River Plain but not beneath the flanking regions. It was concluded that Yellowstone is not caused by a deep mantle plume (Christiansen et al., 2002).

Subsequent work using digital, three-component seismic networks has yielded similar results, with strong, low-V_P anomalies in the upper ~250 km beneath Yellowstone caldera and the eastern Snake River Plain and high-V_P anomalies beneath (Dueker and Sheehan, 1997). Weaker low-V_P anomalies at greater depth are elongated along the incoming bundles of rays (Fig. 5.20). One of these, to the northwest of Yellowstone, is continuous with the shallow anomaly and extends to ~500 km depth. The strength of most of the shallow anomaly is ~1.5%, reaching a maximum of ~3.2%. The deep, weak anomaly generally has a strength of ~0.5%, reaching a maximum of ~0.9%. It

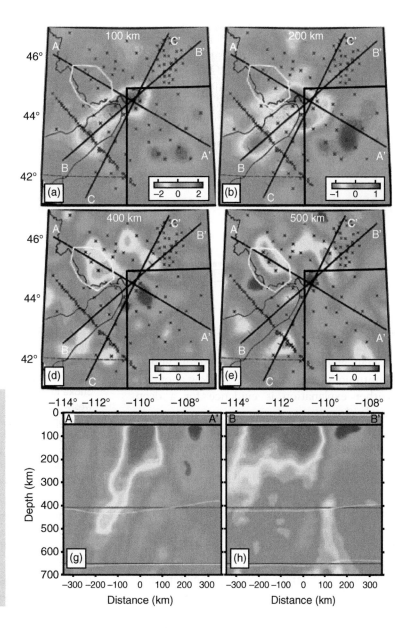

Figure 5.20 Teleseismic tomography V_P structure beneath Yellowstone (from Yuan and Dueker, 2005). White rings on the horizontal sections indicate where the 410-km discontinuity is depressed. In the cross-sections in the lower panels, the 410- and 650-km discontinuities are shown by white lines, and their average depths by black lines (Fee and Dueker, 2004). See Plate 25

was concluded from this work that the deep extension of the shallow low-V_P anomaly represents a plume tail (Yuan and Dueker, 2005).

Transition-zone topography beneath the region shows variations of 35–40 km in the depths of both discontinuities (Fee and Dueker, 2004). These variations are uncorrelated, however, and provide no evidence for any coher-

ent, through-going structures. The 410-km discontinuity is depressed by ~18 km where the deep, weak low-V_P anomaly crosses it, but the 650-km discontinuity is flat.

What can be concluded from these results? No coherent low-V_P anomaly extends from the surface at Yellowstone down into the lower mantle. The weak, low-V_P anomaly that

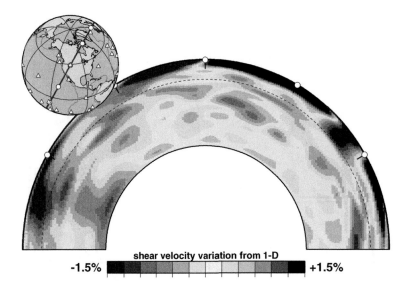

shear velocity variation from 1-D

-1.5% ▮▮▮▮▮▮▮▮▮▮▮ +1.5%

Figure 5.21 Northeast-southwest cross-section through the Yellowstone caldera, showing the mantle tomography model of Ritsema et al. (1999). *"Complex shear wave velocity structure imaged between Africa and Iceland"*, Ritsema, J., van Heijst, H.J. & Woodhouse, J.H. Science **286**: 1925–1928, 1999. Reprinted with permission from Science. See Plate 26

extends down to ~500 km has been interpreted variously as an artifact of smearing of the shallow anomaly along the incoming ray bundle and as a continuous extension of the shallow anomaly. Neither explanation is evidence for a deep mantle plume.

Holistic interpretation of all the seismic results is problematic. A thermal interpretation of the weak, low-V_P anomaly northwest of Yellowstone is consistent with, but not required by, the ~18-km downwarp in the 410-km discontinuity there. However, immediately beneath Yellowstone a strong, high-V_P anomaly is also associated with downwarping of the 410-km discontinuity. There is a troubling lack of correlation between different seismic parameters, not only at Yellowstone, which defies simplistic interpretations in terms of a single parameter.

An anomaly that is confined to the upper mantle, whether thermal or not, cannot be a mantle plume (Section 1.4). In the upper mantle, viscous flow associated with plate motions is most vigorous. There is thus no reason why a thermal diapir arising spontaneously in the upper mantle should be fixed, for example, relative to other melting anomalies with time-progressive volcanic chains in the Pacific Ocean. The plume model for Yellowstone is also ques-

tionable on geological grounds (Christiansen et al., 2002; Humphreys et al., 2000):

- Large-volume basaltic volcanism along the eastern Snake River Plain, extending hundreds of kilometers westward from Yellowstone, has been continuous right up to the Quaternary. The time-progressive volcanic trail, perhaps the strongest argument for a plume, comprises a chain of silicic volcanoes. This chemistry is unique among melting anomalies and must involve substantial melting of the continental crust.

- The eastern Snake River Plain-Yellowstone system formed simultaneously with the onset of Basin and Range extension and volcanism throughout a 2000-km-wide region.

- The eastern Snake River Plain-Yellowstone system lies at the boundary between cratonic lithosphere to the north and thin, hot lithosphere associated with the Basin and Range province to the south. This coincidence is clearly visible in images from whole-mantle tomography (Fig. 5.21).

- A complementary time-progressive volcanic chain – the "Newberry trend" – developed simultaneously and propagated to the northwest, in the opposite direction to the Yellowstone system (Fig. 2.9).[13]

[13] http://www.mantleplumes.org/HighLavaPlains.html

The observations are most consistent with an origin for the eastern Snake River Plain-Yellowstone system in relative motion between the stable cratonic region to the north and the rapidly deforming Basin and Range Province to the south. This interpretation is supported by measurements of surface deformation using GPS surveying, which show differential motion between these two regions (Payne et al., 2008). The eastern Snake River Plain-Yellowstone system probably represents a lithosphere-penetrating shear zone with an opening component, the eastern end of which propagates northeast as Basin and Range extension evolves. Such a model is consistent with the topographic depression of the eastern Snake River Plain. The seismic anomaly in the mantle then developed from the surface downward in response to passive extraction of melt at the surface and replacement by compensating upwelling of volatiles and melt (Section 5.8).

5.5.2 Less well-studied melting anomalies in remote regions

The advantage that an extensive landmass offers in facilitating the deployment of seismic instrumentation over broad areas is not enjoyed by the many melting anomalies that form small islands or archipelagos. In these cases, seismological investigations tend to be limited to the study of shallow, small-scale structure or to lower-resolution studies of extremely large-scale structure using global tomography.

Neither of these techniques can cast light on the scale and vertical extent of the medium-scale structures that can potentially distinguish between a Plate and Plume origin. In order to do that, on-land deployments must be supplemented by large numbers of ocean bottom seismometers, a technically challenging and costly exercise. Such deployments are only beginning to be used to study oceanic melting anomalies.

(a) Hawaii

The limitations imposed on seismology by a restricted land area are nowhere better illustrated than at Hawaii.[14] The Hawaiian archipelago is essentially one-dimensional. The inhabited part is ~550 km long but the largest island is only ~100 km in diameter. Most seismic investigations of deep structure have relied on stations deployed on these islands. Their spatial limitations have precluded them from casting significant light on the deep structure beneath the "Big Island", where the type example mantle plume has been proposed. A broad ocean-bottom seismometer network was recently deployed around Hawaii in order to image deeper into the mantle (Laske et al., 2007).

Teleseismic tomography using data from the Big Island of Hawaii (Fig. 5.22) can only resolve structure down to a depth of ~100 km, the approximate diameter of the island. This depth range includes the ~8-km-thick basalt pile of the island, the ~8-km-thick underlying Cretaceous basaltic oceanic crust, and the upper portion of the ~100-km-thick Cretaceous mantle lithosphere beneath. It cannot image deep into the asthenosphere. A larger-scale experiment used seven three-component digital stations deployed throughout the inhabited islands to study V_P and V_S (Fig. 5.22) (Wolfe et al., 2002). Mantle structure was imaged down to ~350 km depth and found low wave-speeds beneath the islands of Maui and Molokai. The strength of the anomalies was typically ~0.5% in V_P and ~1% in V_S, reaching maxima of 0.6% in V_P and 1.8% in V_S. No significant anomaly was found beneath the Big Island, although resolution was poor at depths >150 km.

At the other end of the scale range, whole-mantle tomography shows Hawaii to lie within part of a low-wave-speed anomaly that occupies much of the mantle beneath the Pacific (Fig. 5.32) (Ritsema, 2005). The strong low-wave-

[14] http://www.mantleplumes.org/TopPages/HawaiiTop.html

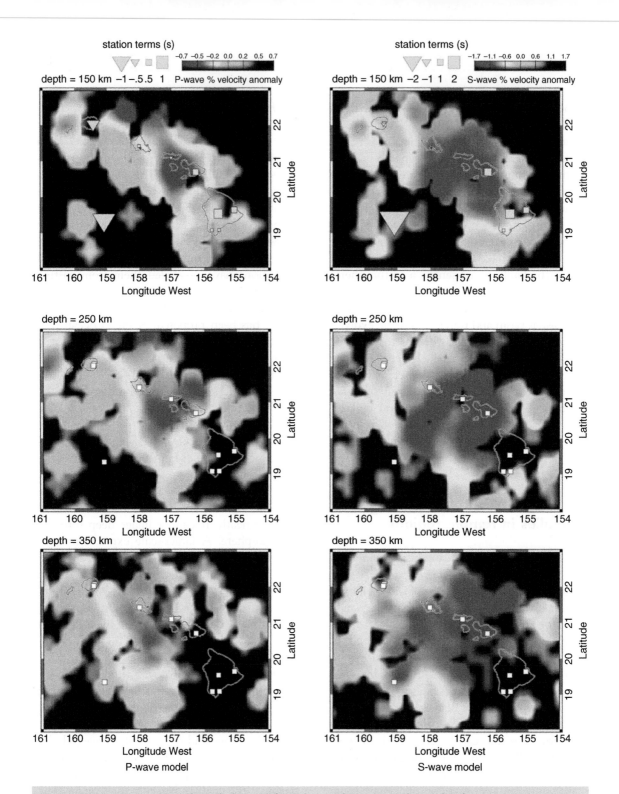

Figure 5.22 Cross-sections through the V_P (left) and V_S (right) tomography models for Hawaii. Regions with few seismic rays are black. The white squares in the middle and bottom rows show the station locations, and the square and triangular symbols in the top row show the station terms, which are used to correct for the effect of the top 100 km, where lateral variations in structure are unconstrained because of the absence of crossing rays. The velocity perturbations are relative variations across the modeled volume; the absolute velocities are unconstrained (Wolfe et al., 2002). See Plate 27

speed anomaly found northwest of the Big Island at ~300 km depth may be the same feature detected beneath Maui and Molokai by teleseismic tomography. Cross-sections show apparent continuity of low-wave-speed anomalies between the surface at Hawaii and the core-mantle boundary beneath a large swath of the Earth ranging from the New Hebrides and Samoa, throughout the south Pacific and north along the East Pacific Rise. These structures have resolutions no better than ~1000 km and are thus unable to test for the existence of a narrow plume. However, the cross-sections in Fig. 5.23 illustrate well the way in which many different

images may be constructed merely by choosing different orientations.

The whole mantle directly beneath the island chain was also studied using a single, large earthquake that occurred in 1973 on the island of Oahu (Best et al., 1975). The travel times and attenuation of multiple-ScS waves – shear waves that reverberate between Earth's surface and the core-mantle boundary – were initially interpreted as showing that mantle wave-speed beneath Hawaii was higher than the average for the southwest Pacific (Katzman et al., 1988), and attenuation lower (Sipkin and Jordan, 1980), the opposite of what is expected for a hot

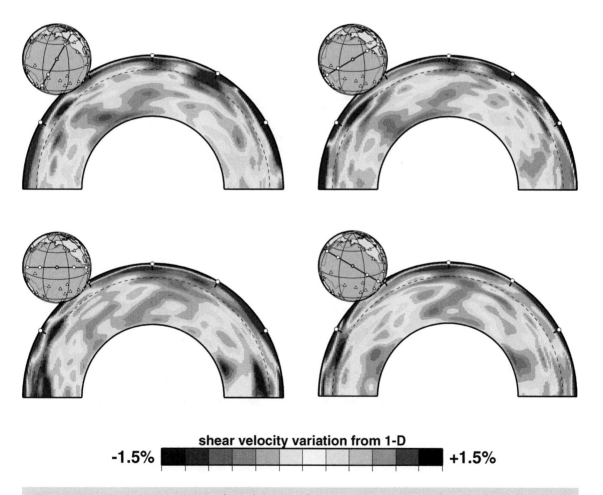

shear velocity variation from 1-D

-1.5% +1.5%

Figure 5.23 Cross-sections passing through Hawaii showing seismic wave-speed structure in the mantle (Ritsema et al., 1999). See Plate 28

plume. A later investigation showed, however, that this experiment also had insufficient resolution to rule out a plume. The volume to which such *ScS* waves are sensitive is ~1000 km wide, much larger than the 100–200 km resolution required to test typical plume models (Julian, 2005).

Various estimates have been made of the depths of the 410- and 650-km discontinuities beneath Hawaii using receiver functions. The results are variable with different recent studies obtaining 405 km and 651 km (Deuss, 2007), and 436 km and 672 km, compared with average global values of 410 km and 653 km (Gu et al., 2009).

Because of the small size and remoteness of the Hawaiian Islands, it has not been possible hitherto to study the structure beneath them seismically in anything approaching the same

detail as at Iceland, Eifel and Yellowstone. As a result there is at present little significant seismic evidence either for or against a mantle plume, or providing strong constraints on Plate models.

(b) Galápagos

The Galápagos melting anomaly has been studied using both explosion- and earthquake seismology.[15] Several seismic-refraction profiles up to 350 km long have been shot on the Cocos, Carnegie and Malpelo ridges and the structure of the ~15–20-km-thick crust has been calculated using tomography (Fig. 5.24) (Sallares et al., 2005). The relationship between crustal thickness and seismic wave-speed is similar to that in Iceland, with a negative correlation

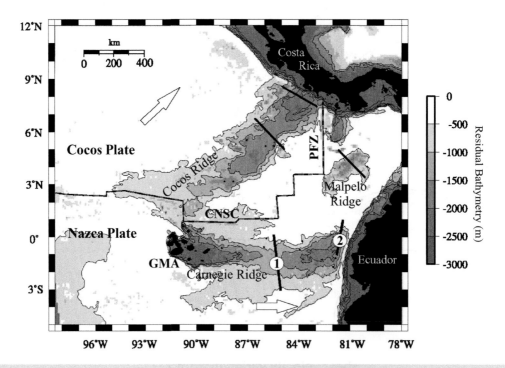

Figure 5.24 Map of the Galápagos area showing the location of explosion seismology profiles (from Sallares et al., 2005). Arrows show the diverging directions of motion of the Cocos and Nazca plates.

[15] http://www.mantleplumes.org/TopPages/GalapagosTop.html

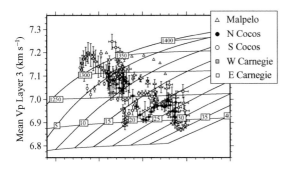

Figure 5.25 Mean compressional-wave-speed of the lower crust as a function of whole crustal thickness for the Carnegie, Cocos and Malpelo ridges, Galápagos region (from Sallares et al., 2005).

between wave-speed and crustal thickness (Fig. 5.7; Fig. 5.25). This supports a similar conclusion that the thick crust formed from melting an unusually fusible source and not a high-temperature one.

In order to study mantle structure immediately beneath the Galápagos islands, data were gathered on a 12-station network of 3-component digital seismic stations installed in 1999 (Hooft et al., 2003). The archipelago is a quasi-circular cluster of volcanic islands some 250 km in diameter. Tomography using such a restricted network is limited to the upper ~300 km and cannot probe as deep as the transition zone.

Surface-wave tomography of the upper ~150 km indicates that V_S is as much as ~5% higher than the global average in the upper 15–45 km of the mantle. The strength of the anomaly decreases to just ~2.5% lower than average at greater depth. Similar kinds of studies give V_S values 5% low for Iceland and 15% low for Afar (Vinnik et al., 2005).

The thickness of the transition zone beneath the Galápagos was studied by calculating ~200 receiver functions (Hooft et al., 2003). The results show a great deal of scatter, with individual measurements suggesting thinning by as much as ~100 km. Such extreme results are implausible and indicate errors so large that no reliable conclusions can be drawn from them.

The seismic results as a whole provide limited constraints on the origin of the Galápagos melting anomaly. The explosion seismology results indicate that high source temperatures are not responsible for the thick crust. The earthquake seismology, on the other hand, has inadequate depth resolution to answer critical questions such as the depth extent of structures.

(c) Azores

The Azores archipelago is similar in some respects to the inhabited part of the Hawaiian island chain.[16] It comprises nine main islands arranged roughly in a line ~600 km long and with the largest island being ~60 km in diameter. Few seismic investigations have been conducted there, however. The crust has not been studied in any detail, and crustal thickness has been estimated only by using low-resolution surface waves. The crust is thought to be ~9–12 km thick, considerably less than the 20–40 km crustal thicknesses estimated for Iceland and the Galápagos.

The results of two independent surface-wave tomography studies illustrate several common pitfalls in seismological studies (Figs 5.26 and 5.27). First, the repeatability is poor. A low-V_S anomaly dipping steeply to the south found in one study and interpreted as a plume (Fig. 5.26, top) is absent in the result of the other study, even though similar methods and data sets were used (Fig. 5.27). Second, large vertical exaggeration causes the figures to give distorted impressions. The cross-sections of Fig. 5.26 have vertical exaggerations of ~10 and give the impression that the low-wave-speed anomaly plunges steeply and is plume-like, when in truth it is sub-horizontal and dips at only about 20°. Third,

[16] http://www.mantleplumes.org/TopPages/AzoresTop.html

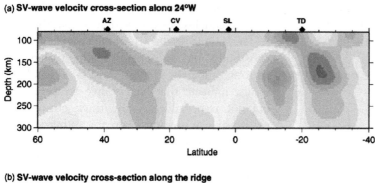

(a) SV-wave velocity cross-section along 24°W

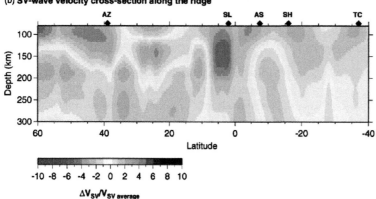

(b) SV-wave velocity cross-section along the ridge

Figure 5.26 Cross-sections of mantle V_S structure from surface-wave tomography: (a) along longitude 24°W passing through the Azores (Fig. 3.19); (b) along the Mid-Atlantic Ridge (Silveira and Stutzmann, 2002). See Plate 29

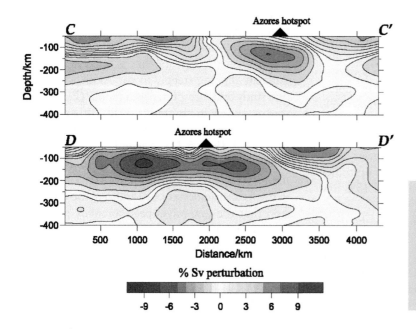

Figure 5.27 Vertical cross-sections through a different surface-wave tomography image: top – across the Mid-Atlantic Ridge; bottom – parallel to the ridge (Pilidou et al., 2005). See Plate 30

wave-speed anomalies of 5–10%, as strong or stronger than those beneath the Azores, occur in many places along the Mid-Atlantic Ridge, and not only at melting anomalies.

Surface-wave tomography cannot resolve bodies smaller than ~1000 kilometers across, and are thus unable to detect narrow structures such as plumes. A land-based teleseismic tomography study of the Azores, using recordings from six seismic stations installed on the islands (Yang et al., 2006), found several low-V_P anomalies. One plunges to the north and generally has a strength of ~1%, peaking at ~1.5% (Fig. 5.28). Determining the V_S structure was impossible,

because the data were too noisy. Thus the interpretation of the V_P anomaly, as well as the other bodies imaged, remains ambiguous.

The low-V_P anomaly plunges at a similar angle to the deeper part of the V_P anomaly beneath Yellowstone, i.e., along the incoming ray bundle. The anomalies beneath the Azores are stronger at depth than at shallow levels, the opposite of what is observed at Yellowstone. It has been suggested that the negative V_P anomaly at the Azores represents a plume plunging to the north as a result of a north-south component of mantle flow. However, shear-wave splitting studies consistently suggest east-west

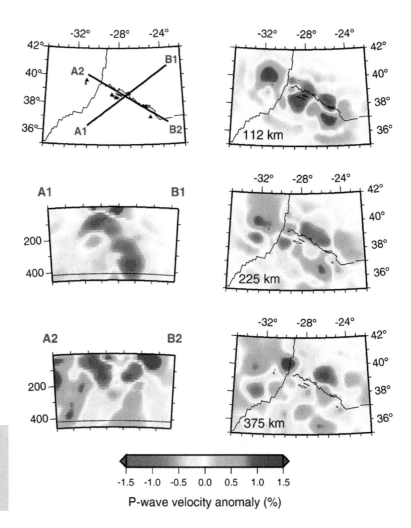

Figure 5.28 V_P structure of the mantle beneath the Azores from teleseismic tomography (Yang et al., 2006). See Plate 31

P-wave velocity anomaly (%)

flow directions (Silveira and Stutzmann, 2002; Silveira et al., 1998). The northward plunge is the opposite of that suggested by some surface-wave tomography, which detected a southward plunging anomaly (Fig. 5.26). Estimates of transition-zone thickness determined from receiver functions vary, with some results suggesting a larger thickness than the global average (Deuss, 2007) and others suggesting a smaller thickness (Gu et al., 2009).

The Azores lie above the middle of an extensive low-V_S anomaly that underlies some 2000 km of the Mid-Atlantic Ridge in the North Atlantic. Significant local variations in V_P occur in the mantle beneath the islands, but it is questionable whether strong wave-speed heterogeneity extends down as far as the transition zone. It is at present not possible to reduce the ambiguity in interpretation of the anomalies between composition, phase and temperature, because too few independent variables have been measured.

The Azores melting anomaly exhibits none of the primary features predicted for mantle plumes. There is no flood basalt with precursory uplift, no time-progressive volcanism in the archipelago, and no evidence for high source temperatures. The amount of melt, as measured by crustal thickness in excess of the ~7–8 km average for mid-ocean ridges, is small. Formation of thicker-than-average crust began at ~20 Ma, and occurred along all three branches of the Azores ridge-ridge-ridge triple junction. Enhanced melt production is expected at such a triple junction because asthenospheric upwelling beneath the ridge branches is focused toward it (Georgen, 2008). Magmatism may have been further enhanced by the long-term tectonic instability and migration of major fault zones that have occurred at the Azores, similar to events in the geological history of Iceland (Gente et al., 2003; Luis and Miranda, 2008; Madeira and Ribeiro, 1990; Searle, 1980). Such an explanation fits the observations better than an *ad hoc* plume that coincidentally impacted at one of the only four major ridge-ridge-ridge triple junctions on Earth.

5.5.3 Structure beneath extinct melting anomalies, and elsewhere

In order to place the results described above in context, it is necessary to compare them with the seismic structure of the mantle away from currently erupting melting anomalies. If similar structures are widespread, the case linking them with active volcanism is weakened. Only a few areas have been studied in sufficient detail for such comparisons to be made.

(a) The Ontong Java Plateau

The long-extinct Ontong Java Plateau (Section 3.5.3) is entirely oceanic and contains no suitable islands on which seismic stations can be installed.[17] However, the mantle beneath the plateau has been studied using surface waves from earthquakes in the subduction zones to the south, recorded at seismic stations on the Caroline Islands to the north (Richardson et al., 2000).

A low-V_S anomaly with wave-speeds up to 5% lower than the global average extends from the near-surface down to ~300 km depth (Fig. 5.29). This anomaly is very similar to that found beneath Iceland. A study of seismic ScS phases (shear waves that bounce between the core and the surface) also revealed that attenuation within this anomaly is exceptionally low compared with the rest of the Pacific mantle (Gomer and Okal, 2003). The combination of low-V_S and low attenuation rules out temperature alone as the cause and requires a chemical interpretation that implies a high-viscosity body (Klosko et al., 2001).

Is this anomaly associated with formation of the Ontong Java Plateau, or is it an unrelated feature over which the plateau has been transported by Pacific plate motion? Since its forma-

[17] http://www.mantleplumes.org/TopPages/OJTop. html

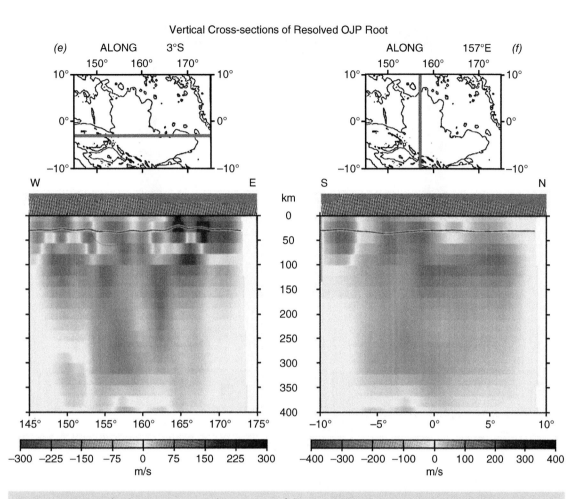

Figure 5.29 Rayleigh-wave tomography structure of the mantle beneath the Ontong Java Plateau (from Richardson et al., 2000). See Plate 32

tion at ~122 Ma, the plateau has drifted several thousand kilometers from where it initially formed. If the mantle anomaly is a remnant associated with formation of the plateau, then it must have been transported with the oceanic lithosphere. Such a process would require a radical revision of the concept of a vigorously convecting upper mantle overlain by an independently drifting lithosphere. Regardless of this issue, the anomaly beneath the Ontong Java Plateau is an example of an anomaly that must be chemical, and not thermal in nature, and yet

is similar to the low-V_S anomaly underlying Iceland. It provides observational evidence that the interpretation of such low-V_S anomalies as hot, buoyant material, is not unique.

(b) The Paraná basalts, Deccan Traps and Ireland

Other low-wave-speed anomalies have been found in the mantle far from currently active melting anomalies. One underlies the ~135 Ma,

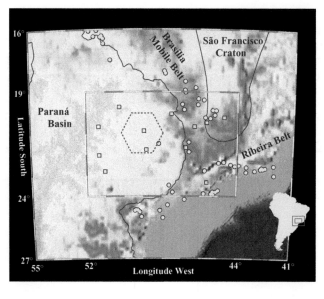

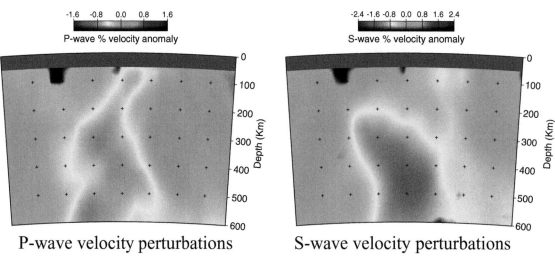

P-wave velocity perturbations   S-wave velocity perturbations

Figure 5.30 Map of the region of Brazil studied using teleseismic tomography, and cross-sections through the structure imaged (from VanDecar et al., 1995). See Plate 33

extinct Paraná flood basalt province on the Brazilian shield (VanDecar et al., 1995). V_P and V_S tomographic images of the mantle there reveal an anomaly ~300 km wide with wave-speeds depressed by as much as 1.5% in V_P and 2% in V_S. This anomaly extends from the near-surface down to the lower limit of resolution at ~500 km depth (Fig. 5.30).

Beneath the ~65 Ma Deccan Traps,[18] a cylindrical, low-V_P anomaly ~250 km in diameter extends from near the surface down to ~200 km depth, where it merges with a broader anomaly with wave-speeds as much as ~2%

[18] http://www.mantleplumes.org/TopPages/DeccanTop.html

lower than beneath surrounding continental regions (Kennett and Widiyantoro, 1999). Both this anomaly and the one beneath the Paraná basalts present the same interpretive difficulties as the anomaly beneath the Ontong Java Plateau. Both the Deccan Traps and the Paraná basalts have been transported several thousand kilometers laterally by plate motion subsequent to their formation. Thus, if the anomalies beneath them are associated with flood-basalt formation, they must have been transported with the plates. This process would require the top several hundred kilometers of the mantle to be dragged along with the plates and to retain low seismic wave-speeds for as long as 135 Ma.

A steeply dipping anomaly beneath Ireland has V_P depressed by as much as 3% compared with the regional average (Wawerzinek et al., 2008). The anomaly extends from near the surface down to ~150 km depth. It has been interpreted as an ancient or modern plumelet associated with a plume currently active beneath Iceland. An alternative view would be that the shallow mantle is inhomogeneous in composition, degree of partial melt, mineral phase and temperature, and that such anomalies are the expression of this inhomogeneity and are not uniquely interpretable in the absence of independent constraints. Where seismic wave-speed variations are described relative to regional averages, approximately half of the mantle will have low seismic wave speeds. It is absurd to expect that every low-wave-speed anomaly is caused by a hot plume.

5.6 Global observations

5.6.1 Whole-mantle tomography

What can whole-mantle tomography contribute to the study of melting anomalies? Seismic studies on very large scales cannot image struc-

tures postulated to be only a few tens, or even hundreds of kilometers in diameter. Thus the method is of limited usefulness when it comes to targeting individual localities. Another limitation is that ray coverage throughout the mantle is extremely non-uniform because most seismic stations are deployed on land, and most large earthquakes are restricted to the Pacific rim and the Mediterranean-Himalaya belt. Consequently, in almost any extensive region of the mantle there will be large parts that are poorly resolved.

A recent effort to address this problem led to the "banana-doughnut" controversy.[19] A method for calculating how seismic waves sample the Earth, which is more advanced than simple ray theory, was included in tomographic inversions. "Finite-frequency" tomography (nicknamed "banana-doughnut" tomography) accounts for the fact that seismic waves do not, in truth, sample the Earth along infinitesimally thin lines (rays), but they sense a finite volume around these theoretical minimum-time paths (Dahlen et al., 2000). This volume has the shape of a hollow banana. A global database of earthquake arrival-time data was inverted, taking this into account. The mantle beneath melting anomalies was studied in particular detail and images of low-seismic-wave-speed anomalies traversing the entire mantle were presented, and interpreted as plumes (Montelli et al., 2004a, b; 2006).

This work caused controversy among seismologists and attracted criticism on several grounds (Boschi et al., 2006; de Hoop and van der Hilst, 2005; Julian, 2005; Trampert and Spetzler, 2006). Although finite-frequency theory is a more correct description of how seismic waves sense the Earth's structure, the practical problems inherent in applying it to real data are such that it produces results that are statistically indistinguishable from those of the conventional, simple ray theory that has been used for many years. The improvement in resolution

[19] http://www.mantleplumes.org/BananaDoughnuts.html

obtained by using finite-frequency theory was smaller than the effects of factors, such as the choice of damping.

Comparisons of the finite-frequency tomography results with those obtained from different ray-theory approaches showed, instead, that the resolution of the most robust features in the Earth's structure such as subducting slabs in the upper mantle is, in fact, worse in the finite-frequency tomography results. The low-wave-speed anomalies in the upper mantle interpreted as plumes, for example, beneath the Indian Ocean, are essentially the same in older inversions that used only ray theory. Those anomalies correlate precisely with the locations of seismic stations on islands and have long been known to be artifacts of poor resolution resulting from the absence of seismic stations in the oceans and the restricted distribution of earthquakes.

The interpretation of wave-speeds is, in any case, hampered by the ambiguity problem. Seismic wave-speeds depend on several physical variables, and seismic tomography images are not maps of mantle temperature. Large-scale mantle tomography can perhaps contribute most to understanding melting anomalies by shedding light on the extent of material exchange between the upper and lower mantle. The key question is whether structures are continuous throughout the mantle, from the surface, through the transition zone, into the lower mantle and on to the core-mantle boundary (see Gu et al., 2001 for a review).

A robust and repeatable feature of whole-mantle tomography images is that continental lithosphere typically has high wave-speeds extending as deep as ~300 km (Fig. 5.31), whereas the oceans are associated with low wave-speeds at these depths. This difference in wave-speeds beneath continents and oceans decreases with depth. A radical change occurs at ~650 km. There, horizontal wave-speed variations become much weaker and there is almost no correlation between the seismic structure above and below – the patterns are quite different. This lack of correlation suggests that large-scale processes above and below the transition zone are largely independent.

What of cold sinkers and hot risers? Subducting slabs are cold, compositionally distinct, and expected to have high seismic wave-speeds. Slab-shaped, high-wave-speed bodies in the western Pacific Ocean are the clearest features seen in tomographic images of the upper mantle. Many of these anomalies become horizontal near the base of the transition zone or a little beneath it and extend horizontally for distances of >1000 km beneath eastern Asia (Fukao et al., 2009; Goes et al., 2000; Hamilton, 2007b; Li et al., 2000a). Tomographic images of the mantle beneath South America and Europe show similar anomalies caused by flat slabs. Some of these slabs may eventually founder and sink to ~1000 km depth, but there is no unequivocal evidence that they sink deeper.

The seismic structure of the mid-mantle is clearly fundamentally different from that of the upper mantle. Rare large-scale zones where high-wave-speed anomalies are continuous throughout the lower mantle can be found, but it is not clear that they represent continuous geological structures. The most widely cited zone of this sort extends from ~650 km beneath North America down to the core-mantle boundary. Its detailed shape varies significantly between different tomography studies. It is slab-shaped throughout some of its range in some, but not all, and it merges with a high-wave-speed region of vast volume, many thousands of kilometers wide, in the lowermost mantle. A slab (the "Farallon slab") is known from plate-motion models to have subducted beneath North America in the past. It has been postulated that it is represented by the observed high-wave-speed anomaly and this has been claimed to be evidence that slabs sink to the core-mantle boundary, delivering near-surface materials that may later be transported back up to the surface by mantle plumes. It is questionable, however, whether this high-wave-speed anomaly is a single, coherent geological body (Gu et al., 2001; Hamilton, 2007b). Such continuity

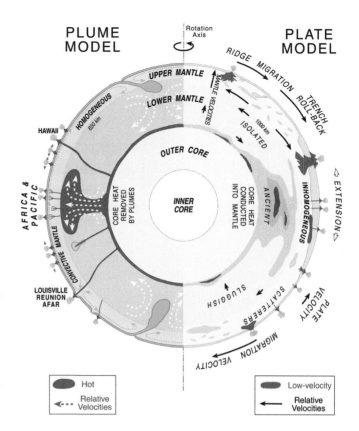

Plate 1 Schematic cross-section of the Earth showing the Plume model (left, modified from Courtillot et al., 2003) and the Plate model (right). The left side illustrates two proposed kinds of plumes – narrow tubes and giant upwellings. The deep mantle or core provides the material and the heat and large, isolated but accessible chemical reservoirs. Slabs penetrate deep. In the Plate model, depths of recycling are variable and volcanism is concentrated in extensional regions. The upper mantle is inhomogeneous and active and the lower mantle is isolated, sluggish, and inaccessible to surface volcanism. The locations of melting anomalies are governed by stress conditions and mantle fertility. The mantle down to ~1000 km contains recycled materials of various ages and on various scales (from Anderson, 2005).

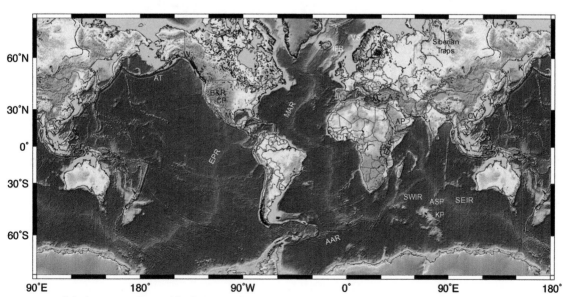

Plate 2 Global topography and bathymetry, from gravity measurements: IFR – Iceland-Faeroe Ridge; VS – Vøring Spur; CV – Changbai Volcano; AP – Arabian Peninsula; W – Wrangellian terrain; AT – Aleutian Trench; EPR – East Pacific Rise; CP – Colorado Plateau; K – Kamchatka; A – Anatolia; M – Mexico; B&R – Basin and Range Province; EAR – East African Rift; AAR – American-Antarctic Ridge; MAR – Mid-Atlantic Ridge; ASP – Amsterdam-St Paul Plateau; SWIR – Southwest Indian Ridge; SEIR – Southeast Indian Ridge; KP – Kerguelen Plateau.

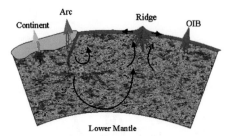

Plate 3 The upper mantle is kept inhomogeneous by continuous melt extraction and lithosphere recycling at subduction zones and beneath continents, which makes plate tectonics work (from Meibom and Anderson, 2004).

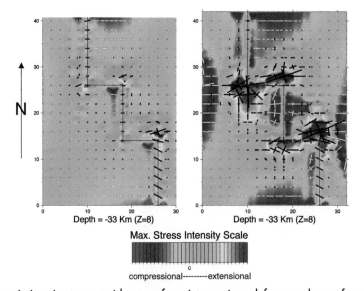

Plate 4 Predicted variations in stress at ridge-transform intersections: left: a weak transform, right: a strong transform. Lines indicate the type and intensity of maximum stress, black: compressional, white: extensional. Regions of strong lithospheric extension, encouraging magmatism, are predicted on the outer corners of the ridge-transform intersections (from Beutel, 2005).

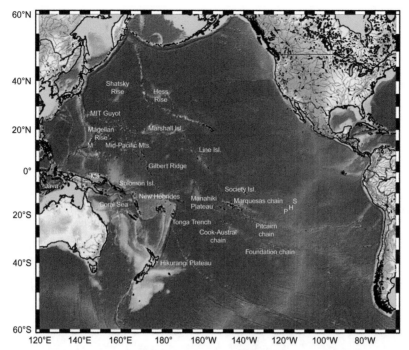

Plate 5 Map of the Pacific showing place names: P – Pukapuka Ridge; S – Sojourn Ridge; H – Hotu Matua Ridge; M – Mariana Trench; Ma – Manus Basin.

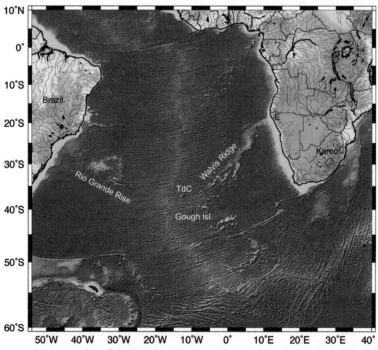

Plate 6 Bathymetry and topography of the South Atlantic region. TdC – Tristan da Cunha Island; E – Etendeka Basalts (basemap from Smith and Sandwell, 1997).

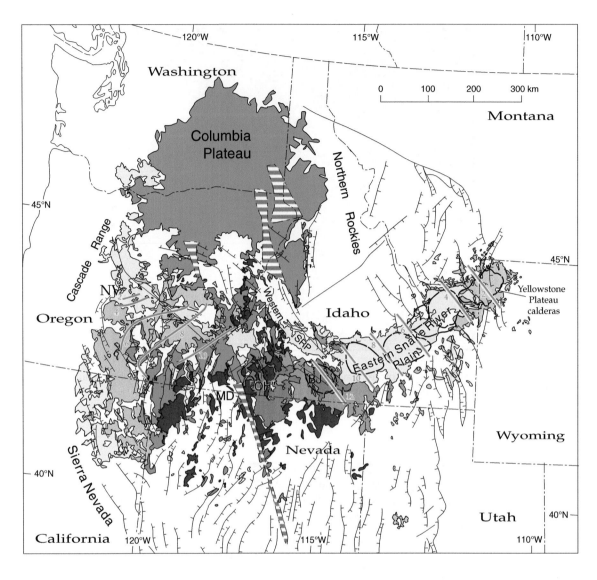

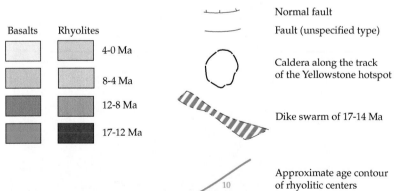

Plate 7 Map of the northwestern United States showing features of the Columbia River Basalts, Snake River Plain and Yellowstone. NV – Newberry Volcano (from Christiansen et al., 2002).

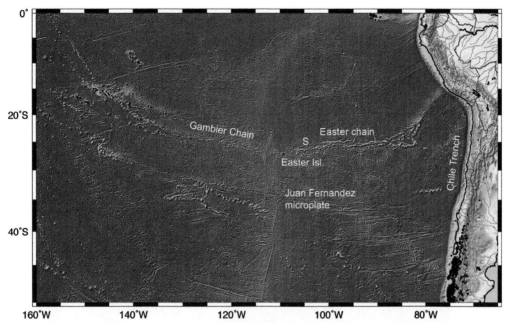

Plate 8 Bathymetry and topography in the region of the Easter volcanic chain. S – Sala y Gomez Island (basemap from Smith and Sandwell, 1997).

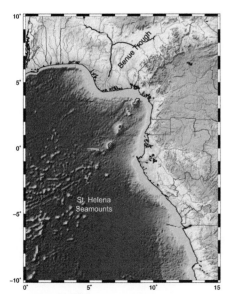

Plate 9 Bathymetry and topography in the region of the Cameroon volcanic line (basemap from Smith and Sandwell, 1997).

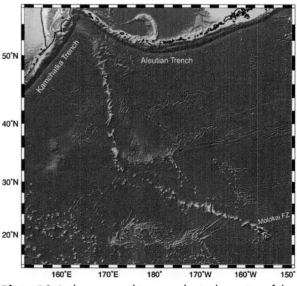

Plate 10 Bathymetry and topography in the region of the Emperor and Hawaii volcanic chains (basemap from Smith and Sandwell, 1997).

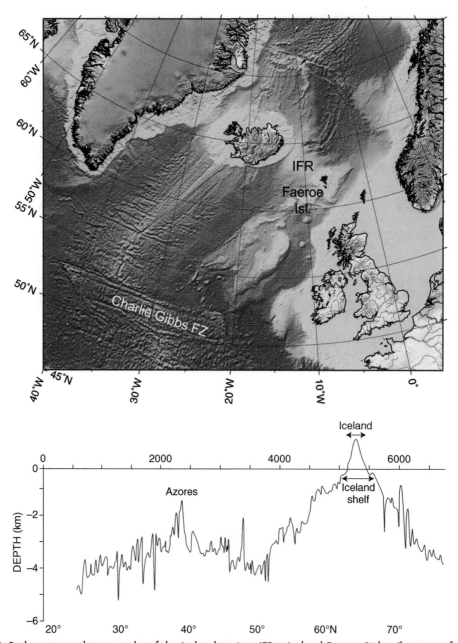

Plate 11 Bathymetry and topography of the Iceland region. IFR – Iceland-Faeroe Ridge (basemap from Smith and Sandwell, 1997). Inset shows bathymetry along a profile along the Mid-Atlantic Ridge. Iceland is not an isolated topographic anomaly but simply the summit of a vast bathymetric high ~3000 km wide that fills much of the North Atlantic (from Vogt and Jung, 2005).

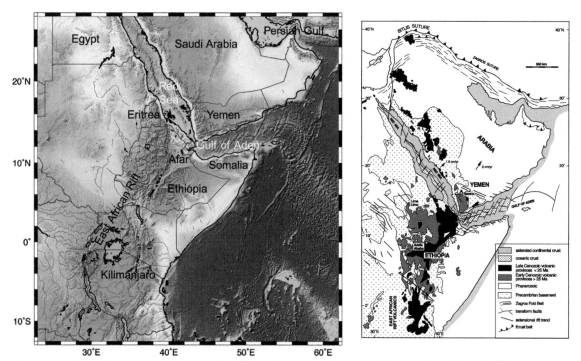

Plate 12 Left: Bathymetry and topography of the Afar region (basemap from Smith and Sandwell, 1997); right: Geological map showing the distribution of Cenozoic volcanism of different ages (modified from Baker et al., 1996).

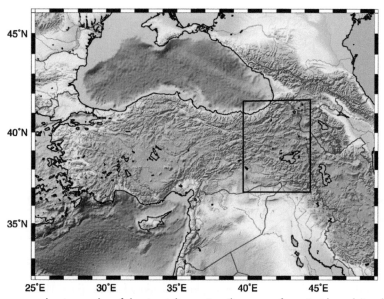

Plate 13 Bathymetry and topography of the Anatolia region (basemap from Smith and Sandwell, 1997). Box indicates the location of recent volcanism.

Plate 14 Sea-floor age based on identified magnetic anomalies and relative plate reconstructions. The age of the Cretaceous "quiet zone" is ~64–127 Ma (from Sandwell et al., 2005).

Plate 15 Satellite gravity field of the North Atlantic with the long-wavelength component corresponding to plate cooling removed. The chevron ridges are clearly seen. JMM – Jan Mayen Microcontinent (from Jones et al., 2002).

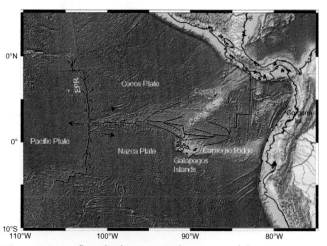

Plate 16 Seafloor bathymetry in the region of the Cocos-Nazca spreading axis and the Galapagos islands, with the main plate boundary features (from Hey et al., 1992) superimposed. The 91° transform lies immediately north of the Galapagos Islands (basemap from Smith and Sandwell, 1997).

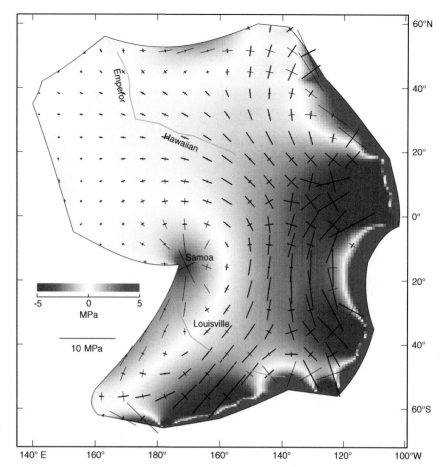

Plate 17 Horizontal stress field for the Pacific plate from cooling and thermal contraction. Horizontal principle stresses are shown by orthogonal lines – thick for compression and thin for tension. Color shows horizontal mean normal stress – red for tension and blue for compression (from Stuart et al., 2007).

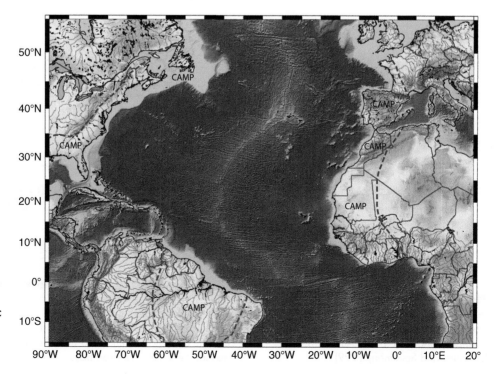

Plate 18 The extent of the Central Atlantic Magmatic Province (CAMP).

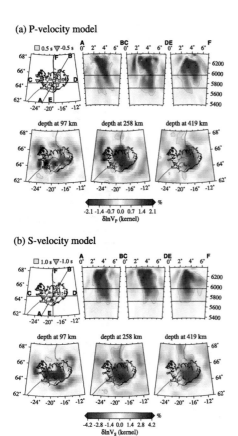

(a) P-velocity model

(b) S-velocity model

Plate 19 Teleseismic tomography images of compressional- and shear-wave-speed perturbations beneath Iceland (from Hung et al., 2004).

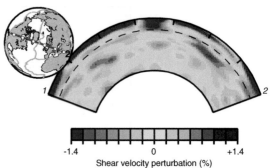

Shear velocity perturbation (%)
-1.4   0   +1.4

Plate 20 Whole mantle tomography cross-section through Iceland, calculated using earthquakes that occurred 1980–1998 and were recorded on several global networks. Over a million surface-wave phase velocity measurements and 50,000 hand-picked compressional- and shear-wave arrival time measurements for a large variety of seismic phases were used. This image is currently the best available for the global transition zone. The results show a low-wave-speed anomaly filling the whole North Atlantic, but extending no deeper than the transition zone (from Ritsema et al., 1999).

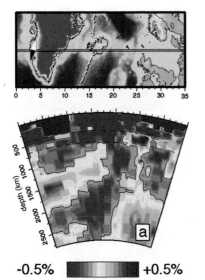

-0.5%   +0.5%

Plate 21 Whole-mantle tomography model purporting to show a mantle plume extending from the surface down to the core-mantle boundary under the North Atlantic (from Bijwaard and Spakman, 1999).

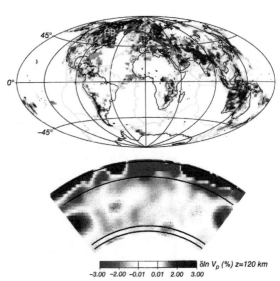

δln V_p (%) z=120 km
-3.00 -2.00 -0.01 0.01 2.00 3.00

Plate 22 Model shown in Fig. 5.14 replotted with the color scale saturated at an anomaly strength of ±3%, and the line of section extended to underlie Canada and Scandinavia. HB – Hudson Bay.

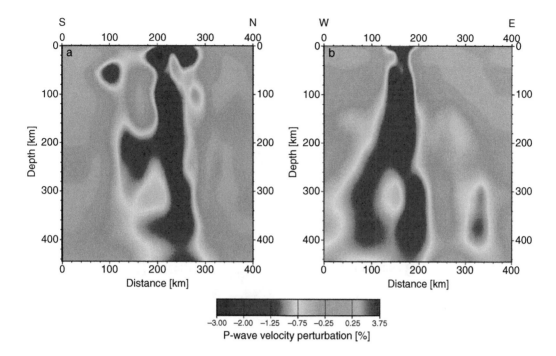

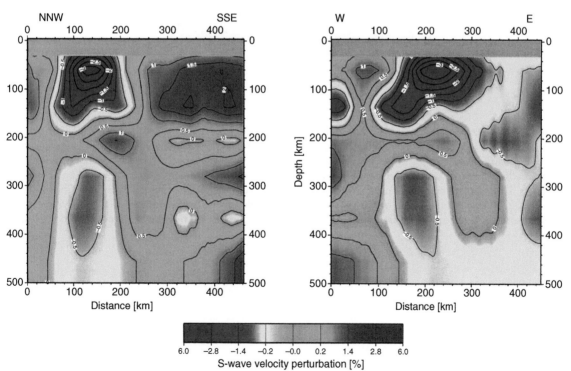

Plate 23 Cross-sections through the V_P (top) and V_S (bottom) teleseismic tomography models from Eifel (from Keyser et al., 2002; Ritter et al., 2001).

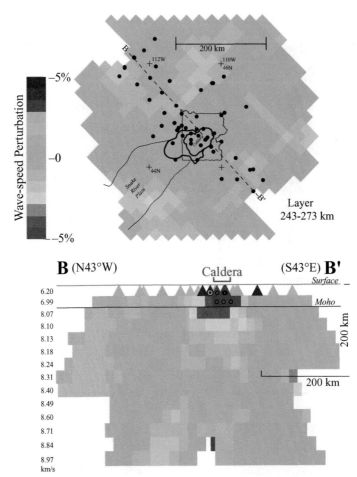

Plate 24 Teleseismic tomographic V_P structure beneath Yellowstone. Top: dots show the seismic stations used, the boundary of Yellowstone National Park, the calderas of the Yellowstone Plateau volcanic field, the edges of the eastern Snake River Plain, and line of cross-section BB′ shown in lower panel. Colors in top panel indicate wave-speed variations in the depth interval 243–273 km, where a deep plume-like structure would be imaged if one exists. Lower panel: cross-section through the model at the northeast edge of the caldera (from Christiansen et al., 2002). "Upper mantle origin of the Yellowstone hotspot. *Bulletin of the Geological Society of America*", Christiansen, R.L., Foulger, G.R. & Evans, J.R. **114**: 1245–1256, 2002. Reprinted with permission from *GSA*.

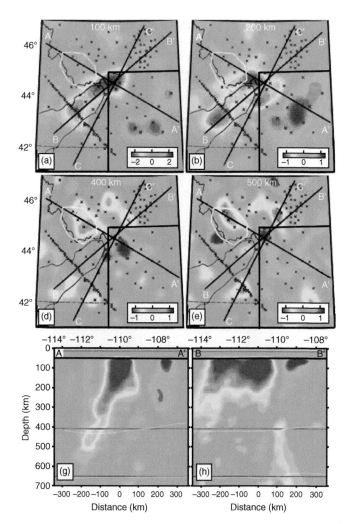

Plate 25 Teleseismic tomography V_P structure beneath Yellowstone (from Yuan and Dueker, 2005). White rings on the horizontal sections indicate where the 410-km discontinuity is depressed. In the cross-sections in the lower panels, the 410- and 650-km discontinuities are shown by white lines, and their average depths by black lines (Fee and Dueker, 2004).

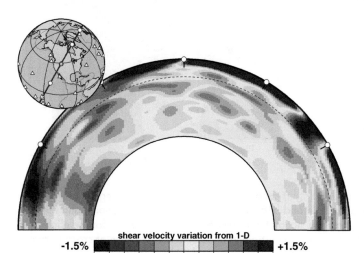

shear velocity variation from 1-D

-1.5%   +1.5%

Plate 26 Northeast-southwest cross-section through the Yellowstone caldera, showing the mantle tomography model of Ritsema et al. (1999). *"Complex shear wave velocity structure imaged between Africa and Iceland"*, Ritsema, J., van Heijst, H.J. & Woodhouse, J.H. Science **286**: 1925–1928, 1999. Reprinted with permission from Science.

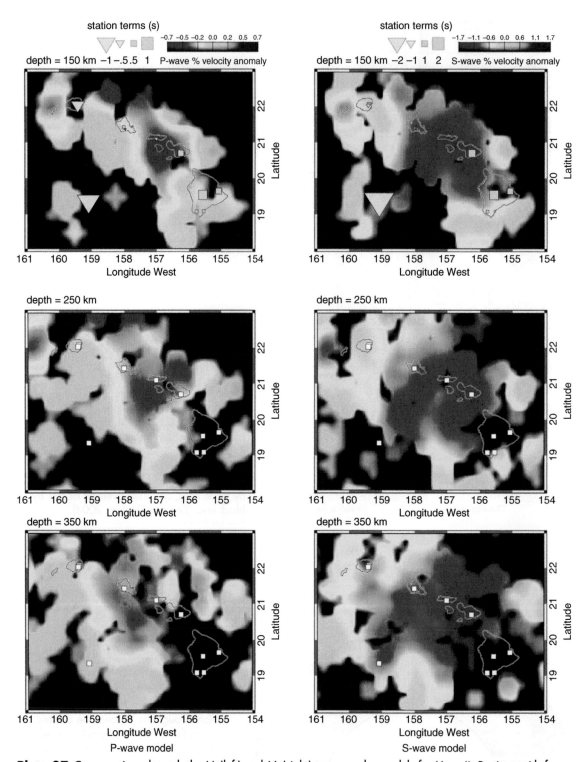

Plate 27 Cross-sections through the V_P (left) and V_S (right) tomography models for Hawaii. Regions with few seismic rays are black. The white squares in the middle and bottom rows show the station locations, and the square and triangular symbols in the top row show the station terms, which are used to correct for the effect of the top 100 km, where lateral variations in structure are unconstrained because of the absence of crossing rays. The velocity perturbations are relative variations across the modeled volume; the absolute velocities are unconstrained (Wolfe et al., 2002).

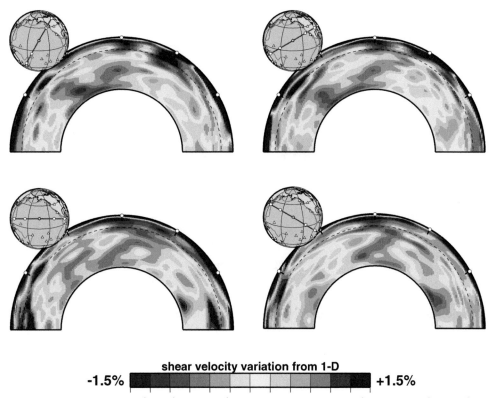

shear velocity variation from 1-D

-1.5% [color scale bar] **+1.5%**

Plate 28 Cross-sections passing through Hawaii showing seismic wave-speed structure in the mantle (Ritsema et al., 1999).

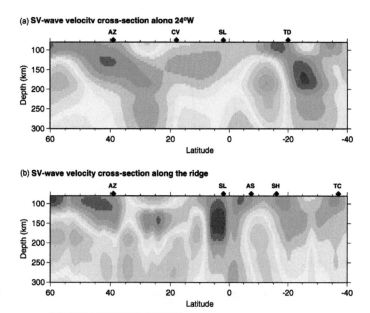

(a) SV-wave velocity cross-section along 24°W

(b) SV-wave velocity cross-section along the ridge

-10 -8 -6 -4 -2 0 2 4 6 8 10

$\Delta V_{SV}/V_{SV\ average}$

Plate 29 Cross-sections of mantle V_S structure from surface-wave tomography: (a) along longitude 24°W passing through the Azores (Fig. 3.19); (b) along the Mid-Atlantic Ridge (Silveira and Stutzmann, 2002).

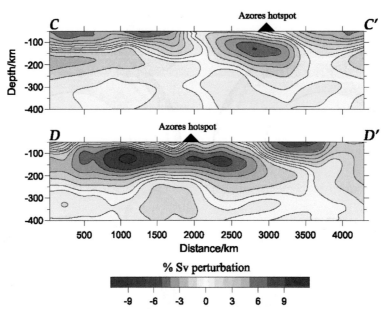

Plate 30 Vertical cross-sections through a different surface-wave tomography image: top – across the Mid-Atlantic Ridge; bottom – parallel to the ridge (Pilidou et al., 2005).

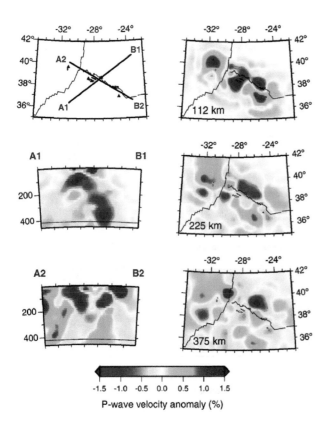

Plate 31 V_P structure of the mantle beneath the Azores from teleseismic tomography (Yang et al., 2006).

Vertical Cross-sections of Resolved OJP Root

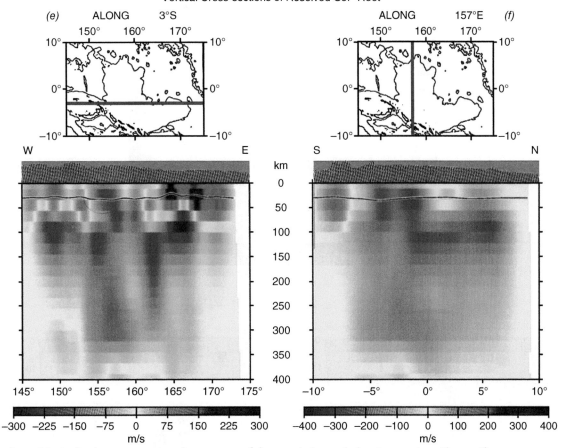

Plate 32 Rayleigh-wave tomography structure of the mantle beneath the Ontong Java Plateau (from Richardson et al., 2000).

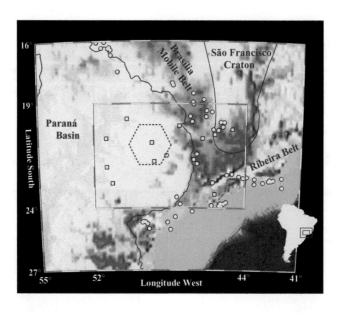

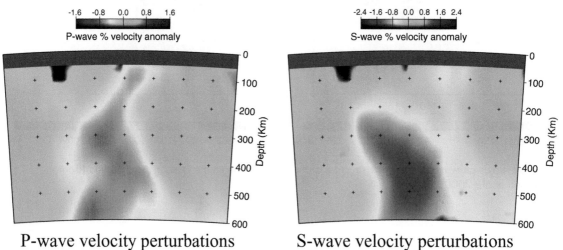

P-wave velocity perturbations   S-wave velocity perturbations

Plate 33 Map of the region of Brazil studied using teleseismic tomography, and cross-sections through the structure imaged (from VanDecar et al., 1995).

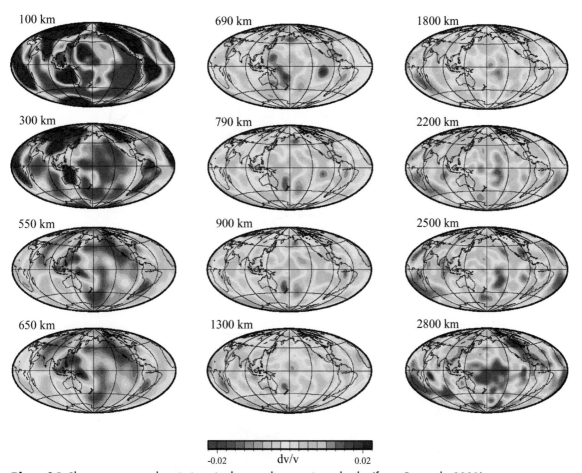

100 km
300 km
550 km
650 km
690 km
790 km
900 km
1300 km
1800 km
2200 km
2500 km
2800 km

-0.02   dv/v   0.02

Plate 34 Shear-wave-speed variations in the mantle at various depths (from Gu et al., 2001).

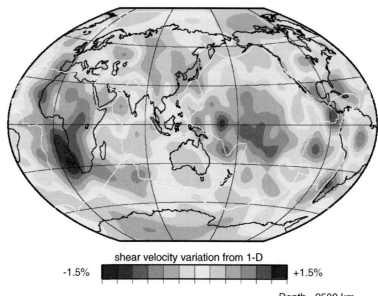

shear velocity variation from 1-D

-1.5% ▮▮▮▮▮▮▮▮▮▮▮ +1.5%

Depth= 2500 km

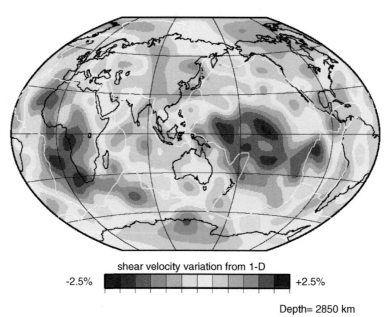

shear velocity variation from 1-D

-2.5% ▮▮▮▮▮▮▮▮▮▮▮ +2.5%

Depth= 2850 km

Plate 35 Maps of shear-wave-speed structure in the lower mantle (Ritsema, 2005).

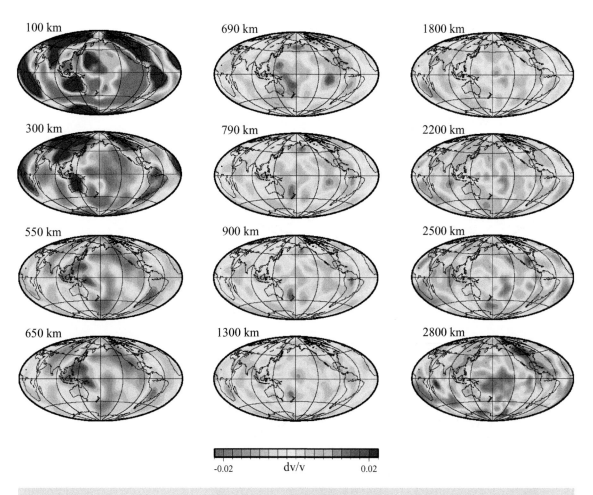

Figure 5.31 Shear-wave-speed variations in the mantle at various depths (from Gu et al., 2001). See Plate 34

throughout the mantle is the exception rather than the rule. Most high-wave-speed anomalies associated with subducted slabs are confined to the upper ~1000 km.

Two vast, low-V_S anomalies occupy much of the lower mantle beneath the Pacific and south Atlantic oceans. They are oxymoronically nicknamed the "superplumes", despite the fact that such vast features are entirely different from what Morgan (1971) originally envisaged (Fig. 5.32). They are usually assumed to be hot and buoyant, but this assumption is demonstrably incorrect. The "superplumes" actually have positive anom-

alies in the elastic bulk modulus, and because temperature affects bulk modulus and V_S similarly, the positive bulk modulus anomalies and negative-V_S anomalies indicate that they cannot be due to the effects of temperature alone. In fact, seismic evidence derived from the Earth's normal modes indicates that these regions are dense, and not buoyant. The relative magnitudes of the anomalies in V_P, V_S and density imply that the "superplumes" have no significant temperature anomalies and are caused primarily by chemical heterogeneity (Brodholt et al., 2007; Trampert et al., 2004). It is a misnomer to call

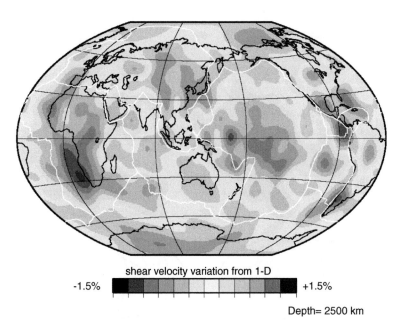

shear velocity variation from 1-D

-1.5% ▮▮▮▮▮▮▮▮▮▮▮ +1.5%

Depth= 2500 km

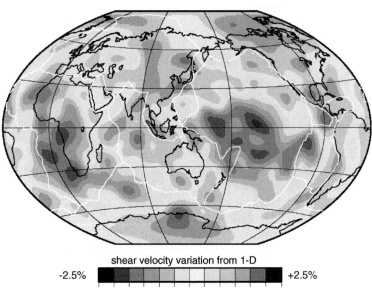

shear velocity variation from 1-D

-2.5% ▮▮▮▮▮▮▮▮▮▮▮ +2.5%

Depth= 2850 km

Figure 5.32 Maps of shear-wave-speed structure in the lower mantle (Ritsema, 2005). See Plate 35

these bodies "plumes" of any kind, and they are more appropriately named "superpiles" (Garnero et al., 2007). The "superplumes" are the strongest evidence yet that overly simplistic interpretation of incomplete seismic data can lead to conclusions that are the exact opposite of the truth.

5.6.2 Global variations in the transition zone

Global studies of the transition zone also are limited in resolution and can only map variations on scales of hundreds or thousands of kilometers. Nevertheless, they may indicate to what degree the 650-km discontinuity impedes the progress of cold sinkers and hot risers, at least on a large scale, and whether or not topography on the transition-zone discontinuities can be used as a thermometer.

Globally, topography on the 410-km discontinuity is uncorrelated with topography on the 650-km discontinuity (Gu et al., 2003; 2009). This suggests that the physical conditions that influence topography on the discontinuities – composition, water content and temperature – are also decoupled between these two depths. On the other hand, the thickness of the transition zone correlates with surface lithospheric structure, and is several kilometers thinner beneath the oceans than the continents. These intriguing observations hint that the influence of the surface lithospheric plates extends deeper than the asthenosphere and into the transition zone, but not through it.

Studies have focused specifically on subduction zones and melting anomalies, places where structures penetrating through the transition-zone have been suggested (Deuss, 2007; Tauzin et al., 2008). Such studies usually seek to detect and measure thermally induced discontinuity topography, i.e., an elevated 410-km discontinuity where cold slabs penetrate and a depressed 410-km discontinuity beneath melting anomalies where high temperatures are postulated. Where slabs subduct, a tendency is observed for the 410-km discontinuity to be elevated, as expected for an Fe-rich petrology, high water content, low temperature or a combination of these conditions. Beneath melting anomalies, the evidence for systematic deepening of the 410-km discontinuity is variable between studies. Some studies report tendencies for depths greater than the global average (Gu et al., 2009)[20] but others find no such tendencies (Tauzin et al., 2008).[21] There is no evidence for

systematic thinning of the transition zone above the Pacific and Atlantic "superplumes".

Formidable barriers exist to using observations from the transition zone discontinuities to infer the physical conditions beneath melting anomalies. Even the most sophisticated experiments cannot determine depth to an accuracy better than 5–10 km. The repeatability of results between independent studies is poor. For example, at Iceland, local studies have concluded that the 410-km discontinuity beneath Iceland is depressed by ~20 km (Du et al., 2004; Shen et al., 1998), but different global studies have found it to lie at the average global depth (Gu et al., 2009) and to be elevated by ~15 km (Deuss, 2007). It is difficult to know what significance to place on such a body of results. Locally, discontinuity topography often varies radically over short distances, and has no correlation with surface features, for example, at Yellowstone and the Baikal region (Fee and Dueker, 2004; Liu, 2006).[22] Most problematically, it is at present not possible to separate the several physical factors that can cause topographic variations. The current weight of evidence suggests that the transition zone hinders the exchange of material between the upper and lower mantle.

5.6.3 The core-mantle boundary

The core-mantle boundary region – the lowermost few hundred kilometers of the mantle, is where classical plumes are predicted to arise (Garnero et al., 2007; Lay, 2005).[23] This region is physically complex and seismically heterogeneous. A crystallographic phase change, from perovskite to post-perovskite, occurs in magnesium silicate at deep lower-mantle pressures, and a temperature jump of ~1000 °C occurs on passing from the mantle to the core.

[20] http://www.mantleplumes.org/TransitionZone2.html
[21] http://www.mantleplumes.org/TransitionZone3.html

[22] http://www.mantleplumes.org/Baikal.html
[23] http://www.mantleplumes.org/DDub.html

There is good repeatability between seismic studies of V_S in this region, but poor repeatability for V_P. The ratio between the two wave-speeds varies much more than in the shallow mantle. Interpretive difficulties are thus compounded, with variations in composition, crystallographic phase, partial melt content and temperature all expected, but too few reliable independent data to distinguish these different effects. Special seismic studies of local regions ranging from ~1000 km in size down to a few tens of kilometers find radical low-V_S anomalies, as strong as 10% in some places, which almost certainly requires partial melt.

Some local regions with very low V_S correlate spatially with flood basalts on the surface, and have been proposed to be the roots of mantle plumes (e.g., Helmberger et al., 1998). However, no such region has been shown to be connected to the surface by a vertically extensive low-wave-speed zone, and structure throughout most of the lower mantle is not correlated with structure either at the core-mantle boundary, the upper mantle or the surface (Dziewonski, 2005). Thus, physical connections linking the core-mantle boundary to the surface remain speculative and unsupported.

5.7 Plume variants

The original Plume hypothesis predicts that continuous, relatively narrow, hot conduits extend from the surface down to the core-mantle boundary. These conduits are expected to manifest themselves as low-seismic-wave-speed anomalies. Some observations of such anomalies have been reported (e.g., Bijwaard and Spakman, 1999; Montelli et al., 2004b; 2006), but they are not repeatable between independent studies and their reliability is questionable (van der Hilst and de Hoop, 2005). The repeatable seismic structures found beneath melting anomalies do not resemble conduit-like features extending deep into the lower mantle. In the cases of some

major melting anomalies, for example, Iceland and Yellowstone, such anomalies have been ruled out to a high degree of certainty by multiple independent studies.

In order to account for these observations, variants of the original Plume hypothesis have been proposed. Such variants fall into four categories:

- Plumes that arise from regions other than the core-mantle boundary;
- Dying, or newly born plumes;
- Discontinuous plumes;
- "Hide-and-seek" plumes, that tilt, bend, or are too narrow to detect.

It has been suggested that plumes may arise from the base of the lithosphere, the 650-km discontinuity, or indeed almost any depth in addition to the core-mantle boundary. No mechanism by which long-lived plumes might arise from the shallower boundaries has been given, however. It is not clear how a plume originating at the base of the lithosphere is envisaged to differ from asthenosphere rising through a lithospheric crack. Furthermore, for an adiabatic mantle, the temperature is expected to drop by ~30 K beneath the 650-km discontinuity, not rise (Section 6.1.2) (Katsura et al., 2004). It unexplained how long-lived volcanism or relative fixity between distant plumes originating at such diverse depths would be maintained, in particular since lateral flow in the mantle is expected to vary with depth as a result of depth-dependent shearing in response to plate motion (Doglioni et al., 2005). Another model assumes that the "superplumes" are hot upwellings and that their upper surfaces comprise thermal boundary layers from which small plumes rise (e.g., Courtillot et al., 2003; Fig. 1.2). This model may be dismissed because the "superplumes" are not thermal upwellings (Trampert et al., 2004).

Dying plumes have been proposed to account for anomalies that are confined to the shallow mantle. Such anomalies would represent the last rising remnants of plumes originally fed from the core-mantle boundary but whose feeder conduits have now expired. Such a model has been suggested for the Iceland, Eifel, Bowie, Hainan (China), Eastern Australia and Juan Fernandez melting anomalies (e.g., Montelli et al., 2006). The inverse case, low-wave-speed anomalies confined to the deep mantle, have been interpreted as young plumes that have not yet risen as far as the surface. Anomalies south of Java, east of the Solomon Islands and in the Coral Sea have been interpreted in this way (Montelli et al., 2006). Scattered, discontinuous low-wave-speed anomalies, or the absence of a through-going structure in the transition zone, have been interpreted as dissociated pulses of plume material, for example, at Yellowstone (Yuen et al., 1998). Branching, or headless plumes have been suggested to explain irregularly shaped or forking low-wave-speed anomalies.

The concept of "hide and seek" plumes expands still further the suite of excuses for the failure of seismology to find plume conduits. It has been suggested that vertically continuous low-wave-speed anomalies do in truth exist but are not observed because they do not directly underlie the surface melting anomalies where studies have been performed. Such plumes are suggested to tilt, or to lie a long way from the melting anomaly and to supply it by lateral flow. It has also been suggested that plume conduits are too narrow – perhaps only a few kilometers in diameter – to be detected by seismology. This suggestion is untestable as a practical matter and also inconsistent with the expectation that conduits would be wider in the high-viscosity lower mantle than in the lower-viscosity upper mantle.

5.8 Discussion

The use of seismology to explore melting anomalies is a vast, multi-faceted subject. Ideally, tomography would be used to look for an under-

lying conduit extending to the surface from the deep mantle. In a simple world, observation of such a feature would support the Plume hypothesis and failure to observe it would support the Plate hypothesis. In reality, things are not so simple. To date only three currently active melting anomalies have been studied in detail – Iceland, Eifel and Yellowstone. These places display several common features:

- Low-wave-speed anomalies underlie them, but all have been repeatedly found to be restricted to the upper mantle.
- The 410-km discontinuity is deflected downward locally, though not always directly beneath the melting anomaly.
- No topography occurs on the 650-km discontinuity that might be consistent with a downward-continuous hot body.

The wave-speed anomalies beneath these three areas are restricted to the shallow mantle. Even taking into consideration the expected reduction with depth of the sensitivity of wave-speeds to temperature (Fig. 5.10), it is clear that the effects that cause the anomalies weaken downward and disappear in or above the transition zone. This observation is not consistent with a bottom-driven mechanism, which would be expected to generate an anomaly that is strongest in its lower part and weaker higher up, when corrected for the depth sensitivity of seismic wave-speeds. The observed seismic anomalies are thus not only inconsistent with plume tails arising from the core-mantle boundary, but they are also inconsistent with variants such as "upper-mantle plumes" or plumes that arise from the 650-km seismic discontinuity (Section 5.7). They are not consistent with causative properties, for example, temperature anomalies, that strengthen downward.

Top-down processes offer a simpler explanation. Persistent surface extraction of melt and volatiles as a result of on-going extension may induce a reaction analogous to "chimney draw".

The escape of gas and smoke from the top of a chimney draws an updraught of replacement material in from below, even in the absence of a heat source at the bottom. In this model, the strong, near-surface seismic anomalies are associated with high degrees of partial melting, including accumulations in magma chambers. Continued extraction of volatiles and melt induces upflow of replacement volatiles and melt from below, which also transports heat upward. The supply region grows downward and outward to a degree related to the surface magmatic rate and the local geological situation including stress, temperature and composition. It is this supply region that is reflected in the deeper, weaker seismic anomalies. This process – expansion of seismic anomalies in volume and strength in response to continued extraction of fluid from the surface – has been observed on a small scale in The Geysers geothermal field, California (Gunasekera et al., 2003).

This model proposes that it is the escape of volatiles and melt at the surface, permitted by extensional stress, which induces development of an underlying melt-delivery system, not vice versa. Highly mobile volatiles (CO_2 and H_2O) may play a critical part in the detailed physical processes at work.

Testing whether thick layers of basaltic crust result from melting at high temperature depends on indirect geochemical arguments (Korenaga et al., 2002). This is not to say that such arguments are wrong, but simply that such a test is further away from direct observation of the primary features of melting anomalies. Nevertheless, in the two cases where this method has been applied (the North Atlantic and the Galápagos) the results suggest that the flood basalts did not form at unusually high temperatures (Sallares and Calahorrano, 2007).

Even more circumstantial are arguments that rely on structures much larger than those expected beneath melting anomalies. Such studies can only give broad indications of the structure and behavior of the Earth, to infer smaller-scale features. For example, whole-mantle tomography may indicate whether the mantle convects as a single layer or two independent layers separated by the 650-km discontinuity. Single-layer convection would be consistent with the view that subducting slabs transport material from the surface down to the core-mantle boundary, whence it is transported back up to the surface in plumes. Layered convection would be consistent with shallow-genesis models for melting anomalies (Hamilton, 2002; 2007b). Current results suggest that the 650-km discontinuity is a barrier to convection, and decouples the upper and lower mantles, at least to a large degree.

A particularly challenging problem is the tendency of different studies to give results that seem to be mutually exclusive. This is common with results from different disciplines, for example, geochemistry and seismology, but it is also common within single disciplines, including seismology. Clearly, such problems arise from either experimental errors or invalid interpretations. Valid interpretive models must be consistent with all reliable results.

It has been suggested, for example, that the southward migration of the Emperor melting anomaly with respect to the Earth's geomagnetic pole resulted from deflection of a plume by mantle flow ("mantle wind"; Fig. 4.11) (Steinberger, 2000). The present-day position of such a plume at the 650-km discontinuity is calculated to be ~100–250 km south of the Big Island. This suggestion is at odds with interpretations of receiver functions that propose that a plume feeding Hawaii penetrates the 650-km discontinuity ~200 km west of the Big Island (Li et al., 2000b). The mantle-wind model predicts that the plume root should lie at the core-mantle boundary ~1300 km north of the Big Island, but this prediction is inconsistent with reports of ultra-low-wave-speed regions at that depth variously 1000 km to the southeast (Montelli et al., 2004b; Russell et al., 1998), 2000 km to the southwest (Bréger and Romanowicz, 1998) and 200 km to the northwest (Ji and Nataf, 1998). The mantle wind hypothesis is also inconsistent with suggestions that the low-wave-speed anomalies beneath long-extinct flood basalts such as the Ontong Java Plateau, the Deccan Traps and the Brazilian shield represent fossil plumes that have traveled with the overlying plates as they drifted for thousands of kilometers.

In the north Atlantic, different receiver function studies find a depressed 410-km discontinuity variously beneath central Iceland and beneath south Iceland. Different studies of the deeper mantle and the core-mantle boundary have been interpreted as indicating that the root of an Icelandic plume plunges to the south (Shen et al., 2002), the east (Helmberger et al., 1998) and the west (Bijwaard and Spakman, 1999). At Yellowstone, seismic observations have been interpreted as indicating that a plume plunges to the northwest (Yuan and Dueker, 2005), whereas convection modeling suggests that one would plunge to the southwest (Steinberger, 2000).

Seismic experiments to study melting anomalies increase continually in scale and sophistication, and the number of data available has mushroomed in recent years. However, such increases will do us little good if the results do not build on earlier work, but only proliferate mutually exclusive models based on selective subsets of the available evidence. As newer, superior techniques provide improved results to replace older, less accurate ones, the problem of disagreement between seismic results can be potentially resolved if the will is there. An innovative new technique that has been proposed is to search for "plume waves" – guided seismic waves analogous to the light that travels in optical fibers. Such waves would be trapped in a continuous, low-wave-speed conduit regardless of its topology, and would be distinctively dispersed (Julian and Evans, 2010). A problem more difficult than seismology is, however, the disagreements between interpretations that involve additional assumptions. Such is the diversity of interpretations, sometimes of the same results, that a skeptic might reasonably ask how much power seismology has, in truth, to constrain the processes responsible for surface melting anomalies. What is a vigorously upwelling plume tail to one eye may be nothing but a wispy smearing of ray bundles or a largely static region of low-degree melt, to another.

No reason why curiously similar anomalies should exist in the upper mantle beneath very different regions has yet been offered. For example, the low-wave-speed anomaly that fills much of the upper mantle beneath Iceland is similar in strength and depth extent to those beneath Eifel and the Ontong Java Plateau. Iceland is a vast, persistently productive volcanic region, in stark contrast to Eifel which has erupted only very small volumes intermittently, and the Ontong Java plateau which has been extinct for ~122 Ma. Beneath western North America, anomalies similar to that beneath Yellowstone also underlie the eastern Snake River Plain, Newberry volcano in Oregon, and southern Utah, where a voluminous melting anomaly does not exist (Roth et al., 2008).

Notwithstanding the technical problems of seismology, the most intractable problem lies in geological interpretation of seismic images. Seismic wave-speed does not correspond directly to petrology, composition, mineralogical phase, degree of partial melt, geological structure or temperature. A single continuous seismic wave-speed anomaly does not necessarily correspond to a single geological feature. It is even less certain that discontinuous regions with similar anomalies correspond to homogenous physical properties. Most importantly, neither seismic wave-speed nor transition-zone-discontinuity topography are thermometers. With the exception of the normal-mode study of the "super-plumes", seismology alone cannot test for the existence of hot bodies, nor can it measure whether or not material is rising. The challenge for the future is to use seismology as a powerful component in multidisciplinary work, and to use it to answer questions that it is capable of answering, rather than to attempt to extract information from it that it is inherently unable to give.

5.9 Exercises for the student

1 How can seismology distinguish between the Plate and Plume hypotheses?

2 How likely are apparently plume-like and slab-like anomalies to occur in a synthetic, random Earth model?

3 Why are the topographies on the 410- and 650-km discontinuities uncorrelated?

4 Why is seismic structure near the core-mantle boundary and the shallow mantle correlated, when neither correlates with mid-mantle structure?

5 Do whole-mantle tomography images show flattened discs of low-seismic-wave-speed material beneath the 650-km discontinuity where plume heads are predicted to have impacted?

6 Map globally the base of the seismic lithosphere.

7 How widespread is the global low-velocity layer?

8 Is there seismic evidence for large, long-lived magma reservoirs beneath thick continental lithosphere?

9 What is the absolute strength and spatial extent of the low-wave-speed body beneath the Iceland region?

10 If the low-wave-speed seismic anomaly beneath the Yellowstone region were interpreted solely in terms of temperature, how would the deduced temperatures vary throughout the body?

11 What kind of teleseismic delays are typically seen across large faults and sutures?

12 Why are the Ontong Java Plateau, Iceland and Eifel – radically different melting anomalies – all underlain by similar mantle seismic wave-speed anomalies?

13 What is the true depth extent of the low-seismic-wave-speed anomaly that tilts to the north from beneath Yellowstone?

14 The low-seismic-wave-speed region beneath the Ontong Java Plateau also has low attenuation. Is this also true for the seismically anomalous regions beneath the Paraná basalts, the Deccan Traps and Ireland?

15 What are the seismic anomalies under the Ontong Java Plateau, the Paraná basalts, the Deccan Traps and Ireland?

16 Compare and contrast the crustal structures, elevations, and histories of vertical motion of the Ontong Java Plateau and the Iceland Plateau. How similar are they?

17 Model the "chimney draw" process numerically.

18 Estimate the CO_2 flux from melting anomalies.

6

Temperature and heat

It ain't what you don't know that gets you into trouble. It's what you know for sure that just ain't so.

–Mark Twain (1835–1910)

6.1 Introduction

Are "hot spots" hot? What is meant by this question? All volcanic areas are hot in the sense that molten rock at high temperature erupts at the surface, and is intruded into the crust. Some volcanoes even overlie persistent magma chambers containing melt and crystal mush as hot as ~1200 °C. Local temperature gradients may be as high as 200 °C/km (Flovenz and Saemundsson, 1993) and boiling hot springs lie at the surface in geothermal areas.

This is not what is meant, however. The term "hot spot" refers to the concept that volcanic islands such as Hawaii form over unusually hot regions in the mantle, and that plate motion carries volcanoes away, creating time-progressive chains of volcanic islands and seamounts.[1] This idea led on later to the mantle Plume hypothesis, which proposed a source for the heat and a mechanism whereby it might be maintained (Morgan, 1971). The term "hot spot" thus emerged to signify a region in the mantle with a higher temperature than surrounding mantle at the same depth.

A source that is anomalously hot is intrinsic to the Plume hypothesis and is its most fundamental requirement. If the volcanic rocks at melting anomalies do not arise from anomalously hot sources, the Plume hypothesis can be ruled out without the need to test other, secondary, predictions. The task at hand is to measure the temperature of the mantle source of lavas at melting anomalies relative to some "normal" reference value. Unfortunately, this is a surprisingly difficult task. The problem is further complicated by the issue that the absence of evidence may not necessarily make a convincing case that high temperatures do not exist.

Let us first consider some general aspects of the thermal state of the Earth. The planet is thought to have accreted hot at ~4.56 Ga by the gravitational accumulation of particles. Heat sources included residual accretional heat, gravitational settling of iron to form the core, differentiation, and the early impact of a Mars-sized body which tore off a large part of the Earth's mantle to form the Moon (the "big whack" hypothesis; Natland, 2006). As a result, the Earth went through a phase when a large part of its mantle was molten and formed a magma ocean that may have extended over much of the Earth and been up to several hundred kilometers deep (Hartmann and Davis, 1975).

[1] http://www.mantleplumes.org/TopPages/ThermalTop.html

Plates vs. Plumes: A Geological Controversy, 1st edition. By Gillian R. Foulger.
Published 2010 by Blackwell Publishing Ltd.

The Earth subsequently cooled progressively and may today be ~200 °C cooler than it was in the Archean (Hamilton, 2007c). Heat is transported around inside the Earth by advection, conduction and radiation. By far the most efficient method of moving heat is by the advection of fluids in the crust and mantle, including melt percolation, intrusions, eruptions, hydrothermal circulation and degassing. Solid-state mantle convection also redistributes heat by advection.[2] This includes warm, rising convection currents, cold, sinking ones, the subduction of cold slabs and delamination and gravitational instability of the lower, cold, thick continental lithosphere and possibly deeper parts of the lower crust. The latter processes cool the mantle from within. As a result of all these processes, combined mantle potential temperature varies both vertically and laterally.

Conduction, in contrast, is a slow process because the thermal conductivity of rocks is low. The manner in which it occurs is mathematically identical to chemical diffusion and thus the conductive redistribution of heat is described as "thermal diffusion".[3] The rate at which bodies surrounded by material with a different temperature equilibrate thermally is related to the square of their thickness. Thin bodies equilibrate relatively quickly, but thick bodies take much longer. A body ~10 km thick would equilibrate in just ~0.1 Ma, but a subducting lithospheric slab ~100 km thick would take ~10 Ma – roughly the time it takes to sink to the base of the transition zone in subduction zones (Toksöz et al., 1971). Bodies with dimensions of the order of thousands of kilometers take times of the order of the age of the Earth to equilibrate thermally by conduction alone. Transport of heat by radiation is minor at shallow depths but may be more important in the deep mantle where radiative heat transfer is higher (Hofmeister and Criss, 2005).

[2] http://www.mantleplumes.org/Convection.html
[3] http://www.mantleplumes.org/HeatTransport.html

6.1.1 Surface heat loss

How quickly is Earth cooling? In order to answer this question, the rates of heat loss and heat production are needed. Instantaneous heat loss from the Earth's surface can be estimated from temperature gradient measurements in shallow boreholes. In the continents, heat loss is typically ~50–100 mW m^{-2} with a median of ~60 mW m^{-2} (Anderson, 2007b). It is commonly assumed to be related to age, or to the time since the last tectonic or igneous event, with low heat flow over old crust such as cratons and higher values where the crust is young. Nevertheless, the scatter in measurements over individual geological provinces is large.

In the oceans, measured values commonly range from ~25–300 mW m^{-2}, with a median value about the same as for continents, or ~60 mW m^{-2} (Anderson, 2007b). Two-thirds of the Earth is covered by oceanic crust and thus its heat loss dominates the planetary budget. Heat flow is greatest at mid-ocean ridges, which typically lie at depths of ~3 km, and less from older sea floor, which is up to ~100 Ma and ~5.5 km deep (Fig. 6.1). Heat flow decreases with the age

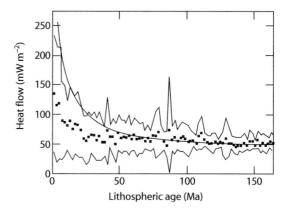

Figure 6.1 Measured heat flow (dots) as a function of lithospheric age, with one-standard-deviation bounds (thick lines). Thin line shows the predicted heat flow from the GDH1 cooling model, which fits the observations best (Anderson, 2007b; Stein and Stein, 1992).

of the ocean floor up to ~50 Ma, as bathymetric depth increases, but is observed, on average, to be approximately constant for older crust. A commonly assumed model assumes the oceanic lithosphere is a cooling plate underlain by asthenosphere maintained at a constant temperature by convection and radioactivity (Parsons and Sclater, 1977; Stein and Stein, 1992). Such a model predicts that heat flow dies off as the square root of lithospheric age, a prediction that fits bathymetry observations but not heat flow (Fig. 6.1). This model also predicts that the lithospheric plates thicken with age by the freezing and underplating of asthenosphere onto their lower surfaces. In this way they are thought to attain a thickness of 50–100 km over ~100 Ma.

Although this model is widely accepted, it fits the heat-flow observations poorly (Parsons and Sclater, 1977; Stein and Stein, 1992). Heat flow falls off rapidly from zero age crust at the crests of spreading ridges to crust of ~20 Ma (Stein and Abbott, 1991). Beyond this, heat flow is approximately constant. The misfit between the heat flow theoretically predicted and observed is generally attributed to hydrothermal circulation, but this has not been independently or experimentally tested. From heat-flow measurements, the total heat flux from the Earth's surface is estimated to be ~32 TW (~12 TW from the continents and ~20 TW from the oceans). If a square-root-of-age cooling model is assumed, an additional ~12 TW must be added. This is because the predicted heat flow from young crust is higher than measured. This increases the total estimate to ~44 TW (Anderson, 2007b; Pollack et al., 1993).[4] It is assumed that hydrothermal circulation removes this from the uppermost oceanic crust and transfers it to the ocean above. However, this is problematic to detect.

What is the origin of the Earth's heat? Radioactivity is predicted to generate a total of ~28 TW. The main contributors are the isotopes ^{40}K, ^{238}U, ^{235}U and ^{232}Th, much of which resides in the continental crust. There is little radioactive material in oceanic crust or in the mantle. The rest of the heat lost from the surface comes from sources such as differentiation of the mantle, earthquakes, tidal friction and secular cooling of the mantle and the core.

The upper mantle may be cooling by 50–100 °C/Ga, and provide 15–35% of the global heat loss. The requirement that the magnetic-field-generating dynamo in the outer core be maintained places a lower bound on the secular cooling of the core. Sufficient heat must be lost to maintain the dynamo but not so much that the core is predicted to solidify too quickly. From this reasoning, the minimum amount of heat that must enter the mantle from the core is ~8 TW, or ~18% of the total lost from the Earth's surface. It has been suggested that much of this is removed by mantle plumes, but some must be transferred by conduction and radiation and contribute to the general background flux of the mantle (Hofmeister, 2007). As a result, any coincidence between the calculated heat flux from the core and that postulated to be lost at melting anomalies cannot be used as an argument in favor of the existence of plumes (Davies, 1988; Sleep, 1990).

6.1.2 Subsurface temperature profiles

Temperature increases with depth in the Earth at rates of typically ~20 °C/km in tectonically stable continental interiors, ~50 °C/km in rift valleys and ~100 °C/km on volcanically active oceanic islands, for example, Iceland. It may rise to as much as ~200 °C/km locally in high-temperature geothermal areas, and rarely and extremely, volcano calderas may contain lakes of basaltic magma open to the surface with temperatures of ~1200 °C.

How does temperature vary at still greater depths? In an object as large as the Earth, progressive compression of minerals with increasing pressure – adiabatic compression – causes

[4] http://www.mantleplumes.org/Energetics.html

absolute temperature to increase with depth, regardless of other effects. This temperature increase is of little interest to the issues of heat transport with which this book is concerned, however, since if material is rapidly displaced vertically and thus subject to pressure change, its absolute temperature simply changes adiabatically and stays the same as its surroundings. The adiabatic temperature gradient (the "mantle adiabat") in the shallow mantle is ~0.4 °C/km, and the total increase in absolute temperature throughout the mantle is ~500–900 °C (Fig. 6.2). The adiabatic gradient changes abruptly where mineralogical phase changes occur in the transition zone, and will change if the amount of radioactive elements varies. It is interesting to note that, because of the negative Clapeyron slope of the mineralogical phase change in olivine at the 650-km discontinuity (Section 5.1), a temperature inversion occurs there (Fig. 6.2) (Katsura et al., 2004).

The most useful measure of temperature is potential temperature (T_P). This is the temperature that material would have at the Earth's surface if it rose adiabatically (i.e., without loss or gain of heat) and without change of state or phase from its original depth (McKenzie and

Bickle, 1988). If temperature in the mantle were purely adiabatic and laterally invariant, T_P would be the same everywhere. However, this cannot be so because dynamic geological processes such as subduction occur, the mantle is radioactively inhomogeneous, and it convects. T_P must vary in the mantle and these variations affect relative buoyancy and thus convection.

The regions of most importance to melting anomalies are the Earth's two major thermal boundary layers – regions throughout which T_P rises steeply with depth. One of these is at the surface and the other is at the core-mantle boundary. The Plate hypothesis predicts that melting anomalies result from processes associated with the surface boundary, whereas the Plume hypothesis predicts that they are associated with processes at the lower one.

The strongest thermal boundary layer is that at the surface. There, T_P rises from ~0 °C to ~1500 °C across a depth interval of ~100–200 km (Fig. 6.3). The region of interest is more accurately referred to as the "conductive layer". The term "thermal boundary layer" strictly speaking applies to a cooling layer overlying an

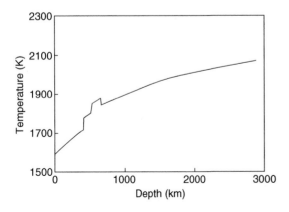

Figure 6.2 Example mantle adiabat assuming a potential temperature (T_P) of 1590 K (1317 °C) throughout the mantle (from Katsura et al., 2004).

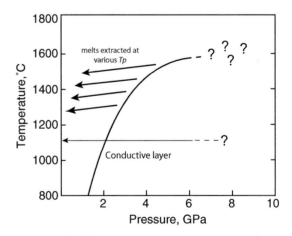

Figure 6.3 Schematic diagram showing the thermal conditions in the shallow Earth. Melts drawn from deeper within the conductive layer have higher T_P than those drawn from shallower layers.

homogeneous fluid, and it does not accurately reflect the complexity of the Earth's lithosphere or the thickness of the layer directly affected by surface processes.

T_P at the base of the conductive layer and beneath it is variable and its average value is not accurately known. It is sometimes assumed to be the same temperature as that of the source of mid-ocean-ridge basalts, but there is no reason why this should be so. In subduction zones, lateral variations in T_P reach extreme values of hundreds of degrees Celsius where cold slabs subduct into warm mantle (Manea et al., 2005). Other processes that cause T_P to vary include melt extraction, which inhibits temperature rises by advecting heat to the surface, thermal disruption of the shallow mantle by continental break-up, the sinking of cold instabilities, and solid-state mantle convection.

The Earth's lower thermal boundary layer lies at the base of the mantle, just above the core. The core is thought to be made largely of molten iron containing subsidiary amounts of nickel, sulfur and lighter elements such as oxygen. It is required to be largely liquid because it does not transmit seismic shear waves. At the core-mantle boundary, T_P increases by ~1000 °C across a depth interval of ~200 km, from ~2000 °C in the lowermost mantle to ~3000 °C in the outer core. The surface area of the core is only ~40% that of the Earth's surface. Thus, if the total heat flux from the core is ~18% that from the surface, the heat flux per unit area from the core is ~72% of that from the surface.

6.1.3 Where is the melt?

How do vertical and lateral variations in T_P affect the ability of the mantle to melt, and if melt is retained inside the Earth, where might it be? The answers to these questions are not dependent on T_P alone, but on homologous temperature. This is the temperature of a material expressed as a fraction of its melting point, on the Kelvin scale. For example, the homologous

temperature of lead at room temperature is 298 K/601 K = 0.50. In other words, whether or not the mantle melts depends on both its ambient temperature and its fusibility.

Rocks are usually composites of several minerals, and the temperatures at which those with the lowest and highest melting points melt are known as the solidus and the liquidus respectively. The mantle is traditionally assumed to be made largely of peridotite. However, it must be inhomogeneous because numerous processes remove melt from it (e.g., at spreading ridges), re-inject near-surface materials (e.g., at subduction zones) and differentiate and transport material within it. Among the many different petrologies expected to exist in the mantle is eclogite, which forms when oceanic crust is subducted to depths > ~60 km (Section 7.2.2).[5] The solidus and liquidus of eclogite are typically ~200 °C lower than those of peridotite, and under some conditions eclogite may be completely molten at a temperature lower than the peridotite solidus (Fig. 6.4) (Cordery et al., 1997).

The presence of volatiles, for example, water and carbon dioxide, is critical. In mantle rocks, water can lower the solidus by several hundred degrees Celsius (Fig. 6.5) (Hall, 1996). The importance of water is particularly well-illustrated at subduction zones. There, arc volcanism cannot be explained by isentropic upwelling of mantle rocks in an extending environment, but almost certainly results from the fluxing of mantle rocks by hydrous fluid rising from the down-going slab.

The effect of CO_2 has been studied in particular detail (Fig. 6.6). This has been done by measuring in the laboratory the solidus of chemical assemblages that reproduce the major-element compositions of various fertile peridotites (the CaO-MgO-Al_2O_3-SiO_2-Na_2O-FeO-CO_2 (CMASNF-CO_2) system). Experiments have been conducted at pressures up to ~8 GPa, corresponding to a depth of ~250 km. At depths >

[5] http://www.mantleplumes.org/Eclogite.html

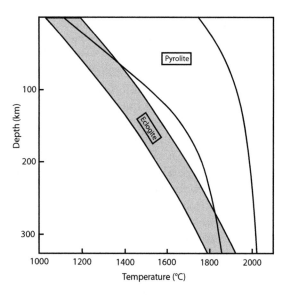

Figure 6.4 Solidi and liquidi of pyrolite (a proxy for peridotite) and eclogite. The eclogite solidus is up to ~200 °C lower than the pyrolite solidus in the depth range 0–250 km. The temperature difference between the liquidus and solidus of eclogite is only ~150 °C for most of the depth range for which data exist (from Cordery et al., 1997).

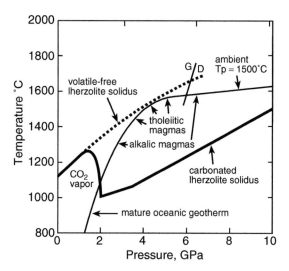

Figure 6.6 Solidus curves for carbonated lherzolite and volatile-free lherzolite. G – graphite, D – diamond (from Presnall et al., 2010).

~60 km, the solidus of mantle rocks drops from ~1300 to ~1000 °C in the presence of CO_2. It thus probably falls below the ambient mantle temperature at these depths, assuming a typical mantle temperature profile. The first-forming melt is carbonatite, but higher degrees of melt have alkalic and tholeiitic compositions (Section 7.2.1). A seismic low-velocity layer, typically ~100 km thick and with its top at ~60 km, is widespread beneath the oceans (Fig. 5.2). In this layer, V_S is reduced by ~5–10%. This can be explained by partial melt, and the solidus curve for CO_2-saturated mantle rocks suggests that the melt may result from fluxing of the mantle by CO_2 (Presnall and Gudfinnsson, 2008).[6]

6.1.4 Temperature and heat

The question of heat is separate from that of temperature. Mantle that is close to melting that

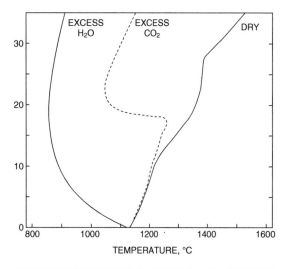

Figure 6.5 The effect of volatiles on the solidus temperature of peridotite (from Hall, 1996).

[6] http://www.mantleplumes.org/Temperature.html

rises adiabatically will, counter-intuitively, both cool and melt. The heat required to melt mantle rocks comprises the specific heat needed to raise the source material to its solidus, plus the latent heat of melting. For a given amount of heat, much more melt will be produced from fusible source rocks such as eclogite than from refractory ones such as peridotite, in particular in the presence of solidus-lowering volatiles. The removal of large volumes of melt from the mantle to the crust extracts large amounts of heat and thus tends to buffer temperature. Ironically, the eruption of large volumes of magma at melting anomalies is thus evidence for a cool mantle beneath rather than a hot one.

6.2 Methods

There are many ways of estimating temperature inside the Earth, but the problems lie in achieving sufficient accuracy and in understanding the results. It is not possible to drill deeper than a few kilometers, and estimates of mantle temperature thus rely on indirect methods. This means that a number of formidable barriers must be faced. Both absolute and relative mantle temperature must be considered. Absolute temperature is likely to be more difficult to estimate than variations in temperature from place to place. This is because absolute temperature is dependent on many factors that are poorly known, such as mantle composition, volatile content, and the physical properties of common rock-forming minerals at conditions corresponding to mantle depths in the Earth. However, absolute temperature is not strictly needed to test the hypothesis that melting anomalies arise from anomalously hot sources. Merely measuring the variation in source temperature between "hot spot" lavas and others presumed to come from "normal temperature" mantle is sufficient. Nevertheless, even this is a difficult problem.

First, to what depth does a temperature estimate apply? The surface conductive layer – the layer throughout which T_P progressively increases with depth – may be up to ~200 km thick. Thus, T_P increases strongly with depth throughout this layer, and if temperature estimates for different regions correspond to different depths, then large differences are expected that do not reflect lateral temperature anomalies. The "normal" temperature of the mantle is commonly taken to be that of the source region of mid-ocean ridge basalts. These are generally assumed to be extracted from the top ~100 km of the mantle, the fraction reducing, probably exponentially, with depth. However, the details are poorly known because calculations depend, among other things, on assumptions concerning source composition. Similarly, in very few cases do robust data exist that constrain the depth of extraction of melt at melting anomalies.

6.2.1 Seismology

Several methods based on seismology have been applied to estimate temperature in the crust and the mantle. In regions subject to strain, the shallow crust deforms by brittle failure, i.e., seismogenically. At depths where the temperature is high enough that the rocks deform plastically, earthquakes do not occur. Thus, the maximum depth of earthquakes in the crust gives an indication of temperature if the petrology, volatile content and strain rate can be estimated (e.g., Hill, 1989).

The thickness of igneous crust, as measured using seismology, is a possible proxy for the amount of melt formed (Section 3.1). Larger melt volumes, and thus greater igneous crustal thicknesses, are expected for higher-temperature melting or more fusible source compositions. These two can be distinguished where details of crustal V_P and thickness are available, for example, from explosion seismology. Igneous crust formed at higher temperatures is expected to be not only thicker but also richer in olivine and MgO and as a result to have higher V_P. Crustal thickness and V_P will thus be correlated for a hot source, but anti-correlated for a fusible source (Section 5.2.1, Fig. 5.7) (Korenaga and Kelemen, 2000).

The most common seismic parameter used to infer temperature is wave-speed, both V_P and V_S. This is unfortunate because it is an invalid approach (Section 5.1.2).[7] Seismic wave-speed does not directly indicate temperature or density. Wave speeds are dependent on several physical parameters – petrology, composition, mineralogical phase, state (liquid or solid) and temperature, with temperature having one of the weakest effects. It is not possible to separate out the effects of these variables with measurements of just one or two wave-speeds. Even if both V_P and V_S are available, it may be difficult to compare them. The errors may be too large to discriminate between different relationships predicted by different interpretations. They may also have been determined using different experimental approaches. For example, V_P may have been mapped using explosion seismology, and V_S using surface waves or receiver functions. The experiments then sample the Earth differently spatially. The sensitivity of seismic wave-speeds to temperature is expected to reduce by a factor of ~3 from the top to the bottom of the mantle (Fig. 5.10).

In some experiments, it is easier to map the ratio V_P/V_S than V_P and V_S separately. V_P/V_S is sensitive to the presence of partial melt, because melt lowers V_S more strongly than V_P. A partially molten region is thus expected to have a high V_P/V_S ratio (Dawson et al., 1999). Seismic attenuation also can be used to estimate whether a region is near its solidus or partially molten and, if the local petrology is known, this can place constraints on temperature (Menke and Levin, 1994). If a region is partially molten, it will absorb seismic energy and be attenuating.

Topography on the transition-zone-bounding seismic wave-speed discontinuities at ~410 km and ~650 km depths has also been used widely in attempts to map temperature variations

(Section 5.1).[8] The 410-km discontinuity is expected to deepen from its global, average value by ~8 km/100 °C where temperatures are elevated, and to shallow by the same amount where temperatures are depressed (Fig. 5.4). At 650 km, the effect of temperature on olivine is expected to result in deflections in the opposite sense. However, these may be cancelled out by the effect of temperature on garnet minerals. Topography on the 650-km discontinuity is currently too poorly understood to make it a useful tool for measuring mantle temperature.

The method is hindered by the substantial difficulty in measuring the transition zone discontinuity depths accurately. Stacking of many tens, or even as many as 100 seismograms, is needed to reduce errors to as little as ~5 km. This is still a substantial error compared with the signals sought, which might amount to no more than a 10–15 km deflection. More problematic is the ambiguity in interpretation. Deepening of the 410-km discontinuity can be achieved not only by high temperatures but also by high Mg contents (i.e., a refractory composition) or dry conditions. Conversely, elevation of the 410-km discontinuity is expected for low temperatures, high Fe (i.e., a fertile composition), hydrous conditions, or a combination of all three. The same problem thus besets both seismic wave-speeds and transition-zone topography; they can be measured, albeit with errors, but they cannot be interpreted directly in terms of temperature.

6.2.2 Petrological and geochemical methods

Several petrological methods have been developed for estimating the temperature of the source where melts formed in the mantle. Attempts have been made to estimate both absolute temperature and temperature variations from place to place. In the case of melting anomalies, the issue of interest is whether their

---

[7] http://www.mantleplumes.org/ TomographyProblems.html

[8] http://www.mantleplumes.org/TopPages/ TransitionZoneTop.html

sources are hotter than "normal" mantle at the same depth. In order to address this question, the temperature of "normal" or "average" mantle is needed.

Petrological geothermometers can only be applied where volcanism samples melt from the mantle. The mid-ocean ridge is a mantle-sampling zone that is both widespread over the Earth's surface and thought to be relatively uniform in process. Temperature estimates from melting anomalies are thus commonly compared with estimates from mid-ocean ridges, made using the same petrological method. Such comparisons are particularly useful in the case of on-ridge or near-ridge melting anomalies, because complications inherent in comparing samples from very different tectonic provenances do not arise. A few of the most noteworthy methods are as follows.

(a) The Global Systematics

An early method proposed for inferring temperature variations was based on the belief that Na_8 and Fe_8 (the Na_2O and FeO contents of basalts, inferred for a reference value of 8% MgO) in basalts from mid-ocean ridges correlate with the axial depth, including near melting anomalies. This correlation was explained by a model whereby higher source temperatures in an adiabatically upwelling mantle result in melting beginning deeper and at higher pressures, leading to a higher extent of melting, higher Fe_8, lower Na_8, a larger melt volume, thicker crust and shallower bathymetry (Klein and Langmuir, 1987; Langmuir et al., 1992) (Fig. 6.7). The geochemical correlations were termed the "Global Systematics". The Na_8 systematics were calibrated against temperature using localities where crustal thickness had been measured seismically, assuming that the crust formed from adiabatic upwelling decompression melting at ridges. It was concluded that melting occurred in the depth range ~40–130 km (corresponding to pressures of ~1.3–4.3 GPa) and the potential temperature beneath ridges varied in the range

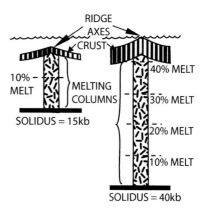

Figure 6.7 The Global Systematics model. Left: Low T_P, deep bathymetry, thin crust, high Na_8 and low Fe_8. Right: High T_P, shallow bathymetry, thick crust, low Na_8 and high Fe_8 (from Langmuir et al., 1992).

~1260–1590 °C. The reasoning inherent in this method requires that the composition of the mantle source of lavas is uniform, but this is now known to be not the case. Furthermore, the "Global Systematics" have been shown by observation to be invalid as the postulated correlations between Na_2O, FeO and axial depth along ridges do not occur (Section 6.5.1) (Niu and O'Hara, 2008; Presnall and Gudfinnsson, 2008).

(b) Mineralogical phase relationships

The main mineralogical phase boundaries for a wide range of pressures and temperatures have been measured in laboratory melting experiments using assemblages of chemicals that replicate the compositions expected for mantle rocks. The results obtained for six-component $CaO-MgO-Al_2O_3-SiO_2-Na_2O-FeO$ (CMASNF) samples in the pressure range ~0.9–1.5 GPa (~ 25–45 km depth) simulate the mineralogical phase relations expected to occur at the natural plagioclase/spinel lherzolite transition where most mid-ocean-ridge basalts are thought to be generated by partial melting. Comparing the compositions of the melts that form in these melting experiments with those of natural samples has shed light on the systematics of mid-ocean-ridge basalt generation under variable temperature, pressure and source-composition conditions (Presnall et al., 2002).[9] Using these results, the relative importance of temperature and compositional heterogeneity in producing the variations seen in mid-ocean-ridge basalts have been investigated.

(c) Olivine control lines

When high-temperature melts also high in Mg cool, olivine is the first crystalline phase to form. Liquid-solid phase relations predict that the first olivine to crystallize will be the highest in Mg (i.e., the most forsteritic) and that the Mg content of subsequent crystals will progressively decrease as the temperature of the melt reduces. Under these ideal conditions, the Mg content of the remaining melt continually reduces as it is removed by the crystals and progressive evolution in composition occurs in a way known as an olivine-controlled liquid line of descent. In order to observe olivine-controlled crystallization with confidence, melts that are sufficiently high in Mg to classify as picrites (i.e., with MgO > 12%) are required.

If samples of the instantaneous melts formed during olivine-controlled crystallization are collected, they can be used to estimate the initial temperature of the original parent liquid. This is done by mathematically modeling the olivine crystal fractionation process in reverse. Olivine is mathematically incrementally added to the most magnesian liquid sampled until a composition is reached that is in equilibrium with the first and most magnesian olivine phenocryst thought to have crystallized out (Fig. 6.8). This is the theoretical parent melt composition. An estimate for the composition of the first olivine phenocryst is thus also needed. The most magnesian olivine sampled, or the most magnesian olivine commonly observed in the study area, are usually used. The final step is to deduce the

[9] http://www.mantleplumes.org/NoRidgePlumes.html

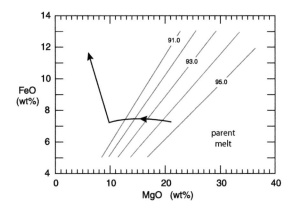

Figure 6.8 Computing parental melt compositions for basalts using the olivine control-line method. Gray lines show the forsterite contents of olivine phenocrysts. Black line shows expected evolution in liquid composition as olivine crystallizes out and MgO reduces. The quasi-horizontal black line segment is the olivine control-line.

temperature at which the calculated melt composition began to crystallize. For this an "olivine geothermometer" is needed. Olivine geothermometers are derived empirically from laboratory measurements of the temperatures at which rock samples melt (Falloon et al., 2007b).

In practice, the olivine-control method for estimating source temperature is dependent on many assumptions and the range of temperatures obtained for a particular locality may be several hundred degrees Celsius. Under such circumstances, the ability of the method to distinguish between the Plate and Plume hypotheses is clearly limited. Why are there such large differences in the estimated temperatures? The problem revolves around several issues:

- It must be assumed that the samples used to construct the olivine-controlled liquid line of descent were once liquids. Only if they are glass is it possible to be confident of this (Presnall et al., 2010). However, high-Mg glass is rare and so crystalline samples are often used, simply because glass is not available. Careful study of some picrite cumulates, for example, from Hawaii (Clague et al., 1995), suggest that the crystals in such cumulates are often scavenged from several different magmas or crystal mushes.[10] The mixing of a partially differentiated mush with a more primitive melt will have the effect of increasing the FeO content for a given MgO content, which will work to give erroneously high apparent temperatures from olivine-control-line analysis. In other words, only if picrite glass is used is it possible to be confident that the results are correct.

- If the starting samples contain plagioclase and pyroxene crystals, the mathematical olivine crystal addition method is not reliable. If plagioclase and pyroxene crystals are present, a point that may also be disputed, this is *prima facie* evidence that the sample does not form part of an array whose compositions were controlled by olivine-only crystallization. Methods are available for correcting for plagioclase and pyroxene crystallization but their robustness is questioned (Falloon et al., 2007a).

- What forsterite composition should be used for the target phenocryst? The higher the forsterite content chosen, the higher the temperature obtained. There are various ways of making the choice, for example, the most-forsteritic crystal, or the most common value, but no way of knowing which, if any, is correct.

- What olivine geothermometer should be used? Many are available (e.g., Beattie, 1993; Falloon et al., 2007b; Ford et al., 1983; Helz and Thornber, 1987; Herzberg and O'Hara, 2002; Putirka, 2005). Different olivine geothermometers use different partition coefficients, which describe how chemical species partition between the solid and liquid phases. They also model the olivine liquidus temperature and crystal composition as a function of different variables. Olivine liquidus temperature is dependent on melt composition and pressure, and the composition of the earliest crystals is dependent on the melt composition, pressure and temperature. Different geothermometers may model different subsets of these variables.

Olivine geothermometers can be tested against one another by comparing the temperatures they predict with temperatures measured by melting samples in the laboratory (Fig. 6 9). They can give

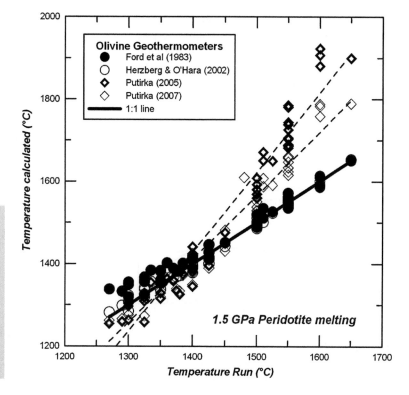

Figure 6.9 Experimental vs. calculated temperatures for 55 peridotite melting and reaction experiments conducted at 1.5 GPa, and the olivine geothermometers of Ford et al. (1983), Herzberg and O'Hara (2002), Putirka (2005) and Putirka et al. (2007) (from Falloon et al., 2007b).

[10] http://www.mantleplumes.org/CoolSamoa.html

differences of >200 °C for the same petrological observations (Falloon et al., 2007b).

■ What choice of reference temperature should be used for mid-ocean ridges? Large ranges are typically calculated for samples from different parts of the global ridge system. Under these circumstances, some investigators take the maximum temperature calculated to indicate the maximum temperature of the mid-ocean-ridge basalt source, whereas others reject the higher temperatures as being influenced by hot, near-ridge plumes. The latter approach assumes a bimodal temperature distribution in the mantle, thereby assuming *a priori* the very hypothesis being tested. Clearly, even if there is agreement about the source temperature of a melting anomaly, if different mid-ocean-ridge reference temperatures are assumed, different conclusions will be reached regarding whether or not a melting anomaly is also a temperature anomaly.

(d) Rare-earth-element modeling

The atomic radii of elements in the Lanthanide-Lutetium (La-Lu) series – rare-earth elements – decrease systematically with atomic number. The smaller atoms are more easily retained in crystal lattices. Thus, if a sample is partially melted, the elements with larger atomic radii systematically partition more readily into the melt (Fig. 6 10).

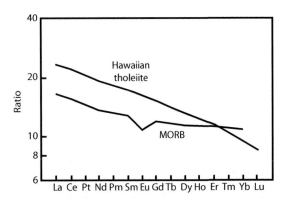

Figure 6.10 Rare-earth-element concentration ratios with respect to depleted mantle for normal mid-ocean-ridge basalt and Hawaiian tholeiite (adapted from McKenzie and O'nions, 1991).

The pattern of rare-earth elements in a melt varies according to which minerals melted and to what degree. Because the stability fields of plagioclase, spinel and garnet are depth dependent, the rare-earth element content can be used to estimate the profile of melting degree and minerals that contribute to the melt. From this, the depth of onset of partial melting may be deduced and, if it is assumed that the melt arises from isentropic upwelling, the temperature of the source can be deduced (McKenzie and O'nions, 1991).

The method involves several critical assumptions, however. These include:

■ The composition of the source, both major and minor elements, and its mineralogy. Usually a composition similar to the primitive mantle (Section 7.1), and peridotite mineralogy are assumed.

■ If melts from different regions are compared, their sources are assumed to have the same composition.

■ The samples analyzed are assumed to represent single melts that formed by fractional melting, and not batch melts, or aggregates from more than one melt.

Rare-earth element modeling in its traditional form yields estimates of source temperature for most basalts. However, because of the assumption of uniform source composition, the differences between these temperatures will be wrong if the compositions of the sources of the different basalts vary. Since the Plate hypothesis postulates variable source compositions, rare-earth element modeling is intrinsically unable to test it.

(e) Komatiites

Komatiites are widely assumed, on the basis of melting experiments, to have been produced at exceptionally high temperatures, possibly >1600 °C, compared with ~1300 °C for mid-ocean-ridge basalts. Komatiites are basaltic rocks that contain 18–30 wt.% MgO, and are thought to derive from liquids that had similarly high MgO contents. This is higher than the 12–18% MgO which defines picrites, and much higher than the 8–10% that typically characterizes mid-ocean-ridge basalts (Le Bas, 2000). Komatiites are widespread in Archean rocks, and as a result have been used to argue that ultra-high-temperature plumes were abundant then (Arndt and Nisbet, 1982). They are rare in the Phanerozoic, however, the only example being from Gorgona Island, 50 km west of the coast of Columbia in Central America (Echeverría, 1980; Kerr, 2005). The Gorgona komatiites erupted at ~88 Ma, and have been variously attributed to plumes postulated to underlie the Galapagos and the Caribbean.[11]

Komatiites might, on the contrary, be produced at low pressures in subduction zones under hydrous melting conditions (Parman et al., 2001). Some komatiites are high in SiO_2, suggesting low pressure melting, and contain hydrous minerals such as amphibole and large concentrations of water in melt inclusions. These komatiites have a large compositional overlap with boninites, which are produced in island arcs, and have similarly high SiO_2 at high MgO contents because they are high-degree melts produced at shallow depths. Large extents of melting at shallow depth are possible beneath arcs because of the high water levels that result from dehydration of the subducting slab and fluxing of the mantle wedge above (Crawford et al., 1989). The origin of komatiites, and whether they provide evidence for melting at high temperatures or simply low-pressure, water-fluxed conditions in subduction zones, is currently a subject of debate (Parman and Grove, 2005; Parman et al., 2001; Puchtel et al., 2009).[12]

6.2.3 Ocean floor bathymetry

In general, the depth of the ocean floor increases with age and this is attributed to the progressive cooling of the lithosphere as it drifts away from the mid-ocean ridge where the crustal part originally formed. This model predicts that, in the absence of non-thermal effects, where the sea floor is shallower than expected for its age, the lithosphere is hotter than normal. On the basis of these assumptions, ocean-floor depth has been used to estimate temperature. Two approaches have been used:

1 The spatial variation in present-day depth of the sea floor has been interpreted as indicating regional variations in lithosphere temperature. Conductive cooling models have been applied to account for bathymetric anomalies such as the swell on which the Hawaiian chain lies, which decays in amplitude as it ages (Ito and Clift, 1998; Sleep, 1987).

2 The patterns of uplift or subsidence through time of single regions have been interpreted in terms of heating and cooling of the lithosphere. The progressive subsidence of the sea floor at a single locality is matched to theoretical subsidence curves to deduce the original temperature of the asthenosphere (Clift, 1997).[13] If the lithosphere formed over asthenosphere that was hotter than normal, unusually rapid subsidence is expected when the region is transported over normal-temperature asthenosphere by plate motion. Subsidence curves can be estimated, for example, from the sedimentary sections of cores drilled into the ocean floor (Fig. 2.2). It is relatively simple and straightforward to estimate the variation in temperature beneath oceanic plateaus during their lifetimes, as subsidence of the ocean floor tends to be simple compared with the vertical motions of continental areas, which are complicated by effects such as tectonism and erosion.

Significant problems include estimating accurately the variation in depth of the sea floor with time. Correction must be made for sediment loading and the sea floor must be restored to the

[11] http://www.mantleplumes.org/TopPages/CaribbeanTop.html
[12] http://www.mantleplumes.org/Komatiites.html

[13] http://www.mantleplumes.org/SedTemp.html

estimated depth in the past for each dated interval, assuming local isostasy. The nature of the sediments and fossils in sedimentary cores are then used to determine water depth of deposition. Benthic foraminifera are, in particular, sensitive to depositional environment. Water depth estimation is generally more accurate for shallow (<200 m) water and can have errors of ~1000 m for greater depths. Variations in sea level introduce additional uncertainty of ~100 m, which cannot currently be corrected for because the timing and magnitude of these variations are poorly known.

Modeling of this kind assumes that all the anomalous depth can be attributed to temperature rather than composition but this may not be the case. Sources of compositional buoyancy include the light residuum that remains following melt extraction (Sections 2.3 and 2.5.3) (Phipps-Morgan et al., 1995). Shallow bathymetry at currently active melting anomalies can be explained naturally either by anoma-

lously hot asthenosphere, or by the residuum left over following melt extraction.

6.2.4 Heat flow

Heat flow from the ocean floor is measured by lowering a thermistor probe from a research ship into sediments on the sea bed (Fig. 6.11). Thermistor probes typically feature a slender lance that may be only a few millimeters in diameter and up to several meters long. It can easily penetrate up to several meters into soft, ocean-bottom sediment, and reach thermal equilibrium within a few seconds.

The vertical temperature profile is measured by a string of thermistors, mounted along the lance, that have a precision of a few hundreths or thousandths of a degree Celsius (Beardsmore and Cull, 2001). Temperature measurements are made at intervals of a few seconds, for a period lasting several minutes. The measured tempera-

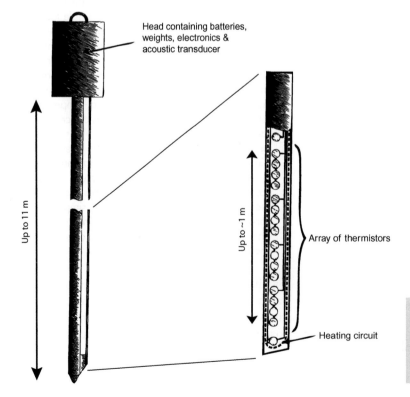

Head containing batteries, weights, electronics & acoustic transducer

Up to 11 m

Up to ~1 m

Array of thermistors

Heating circuit

Figure 6.11 Schematic diagram of a "violin-bow" heat probe for ocean-floor heat-flow measurements (from Beardsmore and Cull, 2001).

tures are then extrapolated to their long-term equilibrium values, assuming asymptotic behavior. Heat flow is calculated using the thermal conductivity of the sediments, which is obtained from core samples or by using a heater in the probe.

Optimum sites for heat-flow measurements are relatively flat, which minimizes the effects of hydrothermal circulation, and blanketed with uniform, thick (>100 m) sediment. An economical survey procedure is the "pogo" method, which involves lowering the probe to the sea floor to make the first measurement, then raising it slightly above the sea floor and towing it to other nearby sites (Stein and Von Herzen, 2007). In this way, a set of closely spaced measurements can be made in a few hours before sailing to the next site. The final heat-flow value for a single site is calculated by combining the measurements made at up to several tens of localities, which suppresses the error by a factor of approximately the square root of the number of measurements.

Although measuring heat flow is perhaps the most direct way of testing for high mantle temperature, a number of hurdles stand between the measurements and their correct interpretation. First, individual measurements must be compared with some global average for lithosphere of the same age. For this, a reliable global model must be used. Many older conclusions of anomalously high heat flow near melting anomalies were based on comparisons with models that underestimated global heat flow (Stein and Stein, 1992).

Second, local geological conditions may either elevate or depress heat flow, so measurements do not reflect the heat being conducted through the lithosphere. Recent magmatism may elevate heat flow, and hydrothermal circulation may reduce it by dispersing heat efficiently, so a temperature anomaly would not be detected even if one existed (Harris et al., 2000).[14] Hydrothermal flow is unlikely, however,

to mask the broad heat-flow anomaly expected for lithosphere substantially reheated by plumes. Careful study of the volcanic, sedimentary and bathymetric context of measurement sites is necessary in order to assess whether volcanism or hydrothermal circulation affects individual measurement sites (Björnsson et al., 2005).

Last, heat flow measurements are relatively insensitive to mantle temperature, particularly beneath thick lithosphere. For asthenosphere several hundred degrees Celsius hotter than normal, and lithosphere several tens of kilometers thick, of the order of 100 Ma is required for a surface heat-flow anomaly of a few mW m^{-2} to develop. Such a heat-flow anomaly is at the lower limit of significance, even for the most accurate heat-flow measurements.

6.2.5 Heat loss from intrusions and eruptives

When flood basalts erupt, large amounts of heat are lost to the atmosphere in relatively short times (a few Ma). It has been argued that high source temperatures are required to supply this heat so rapidly. The latent heat of fusion of basalt is $\sim 4 \times 10^5$ J kg^{-1}, and the specific heat capacity is ~ 1200 J kg^{-1} K^{-1}. The amount of heat required to melt basalt is thus approximately the same as is required to heat it by $\sim 300\,°C$. The argument that a high source temperature is required is based on the assumptions of source homogeneity and melt production on the same time-scale as eruption. These constraints vanish if source composition varies and if the melt accumulated over a time period much longer than eruption (Section 3.3).

6.3 Predictions of the Plume hypothesis

High temperature is a required characteristic of mantle plumes. In order to rise through thermal buoyancy, an anomaly of at least 200–300 °C is

[14] http://www.mantleplumes.org/Heatflow2.html

required, even for the weakest upper-mantle plume (Courtney and White, 1986; Sleep, 1990; 2004). These temperature anomalies are expected to be in excess of the general T_P of the regional mantle as a whole.

The hottest part of a plume is predicted to be the tail and the centre of the head just above it. Most of the plume head is expected to be cooler, as it is predicted to entrain large amounts of normal-temperature ambient mantle as it rises and overturns. The temperature of the plume head is thus expected to reduce from a maximum at its center over the plume tail, to approach ambient mantle temperature at its periphery (Campbell, 2006; Davies, 1999). Indicators of high temperature are thus predicted to be strongest near the centers of flood basalts and over currently active melting anomalies, and weaker in the peripheries of flood basalts.

Several seismological indicators of high temperature are expected (Section 5.3). A thick basaltic crust is expected, indicating high magmatic productivity, both at the flood basalt provinces postulated to represent plume heads, and at currently active melting anomalies. A positive correlation is expected between crustal thickness and V_P, in keeping with the expectation that high-temperature melt is higher in olivine and MgO (Fig. 5.7) (Korenaga et al., 2002). For a currently active plume, a hot tail extending from the surface down to the core-mantle boundary is predicted, and this is expected to have low V_P and V_S. For a plume stem ~300 °C hotter than ambient mantle, the strength of the anomaly is expected to be ~2–3% in V_P and 3–4.5% in V_S throughout the body at shallow depth, reducing to ~0.5–0.75% in V_P and 1–1.5% in V_S in the deep mantle (Fig. 5.10) (Karato, 1993). The high temperatures in plumes are predicted to make them attenuating of seismic energy. For a plume with a 200–300 °C temperature anomaly, a downward deflection of ~15–25 km is expected of the 410-km seismic discontinuity that forms the upper boundary of the mantle transition zone. Such anomalies are not expected beneath long-extinct flood basalts,

which have been transported away from their associated plume stems.

Petrological indications of high temperature are expected to be strongest where eruptions are postulated to have occurred from plume stems and lower in peripheral regions. Picrite glass is evidence for high-MgO, high-temperature melts. It is often stated that picrite cumulates are indicators of high temperature, but this is not safe as it cannot be proven that they represent original melts and do not contain xenocrysts. If abundant cumulate picrites are present, and derive from a hot source, there is no reason why picrite glass should not also be present.

The arrival of a hot plume head is predicted to cause uplift ~10–20 Ma prior to the onset of flood volcanism (Section 2.2) (Farnetani and Richards, 1994; Campbell and Griffiths, 1990). Subsequent plate motion is expected to transport the plume-head flood basalt, away from the relatively stationary mantle thermal anomaly associated with the plume stem, and over mantle with normal temperature. The flood basalt is thus expected to subside unusually rapidly. The predicted pattern of subsidence is expected to be most clearly observable in sedimentary cores from the sea floor because the subsidence behavior of oceanic lithosphere is relatively simple and well understood (Clift, 1997; Stein and Stein, 1992).

Significant surface heat flow anomalies are predicted around currently active plume stems. A strong plume will reheat the bottom third to half of the lithosphere to asthenospheric temperatures throughout a zone several hundred kilometers wide (von Herzen et al., 1982). For a plume impinging on a rapidly moving plate with substantial thickness, a surface heat-flow anomaly is predicted that is zero immediately above the plume, because some time is required for the heat to conduct through the lithosphere to the surface. The maximum heat-flow anomaly is predicted to occur some distance along the volcano chain where sufficient time has elapsed for the heat to conduct to the surface (Fig. 6.12).

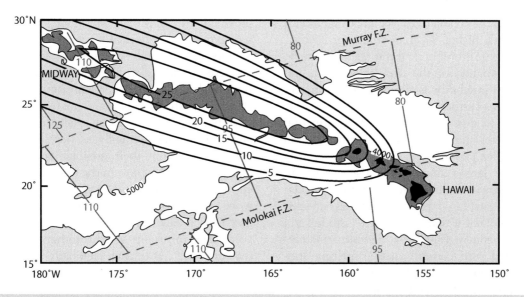

Figure 6.12 Predicted heat flow anomalies for a plume that reheats the bottom of the lithosphere at Hawaii. Heat flow anomalies up to ~25 mW m^{-2} are predicted. Black dots and lines indicate heat flow measurements (from Stein and Von Herzen, 2007).

6.4 Predictions of the Plate hypothesis

A high-temperature mantle source is not required by the Plate hypothesis, which attributes melting anomalies to variations in lithospheric stress and source composition. The Plate hypothesis predicts that the source of melt erupted at the surface is largely within the Earth's surface conductive layer, which comprises the outer ~100–200 km. Within this layer, T_P generally increases with depth and thus material drawn from deeper within it will have a higher T_P than material drawn from shallower levels (Fig. 6.3). There may thus be some variation in the source temperature of melts erupted at the surface, but this will be largely related to depth of extraction. The maximum T_P observed is expected to approach that of the mantle in general below, not a temperature 200–300 °C higher, as predicted by the Plume hypothesis. Lateral variations in the T_P of the mantle in general are also

expected. The mantle beneath the conductive layer is not viewed as being essentially isothermal, with a uniform T_P everywhere except for discrete, isolated high-T_P anomalies of several hundred degrees Celsius. It is viewed as being laterally variable.

Mid-ocean-ridge basalt is thought to be sourced from depths of a few tens of kilometers. Melting anomalies at parts of the mid-ocean-ridge system with high fertility or complex tectonics are expected to be sourced from mantle at similar depths and temperatures to mid-ocean-ridge basalts away from melting anomalies. Exceptionally large volumes of excess melt are attributed to a more fusible source composition. Temperatures may be slightly elevated at plate boundary junctions, for example, ridge-transform intersections and triple junctions, if lateral flow draws source mantle from deeper within the conductive layer. Processes such as rifting continents or intraplate extension, which may fracture lithosphere 100–200 km thick, are expected to draw material up from near the bottom of the conductive layer and thus produce melts with higher T_P than shallower sources. Catastrophic lithospheric thinning, for example, by delamination or the detachment of a gravitational instability, is also predicted to trigger the upflow of asthenosphere from relatively large depths, i.e., near the base of thick continental lithosphere, and so may also elevate material with higher T_P.

Where large-volume magmatism results in thick layers of basalt, a zero or negative correlation between crustal thickness and V_P is expected as melt volume is related to source fusibility and not temperature (Fig. 5.6). Seismic structures associated with currently active melting anomalies are predicted to be confined largely to the upper mantle. Low seismic wave-speeds associated with hot, quasi-cylindrical structures extending from the surface down to the core-mantle boundary are not expected. In the case of some observations predicted by the Plume hypothesis, the Plate hypothesis is equivocal. Seismic anomalies in the upper mantle are not necessarily expected to have high attenuation, though this might occur if they result from partial melt or some compositional variations, for example, eclogite. Similarly, deflections on the transition-zone-bounding discontinuities at 410- and 650-km depth may or may not occur, depending on whether there are lateral variations in composition or volatiles at those depths.

Petrological evidence is expected to indicate relatively uniform temperatures and shallow depths of origin of melts. Petrological indications of relatively high T_P are expected only where magmas were sourced from unusually large depths in the surface conductive layer. Picrite cumulate rocks may occur but picrite glass, an indicator of high source temperature, is expected to be unusual because only rarely are melts sourced from depths of 100–200 km. The heat advected out of the mantle, and the volumes of melt observed, at melting anomalies in strongly extending regions such as mid-ocean ridges, are expected to correspond to the heat provided by isentropic upwelling of fusible source material at normal temperatures.

Flood basalts and oceanic plateaus are not predicted to have a strong radial thermal structure, with a hot centre, as predicted by the Plume hypothesis. They are not expected to have been emplaced over unusually hot mantle and thus subside normally as they are transported away from their original locations by plate motions. Heat flow in the neighborhood of currently active melting anomalies is expected to be normal for lithosphere of that age. Heat flow will be elevated in association with recent volcanism, intrusions and hydrothermal activity, but it will not be high regionally.

6.5 Observations

6.5.1 Mid-ocean ridges

One of the earliest studies of mantle T_P beneath mid-ocean ridges assumed that melt is formed by isentropic upwelling of a standard peridotite lithology (McKenzie and Bickle, 1988). This work assumed that crustal thickness was a direct measurement of melt thickness, and a direct proxy for average mantle T_P. Using these assumptions, a T_P of 1280 °C was calculated for the mantle beneath mid-ocean ridges where crustal thickness is 7 km (Fig. 3.1). The parent melt that formed in the mantle was calculated to contain up to ~11% MgO. The same modeling approach was applied to postulated hot mantle plumes that generate a melt thickness of ~27 km. A T_P of ~1480 °C, generating magma containing up to ~17% MgO, was calculated to be required. This early work suggested a temperature difference of ~200 °C between mantle plumes and mid-ocean ridges and was influential in setting the scene for later work. It was, however, entirely dependent on the assumptions that mantle composition is uniform, that crustal thickness is essentially a measure of T_P, and that the T_P of the source of mid-ocean-ridge basalts is the same as that of the mantle as a whole, with the exception of plumes.

Global systematics geochemistry suggests variations in mantle T_P of up to ~330 °C along the mid-ocean ridge system. However, there are awkward discrepancies in the results. The geochemistry of the Azores and Galapagos regions, which are widely assumed to be influenced by high-temperature plumes, does not conform to the predictions. On the other hand, the geochemistry of the Mid-Atlantic Ridge around the Iceland region does. As a result, it was suggested that the Global Systematics should be applied only to mid-ocean ridges away from melting anomalies (Langmuir et al., 1992). The inconsistencies suggest that either this geothermometer does not work, or that melting anomalies are not hot, or both.

This illogical state of affairs led to a re-appraisal of the Global Systematics and an effort to develop a more self-consistent model for the major-element composition variations in mid-ocean-ridge basalts (Niu and O'Hara, 2008; Presnall and Gudfinnsson, 2008). The Na_8 and Fe_8 contents of basalts from the Mid-Atlantic Ridge from ~55 °S to ~73 °N are anti-correlated as predicted for the entire data set. However, they are positively correlated for individual subsets of the data selected on the basis of tectonic provenance (Fig. 6.13). The overall anti-correlation of the data results from the progressive offset of the individual, positively correlated subsets (Fig. 6.14). In addition, the predicted correlations of Na_8 and Fe_8 with axial depth hold only for the Iceland region and not for other sub-regions of the Mid-Atlantic Ridge (Fig. 6.15). Most parts of the global mid-ocean ridge system show similar results.

The observations are well-explained, on the other hand, by simple melting of fertile peridotite under rather uniform temperature and pressure conditions, in the presence of minor compositional variations between sub-sections of the ridge with contrasting tectonic provenances (Presnall and Gudfinnsson, 2008). Laboratory melting experiments using CMASNF

samples (Section 6.2.2) show that there is a positive correlation between Na_8 and Fe_8 for melts produced in the depth range 25–45 km (pressure range of ~0.9–1.5 GPa). Small variations in composition (namely, variations in FeO/MgO content) cause the "local trends" to be arrayed in Na_2O-FeO space such that the trend of the entire data set (the "global trend") shows broad anti-correlation between Na_8 and Fe_8 (Fig. 6.14). The data are thus consistent with melting of fertile peridotite under largely uniform pressure and temperature conditions, but with minor variations in composition from region to region. This re-interpretation explains the observations consistently for ridges everywhere. It also implies that there is little or no elevation of T_P at near-ridge and ridge-centered melting anomalies, including Iceland, the Azores and the Galapagos.

Olivine control line analysis has also been applied to mid-ocean ridges. In particular, it has been used to obtain a reference mantle T_P with which to compare the results from melting anomalies (Table 6.1). T_P obtained for ridges in general varies from 1243–1475 °C (Table 6.1).

Why do the results vary so widely? First, there are no picritic glass samples from the mid-ocean ridge system and thus no samples high enough in MgO to be confidently claimed to provide evidence for olivine-only crystallization. As a result, cumulate picrite rocks are used, and it is disputed whether this is a valid approach. Second, plagioclase and pyroxene crystals are almost always found in mid-ocean-ridge basalts. This also potentially invalidates the analyses as multi-crystal fractionation is even more difficult to model than olivine-only crystallization. Third, different investigators have used different olivine geothermometers, which in some cases can entirely explain the radically different temperatures calculated (Falloon et al., 2007b). At present, the results from the olivine control line method vary so widely that, taken as a whole, they do not provide a useful guide to the T_P of the source of mid-ocean-ridge basalts. Resolving this problem is a matter of importance.

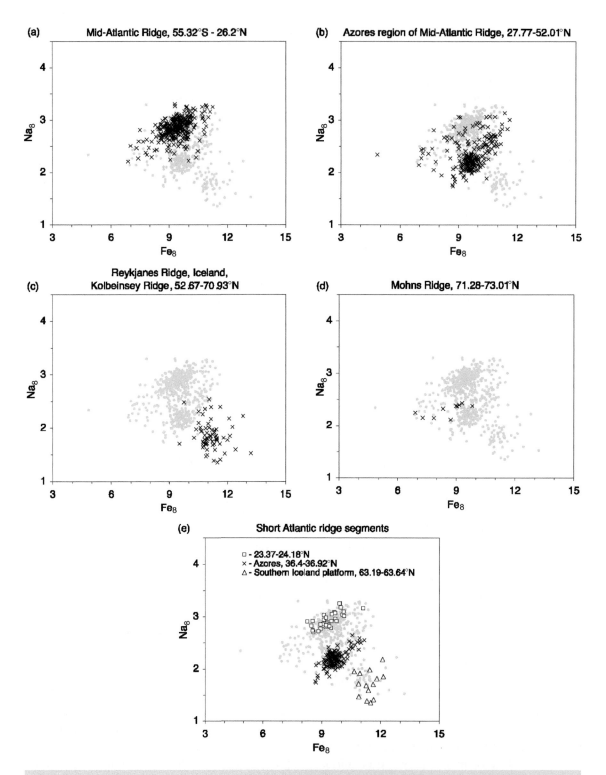

Figure 6.13 Black dots – Na_8 and Fe_8 for different segments of the Mid-Atlantic Ridge against a background of all the remaining samples (gray dots), from the southernmost end of the ridge at the Bouvet triple junction to the Mohns Ridge in the far north (from Presnall and Gudfinnsson, 2008).

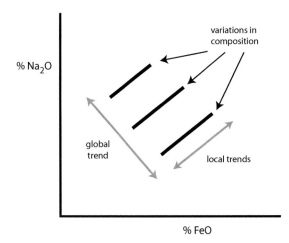

Figure 6.14 Variations in %Na_2O vs. %FeO predicted by laboratory experiments to occur in melts from mantle peridotite with small regional compositional variations, under broadly constant pressure and temperature conditions.

Rare-earth-element modeling has been used to calculate a generalized T_P of 1300 °C for the source of normal mid-ocean-ridge basalt (N-MORB; Section 7.2.7) (McKenzie and O'nions, 1991). This method assumes a uniform peridotite source composition, and thus if this is not the case, as is suggested by mineral phase relations (Presnall et al., 2002), the approach is not applicable.

6.5.2 The North Atlantic Igneous Province

The North Atlantic Igneous Province is where the most extensive and diverse work has been done on the temperature of the mantle source (Fig. 6.16). Crustal seismology, including V_P, V_S, attenuation and V_P/V_S, indicates that the ~40-km-thick crust there is relatively cool and does not exceed ~900 °C down to its base at 30–40 km (Foulger et al., 2003). This interval is, in fact, cooler than at the same depths beneath the East Pacific Rise, a curious result pointed out in a paper ironically titled *Cold Crust in a Hot*

Spot (Menke and Levin, 1994). This result is supported by the crustal structure of the Greenland-Iceland ridge, which is thought to be an offshore continuation of Icelandic-type crust (Korenaga et al., 2002). There, crustal thickness and V_P correlate negatively, suggesting that the thick crust results from melting of a source with enhanced fusibility and not high temperature (Fig. 5.11). Finite-element modeling has shown that a T_P anomalously high compared with that of the source of mid-ocean-ridge basalts is not required to produce the voluminous volcanic margins that formed when the Eurasian continent broke up and the North Atlantic opened at ~54 Ma (van Wijk et al., 2001; 2004).[15]

The temperature of the mantle has also been estimated from seismic wave-speeds, measured using teleseismic tomography and the receiver function method. Several teleseismic tomography experiments have revealed low-V_P and -V_S anomalies of typically ~2% and ~4%, respectively throughout a region several hundred kilometers wide, extending from near the base of the crust down into the transition zone (Section 5.5.1) (Foulger et al., 2001). However, receiver function studies require much stronger anomalies in the mantle low-velocity zone in the depth range ~80–120 km, up to 10% in V_S (Fig. 5.2) (Vinnik et al., 2005). If, due to temperature alone, such a strong seismic anomaly would require temperatures elevated by ~600 °C above the background mantle T_P, an interpretation that can be ruled out because it is entirely at odds with all other observations. This strong V_S anomaly can only be explained by partial melt. If partial melt is required, it may be responsible for all of the low-wave-speed anomaly beneath the Iceland region. The more general regional decrease in V_P of ~2% could alternatively be explained by a decrease of 5–10% in the Mg/(Mg+Fe) content of olivine in the mantle (Fig. 5.5). The seismic observations permit elevated temperature but they do not require it.

[15] http://www.mantleplumes.org/VM_DecompressMelt.html

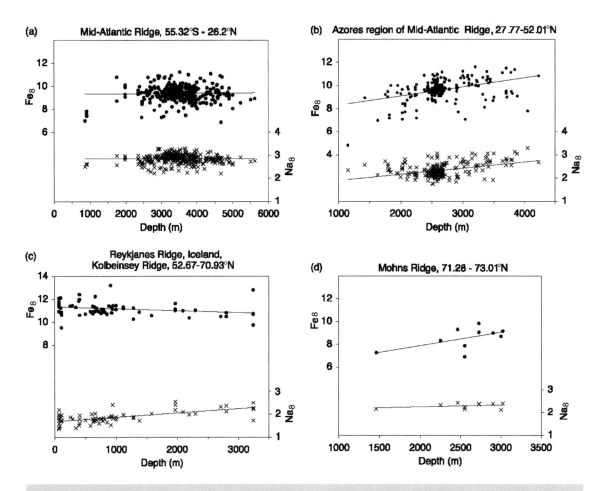

Figure 6.15 Na_8 and Fe_8 vs. depth for different sections of the Mid-Atlantic Ridge (from Presnall and Gudfinnsson, 2008).

The seismic anomaly detected using tomography weakens with depth more rapidly than expected, simply for reducing temperature sensitivity with depth (Figs 5.10 and 5.16). Most importantly, it does not continue through the transition zone. Thus, if the seismic anomaly in the mantle beneath Iceland is partly or wholly due to temperature, then its magnitude must decrease with depth and reduce abruptly to zero at the top of the lower mantle. The 410-km discontinuity is at normal depth beneath most of Iceland. It is depressed by ~20 km locally, although there is disagreement between studies

on the exact location of this depression (Du et al., 2006; Shen et al., 2002). It can be explained by ~200 °C higher T_P, a few mole % higher magnesian olivine, relatively dry conditions, or a combination of all these effects.

In summary, the seismic data constraining crustal structure suggest normal or even unusually low temperatures. Data from the mantle require at least some partial melt and permit, but do not require, normal temperatures. If interpreted in terms of high temperatures, then these do not extend down into the lower mantle. The entire suite of seismic observations are most

Table 6.1 Estimates of mantle source T_P.

	Global	"normal" mid-ocean ridges	Hawaii	Iceland	West Greenland	Gorgona	Réunion
Olivine control lines (Herzberg et al., 2007)		1280–1400	1550	1460		1460	
Olivine control lines (Green et al., 2001)		1430	1430				1323
Olivine control lines (Falloon et al., 2007a)		1243–1351	1286–1372	1361			
Olivine control lines (Putirka, 2005)		1453-1475	1688	1637			
Olivine control lines (Clague et al., 1991)			1358				
Olivine control lines (Larsen and Pedersen, 2000)					1515–1560		
Global systematics (Klein and Langmuir, 1987; Langmuir et al., 1992)		1300–1570					
Geophysics (Anderson, 2000a)	1400–1600						
Plate velocities (Kaula, 1983)	±180K						
Mineralogical phase diagrams (Presnall and Gudfinnsson, 2008)		1250–1280		1250–1280			

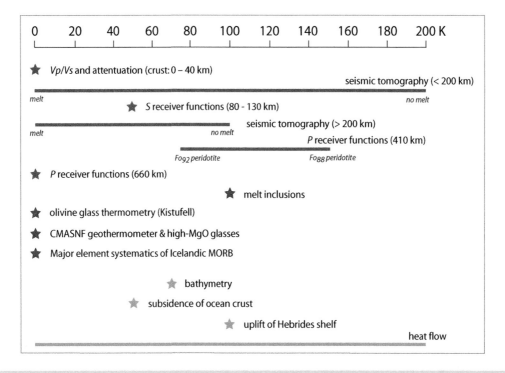

Figure 6.16 Temperature-anomaly estimates for the Iceland region. Bars correspond to temperature ranges and stars to individual estimates (from Foulger et al., 2005a).

consistent with a compositional anomaly and partial melt in a fusible mantle at normal temperatures.

Petrological methods have also been used to estimate the T_P of the source of basalts erupted in the North Atlantic Igneous Province. Iceland has the thickest crust of any portion of the global mid-ocean ridge system. Being an end member in this respect, it was one of the key regions cited in support of the Global Systematics, which predicts an inverse correlation of Na_8 with crustal thickness and, by implication, with mantle T_P (Klein and Langmuir, 1987; Langmuir et al., 1992). However, when this hypothesis was developed, the "thin/hot" crustal model for Iceland was in vogue (Section 5.5.1) and a thickness of 14 km was assumed for Iceland. The crust beneath Iceland is now thought to be up to 42 km thick, and inconsistent with the predictions of the Global Systematics hypothesis.

This in turn invalidates the conclusion that mantle T_P is high under Iceland, derived from this method.

Na_8 and Fe_8 are essentially uncorrelated for the Mid-Atlantic Ridge spanning Iceland, which is also in disagreement with the prediction of the Global Systematics (Fig. 6.13). In Iceland itself, basalts are high in Na_8, the opposite of what is predicted, and highest above the thickest crust in central Iceland (Foulger et al., 2005b). There, primitive lavas have Na_8 similar to basalt glasses from the East Pacific Rise where the crust is only 7 km thick. In common with other parts of the mid-ocean-ridge system, mineralogical phase relationships can explain the Na_8 and Fe_8 systematics in terms of melt extraction under broadly uniform pressure and temperature conditions of ~0.9–1.5 GPa (~25–45 km depth) and 1250–1280 °C, in the presence of moderate compositional variations. A source enriched in fusible components produces larger volumes of melt for the same mantle potential temperature. The ridge spanning Iceland taps a source strongly enriched in FeO and unusually low in MgO/FeO, consistent with enrichment with recycled crust (Presnall and Gudfinnsson, 2008). This is also consistent with an interpretation of the majority of the seismic tomography anomaly as an increase of 5–10% in the Fe/(Mg+Fe) content of olivine.

As is the case for mid-ocean ridges, estimates of temperature beneath Iceland from olivine-control-line modeling vary widely. Different investigators have calculated values of T_P ranging from 1361 °C (compared with 1243–1351 °C for mid-ocean ridges) to 1637 °C (compared with 1453–1475 °C for mid-ocean ridges) (Falloon et al., 2007a; Putirka, 2005).[16] These extreme results imply minimum T_P anomalies for Iceland of 10 or 162 °C, respectively.

The method has also been applied to rocks from West Greenland. Part of the volcanic margin that formed there during continental break-up at ~60 Ma comprises ~22,000 km³ of picritic cumulate rocks, the largest accumulation of such rocks known in the Phanerozoic.[17] These lavas have been attributed to the head of an Icelandic plume postulated to have impacted beneath north-central Greenland. Melt temperatures of 1515–1560 °C have been calculated for parental picritic liquids in equilibrium with olivines with compositions up to $Fo_{92.5}$ (Larsen and Pedersen, 2000). The same method has been used to suggest that the source temperature of the postulated plume subsequently cooled to ~1460 °C (Herzberg and Gazel, 2009). An alternative analysis concluded that the olivine control method could not be applied to these cumulate rocks, and that a more appropriate T_P, calculated using a spinel-based geothermometer, is 1340 °C (Poustovetov and Roeder, 2000).[18]

[16] http://www.mantleplumes.org/MantleTemp.html
[17] http://www.mantleplumes.org/GreenlandHot.html
[18] http://www.mantleplumes.org/Greenland.html
http://www.mantleplumes.org/GreenlandReply.html

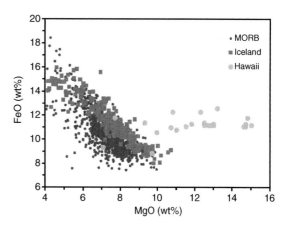

Figure 6.17 Plot of wt% FeO vs. wt% MgO for basalt glass from mid-ocean ridges, Iceland and Hawaii (G. Gudfinnsson, unpublished).

The disagreements notwithstanding, a higher source T_P in plume heads compared with plume tails, is the opposite of what is predicted by the Plume hypothesis (Section 6.3).

In common with mid-ocean ridges, no picrite glass samples are known from Iceland (Fig. 6.17). The most magnesian glass thus far discovered there has only 10.6% MgO (Breddam, 2002), and as for the mid-ocean-ridges, none are thus high enough in MgO to indicate olivine-only crystallization. Small melt inclusions with MgO up to 16% are found in olivine crystals from Iceland, but these are unreliable as they may have gained MgO through re-equilibrating with the surrounding olivine crystal. As a result, all olivine-control-line estimates of T_P for Iceland rely on cumulate picritic rocks.

The same uncertainties thus apply as for elsewhere. The cumulate picrites may represent aggregates of different melts, thus entirely invalidating the olivine-control-line approach. Plagioclase and pyroxene crystals usually co-exist with olivine in Icelandic and Greenland picrites. Their crystallization should also be modeled, but they are often ignored. Many different olivine geothermometers have also been used, resulting in large variations in estimated T_P, even if the same rock samples are used

(Falloon et al., 2007b). In addition, the results have been compared with very different mid-ocean-ridge reference temperatures. In short, T_P estimates for Iceland and Greenland using this method vary so widely that they contribute little to determining whether the North Atlantic Igneous Province resulted from a hot source or not.

Rare-earth element estimates of T_P for Iceland made assuming a peridotite mineralogy suggest anomalies of ~200 °C higher than estimates for mid-ocean ridges, made using the same assumptions. The rare-earth element patterns can, however, also be modeled by reasonable alternative source compositions, for example, a mixture of abyssal gabbro and ocean-island basalt, with no temperature anomaly (Foulger et al., 2005b).

Asthenosphere temperature around Iceland has also been estimated from the bathymetry and the subsidence history of the North Atlantic margins following continental break-up at ~54 Ma (Clift, 2005; Clift et al., 1998; Ribe et al., 1995). Assuming the shallow bathymetry of the North Atlantic to result from hot material delivered by a plume stem to the ridge under Iceland, a mantle T_P anomaly of ~70 °C is calculated. Similar T_P anomalies (<100 °C) are found from modeling the pattern of subsidence at several sites on the volcanic margins of east Greenland, Norway and Britain. There, subsidence of the oldest oceanic lithosphere is more rapid than expected for normal sea floor, consistent with asthenosphere warmer than normal when they formed, and cooling rapidly with time. Enigmatically, subsidence of the Iceland-Faeroe Ridge is no different from normal ocean crust, suggesting that it was emplaced over normal-temperature asthenosphere.

Heat-flow measurements from the ocean floor north and south of Iceland show no significant anomaly compared with global average models (DeLaughter et al., 2005; Stein and Stein, 2003). However, given the large errors in the data, and the insensitivity of surface heat flow to mantle temperature, an anomaly of 100–200 °C beneath the North Atlantic probably cannot be ruled out. Interestingly, in the Iceland region, heat flow is lower west of the Mid-Atlantic Ridge than east of it, the opposite of what would be expected for an eastward-migrating mantle plume.

In summary, the seismic, petrological, bathymetric and heat-flow observations from Iceland and the North Atlantic are all either consistent with normal T_P or elevations of only a few tens of degrees Celsius, or they require them. There is still debate in the case of some methods. However, normal mantle T_P is consistent with modeling results that show that the volcanic margins that formed when the North Atlantic opened do not require elevated temperatures (van Wijk et al., 2001; 2004).

6.5.3 Hawaii

Hawaii is unique because is the only melting anomaly on Earth where picrite glass has been found. This glass is rare, however, and is only known from a few terrestrial samples and small sand grains <300 μm in size dredged from the sea floor. The latter were discovered in 1988 in a core retrieved from 5500 m depth on Puna Ridge, the submarine extension of the East Rift Zone of Kilauea volcano (Clague et al., 1991). They have up to 15.0 wt% of MgO and define an undisputed olivine control line (Fig. 6.17). These data are free of the doubt that plagues the crystalline samples that are used elsewhere, that the samples include xenocrysts and do not represent a single melt.

The olivine-controlled glass trends, along with the compositions of alkalic and tholeiitic basalts from Hawaii, have been studied in the context of mineralogical phase relations over a range of pressures, derived from laboratory melting experiments (Section 6.2.2). The results suggest that Hawaiian tholeiitic basalts are extracted from depths of ~150–200 km (~4.9–6.5 GPa) at a T_P of ~1450–1550 °C, and that the high-MgO glasses found on Puna Ridge experienced almost no olivine crystallization on their way to the surface. However, they also suggest that all the melts at Hawaii are produced under a similar geotherm to mid-ocean-ridge basalts (Presnall et al., 2009). This implies that both Hawaiian and mid-ocean-ridge basalts are extracted from within the surface conductive layer – mid-ocean-ridge basalts from a few tens of kilometers depth and Hawaiian basalts from near its bottom, in the seismic low-velocity layer.

Olivine control line analysis has been performed by several investigators but, despite the unique quality of the data, there is still enormous variation in the mantle T_P deduced (Table 6.1). Estimates range from 1286 °C (Falloon et al., 2007a) to 1688 °C (Putirka, 2005), and the temperature differences estimated between the Hawaiian source and that of mid-ocean ridge basalts ranges from essentially zero (Green and Falloon, 2005; Green et al., 2001)[19] to 270 °C (Herzberg et al., 2007). The variation in these results can be explained almost entirely by the differences in performance of the olivine geothermometers used (Falloon et al., 2007b). Thus the question of whether there is petrological evidence for plume-like high temperatures beneath Hawaii reduces to the question of which olivine geothermometer performs best (Fig. 6.9).

Rare-earth-element modeling suggests a T_P of ~1500 °C for the mantle beneath Hawaii (McKenzie and O'nions, 1991). Again, this estimate assumes a standard peridotite composition. There is presently no consensus about the composition of the source rock of Hawaiian basalts. Peridotite, pyroxenite and eclogite have all been advocated (e.g., Kogiso et al., 2003; Sobolev et al., 2005). Temperature estimates from rare-earth-element modeling are thus unreliable.

Temperature beneath the Hawaiian chain has also been extensively investigated using heat flow. The Hawaiian island chain is the top of a 200-km-wide topographic ridge, which in turn rests on a ~1200-km-wide bathymetric swell.

Plausible explanations for the swell that are linked to the volcanism are buoyant residuum left over from the extraction of melt from the mantle (Section 2.5.3), or thermal buoyancy from replacement of the lower part of the lithosphere with hot material emplaced by a mantle plume (Crough, 1983).

The Pacific plate near Hawaii is moving at ~10.4 cm a^{-1} and the ~90–100 Ma lithosphere there is ~50–100 km thick. The heat flow anomaly from a plume is predicted to gradually increase from zero at the Big Island to a maximum of ~25 mW m^{-2} after ~15–20 Ma, at a distance of ~1800 km along the volcanic chain. The anomaly is predicted to peak along the axis of the chain and to decrease to approach zero at a distance ~600 km transverse to it (Fig. 6.12) (von Herzen et al., 1989).[20]

Such a pattern of heat flow is not observed. Heat flow perpendicular to the chain ~1800 km northwest of the Big Island does not resemble the predicted pattern (Fig. 6.18). Parallel to the chain, the many heat-flow measurements available show a large scatter which, when averaged, indicate similar heat flow to lithosphere of comparable age elsewhere. Heat flow is essentially constant from 800 km southeast of the Big Island to 800 km to the northwest (Fig. 6.19). At ~1800 km northwest of the Big Island, the anomaly is at most a few mW m^{-2} high, and statistically insignificant.

The amount of thermal lithosphere thinning and the surface heat flow anomaly predicted are model-dependent. Some examples assume that hot material with a temperature anomaly of 250 °C replaces the lower lithosphere, as shown in Fig. 6.19. There is no evidence to support significant thermal thinning, but the scatter in the data is too large to rule out small amounts of thinning at large depths. Hydrothermal circulation might mask the thermal signal within ~150 km of the major islands and seamounts, but this cannot explain the absence of a regional heat flow anomaly.

---

[19] http://www.mantleplumes.org/MantleTemp.html

[20] http://www.mantleplumes.org/Heatflow.html

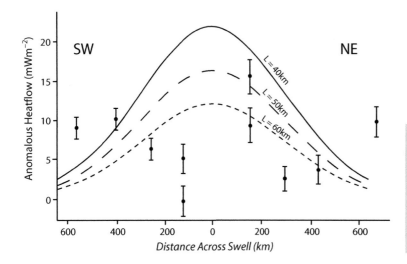

Figure 6.18 Heat-flow measurements perpendicular to the Hawaiian chain compared with predictions for lithosphere thicknesses after thermal thinning of 40, 50 and 60 km (DeLaughter et al., 2005).

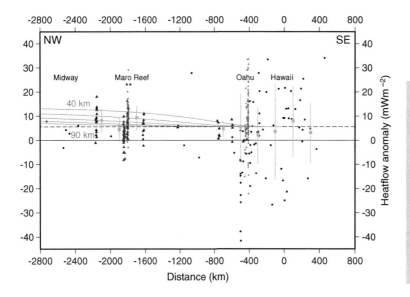

Figure 6.19 Heat-flow data along the Hawaiian volcano chain. The heat flow anomalies are relative to the GDH2 average oceanic model (solid gray line) and to the average heat flow for lithosphere incoming from southeast of Hawaii (dashed gray line). Predicted heat flow (curves) for different depths of lithosphere reheating (from DeLaughter et al., 2005).

6.5.4 Oceanic plateaus

The temperature of the lithosphere has been estimated at a number of oceanic plateaus, ridges and seamounts by reconstructing their sub-sidence histories from the sediments that cap them (Fig. 6.20) (Clift, 2005). Sections through the sediments were obtained from Deep Sea Drilling Project and Ocean Drilling Program drill cores. The results can be divided into four groups (Fig. 6.21).

1 Sites which subsided faster than average oceanic crust, consistent with formation over mantle hotter than normal, and subsequent transportation, by plate motion, over cooler mantle (several sites in the North Atlantic).

2 Sites which subsided at an average rate, consistent with formation over normal-temperature mantle (the Iceland-Faeroe Ridge, several sites along the Ninetyeast Ridge, Shatsky Rise, the Kerguelen Plateau and Walvis Ridge).

3 Sites which subsided more slowly than average, consistent with formation over mantle colder than

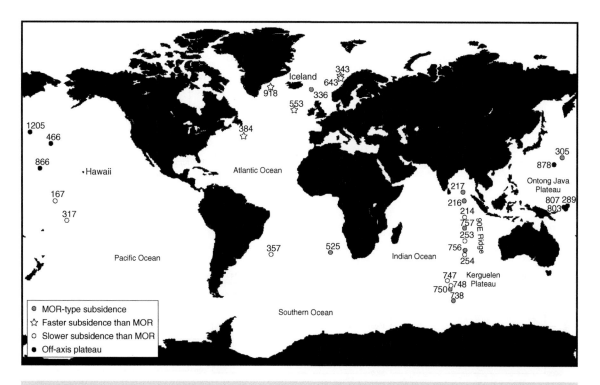

Figure 6.20 Deep Sea Drilling Project and Ocean Drilling Program drill sites where the temperature of the mantle has been estimated using subsidence histories from sedimentary cores. Faster subsidence than mid-ocean ridges (MORs) indicates higher initial mantle temperatures, and slower subsidence than MORs indicates lower initial mantle temperatures (from Clift, 2005).

normal (Magellan Rise, Manihiki Plateau, several sites along the Ninetyeast Ridge, Kerguelen Plateau and the Rio Grande Rise).

4 Sites where the volcanism of interest occurred away from a spreading ridge, where the subsidence could arguably be compared either with that expected for crust of the age of the local sea floor, or with zero-age crust. The latter case would assume that the crust was thermally reset by the magmatism (several sites on the Ontong Java Plateau, the mid-Pacific Mountains, Nintoku Seamount (Fig. 2.13), Hess Rise and MIT Guyot).

The only sites where subsidence patterns provide evidence for crustal formation over anomalously warm mantle are the volcanic margins around the North Atlantic. However, the observations there are consistent with rather small temperature anomalies of no more than ~100 °C through-out a layer ~100 km thick. They preclude thicker layers with temperatures of several hundred degrees Celsius and are thus inconsistent with a plume of the size and temperature that would be required to generate the ~2400-km-wide flood basalt province of the North Atlantic.

All other oceanic melting anomalies studied have subsidence histories compatible with emplacement over normal-temperature or anomalously cool mantle. The relatively large errors in some water depth estimates mean that in some cases a mantle source warmer than average cannot be ruled out. However, with the exception of sites in the North Atlantic, the data do not require it. Upper bounds can be placed on the thicknesses and temperatures of postulated thermal sources and in some cases, for example, the Ontong Java Plateau – the largest volcanic

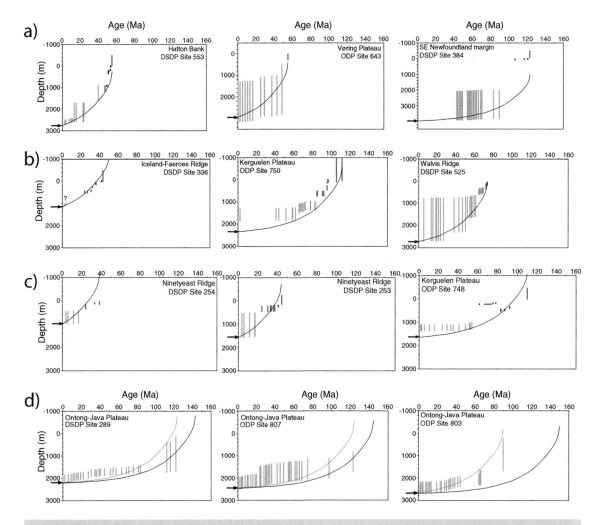

Figure 6.21 Reconstructed depth-to-basement histories for a selection of sites that subsided: (a) faster than average; (b) at an average rate; and (c) more slowly than average. Vertical bars show uncertainties in water depth estimates from sediments or microfossils. Black bars indicate high confidence, and gray bars low confidence. Solid curves show the predicted subsidence history for normal oceanic crust. The end points of these curves are generally known accurately. (d) shows sites where volcanism occurred away from a mid-ocean ridge. The light gray curves show the predicted curve for crust of the same age as the volcanic rocks (equivalent to total thermal resetting of the lithosphere), and the dark gray curves show predicted subsidence for crust of the age of the host crust (from Clift, 2005).

province on Earth – the subsidence data can rule out thermal anomalies of the size and magnitude required to produce the magmatic volumes observed (Ito and Clift, 1998).

Temperature beneath the Ontong Java plateau has also been estimated using seismology (Section 5.5.3). A low-V_S body, similar to that beneath Iceland in lateral extent and strength, has been imaged beneath the plateau using tomography (Fig. 5.29). V_S is as much as 5% lower than the global average. This is consistent with a high-temperature anomaly of up to ~250 °C, but could also be explained by low-degree (<1%) partial melt or by a compositional anomaly such as a reduction of ~5% in the Mg/(Mg + Fe) content of olivine (Table 5.1). In order to reduce the ambiguity in interpretation, seismic attenuation was measured. Attenuation within the body is exceptionally low compared with the rest of the Pacific mantle. The combination of low-V_S and low attenuation rules out a thermal anomaly and requires a viscous, chemical body (Gomer and Okal, 2003), a result consistent with the lack of rapid post-emplacement subsidence.

Other methods used to investigate the source temperature of Ontong Java Plateau basalts have yielded much higher temperatures, at odds with the low temperatures apparently required by the subsidence and seismic observations. Olivine control line analysis applied to cumulate rocks with up to 10.9% MgO yielded estimates of source T_P of 1500 °C, corresponding to a temperature anomaly of 100–220 °C compared with mid-ocean ridges (Herzberg et al., 2007). This analysis is, however, subject to the same difficulties as those from Iceland and elsewhere, which render the results unsafe.

A high-temperature source has also been advocated on the basis of high levels of incompatible elements, that are consistent with a degree of partial melt of ~30% in a near-primitive, peridotite source (Fitton and Godard, 2004). In order to produce such an unusually high degree of partial melting by isentropic upwelling, a T_P of at least 1500 °C is required. Nevertheless, the melt volumes can be explained without the need for high temperature, simply if a more fertile composition is assumed (Korenaga, 2005). The plateau erupted near, possibly at, a very-fast-spreading mid-ocean ridge that became extinct almost immediately afterwards. Eclogite blocks, originating from subducted oceanic crust recycled in the upper mantle, and entrained in the ridge upwelling, can explain the geochemistry and the volumes observed without a high-temperature anomaly (Section 3.5.3).

6.5.5 Swells

Areas with relatively high elevations but no obvious link to tectonism at a plate boundary occur both on land and at sea. Thermal models have been suggested for all these, and investigated using heat flow measurements. Hoggar, in North Africa, is an example of a swell on land. It has been volcanically active from ~35 Ma almost to the present (Liégeois et al., 2005). It comprises a topographic swell ~1 km high and ~1000 km in diameter, capped by ~14,000 km^3 of mafic lavas. It has been attributed to a mantle plume (e.g., Sleep, 1990), an hypothesis that has been tested using heat flow measurements. Heat flow inside the Hoggar swell is normal (~50 mW m^{-2}) for the age of the lithosphere, and lowest at the highest basement elevations. A large heat flow anomaly occurs north of Hoggar, however, below the Saharan basins. This pattern is the opposite of what is expected for a hot source underlying the centre of the swell.

A plume cannot be ruled out at Hoggar using heat-flow data, because 35 Ma is insufficient time for heat conduction through the lithosphere to have caused a measurable surface heat flow anomaly. For example, a plume head 100 km thick and 250 °C hotter than the surrounding mantle, emplaced beneath 120-km-thick lithosphere, would only produce a heat flow anomaly of 0.2 mW m^{-2}. 100 Ma would be required to increase heat flow by 2 mW m^{-2}. The standard deviation of heat flow measurements on land may be at least ~10 mW m^{-2}, so such small anomalies are irresolvable. Nevertheless, the high heat flow north of Hoggar is apparently unexplained.

Cape Verde, Bermuda, Crozet and Réunion are examples of oceanic swells where elevated heat-flow measurements have been reported and cited in support of high-temperature plume models. Ironically, elevated heat flow is absent at Hawaii, and this has been explained as hydrothermal circulation removing the expected signal (Harris et al., 2000). Because of this inconsistent interpretive approach, heat-flow

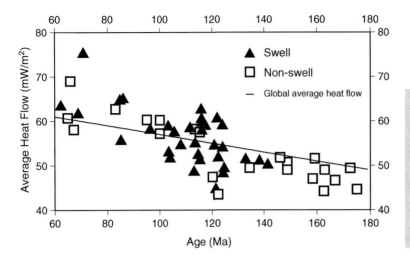

Figure 6.22 Pogo heat-flow data vs. crustal age for the Hawaii, Réunion, Crozet, Cape Verde and Bermuda swells (triangles) and non-swells (squares). There is no significant evidence for elevated heat flow over swells (from Stein and Von Herzen, 2007).

measurements from swells were recently reappraised using a correct global model and reviewing evidence for local recent magmatism and hydrothermal circulation (Stein and Von Herzen, 2007). Very few reliable measurements indicate elevated heat flow (Fig. 6.22). Of these, most are likely to be affected by hydrothermal circulation or recent magmatism. The majority of the remainder are small and within one standard deviation of the measurements, which is typically ~5 mW m^{-2}. The patterns of heat flow anomalies do not have the Gaussian-shaped distributions, centered on the swells, predicted by lithospheric reheating models. The results suggest that recent magmatism may raise heat flow at a few isolated sites, but there is little evidence for broad regions of elevated heat flow over swells. It would not be possible to predict the locations of surface melting anomalies from global oceanic heat flow values.

It has been suggested that a large region of the south Pacific sea floor, which includes the Darwin Rise and the South Pacific "Superswell", is linked to a broad thermal upwelling – a "superplume" (Fig. 6.23) (Larson, 1991; McNutt and Judge, 1990). Individual volcanic chains in the region, such as the Cook-Austral, Marquesas, Pitcairn, and Society seamount chains (e.g., Courtillot et al., 2003), have been attributed to

small "plumelets" that rise from the top of this postulated "superplume" (Fig. 1.2). Volcanism in the Cretaceous is postulated to have uplifted the Darwin Rise (Menard, 1964). In this model, the "Superswell" is the waning phase of that activity (Menard, 1984).

One problem with this theory is that the Darwin Rise is, in fact, not anomalously shallow, but merely a region of abundant seamounts emplaced on sea floor of normal depth. The "Superswell" region is, in contrast, probably up to ~500 m shallower than Pacific sea floor of comparable age (Fig. 6.24, top). The exact amount of shallowing is difficult to estimate because corrections must be made for the volcanic constructs and sediments that blanket the area (Levitt and Sandwell, 1996). The shallowing could be explained by a lithosphere reduced to a thickness of only ~60 km by thermal erosion, with its removed lower part replaced by warm asthenosphere. Heat flow ~15–25 mW m^{-2} higher than the global average for sea floor of comparable age would then be expected. However, heat flow in the "Superswell" region is normal everywhere (Fig. 6.24, bottom) (DeLaughter et al., 2005; Stein and Abbott, 1991; Stein and Stein, 1993).[21] A more likely

[21] http://www.mantleplumes.org/Superswell.html

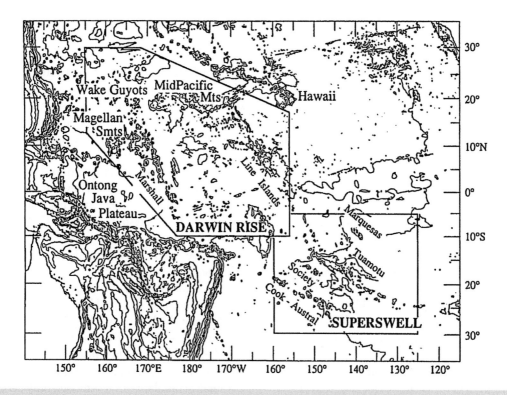

Figure 6.23 Map of the south Pacific showing the Darwin Rise and the "Superswell" (from Stein and Stein, 1993).

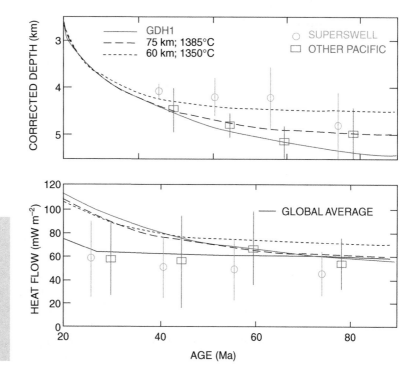

Figure 6.24 Top: Bathymetric depths from the "Superswell" and other regions of the same age in the Pacific for comparison. Bottom: heat flow observed, and predicted by thinning models (Stein and Stein, 1993).

explanation for the shallow bathymetry of the "Superswell" is the presence of widespread volcaniclastic sediments on the ocean floor, produced at the many seamounts and islands in this part of the Pacific Ocean (Levitt and Sandwell, 1996).

A second problem is that the "superplumes" – vast seismic low-wave-speed bodies that occupy much of the lower mantle beneath the "Superswell" and the South Atlantic ocean are not thermal but dense, chemical bodies (Section 5.6.1) (Trampert et al., 2004). There is thus no evidence that they comprise a heat source, nor are capable of generating "plumelets" or fuelling surface volcanism. The available evidence is most consistent with the "Superswell" being supported by a compositional anomaly from residuum left by the volcanism, or by mechanical weakening of the plate as a direct result of the volcanism. There is no evidence that it is elevated by the buoyancy of an unusually hot mantle.

6.6 Variants of the Plume hypothesis

High temperature is a simple, one-dimensional prediction that does not lend itself to complicated variants. Nevertheless, variants have been suggested to accommodate unexpected observations or to explain why predicted observations are not made. These variants can be grouped into the following suggestions.

- High temperatures do indeed exist but evidence for them is difficult to observe and has not yet been found.

- Much or all of a plume has only a small temperature anomaly.

- That plumes exist, but they are not hot.

The temperature anomalies traditionally expected in plumes, compared with a "reference" or "average" mantle value, are typically 200–300 °C. The parental melts that form in the mantle at

such high temperatures are expected to have MgO contents >12% and to have picritic or komatiitic compositions. Thus, if such melts were sampled at the surface, they would provide evidence for the predicted high temperatures.

Picritic cumulate rocks are commonly found at melting anomalies, but it cannot be assumed that they represent melts. They are aggregates that may include olivine crystals from more than one cooling and crystallizing melt. Only picrite glass is proof of a parental picrite melt, and to date these have only been found at Hawaii (Clague et al., 1991).

The total absence of picrite glass at any melting anomaly except Hawaii is commonly attributed its high density. The predicted primary MgO-rich melts are proposed to be trapped in sill-like complexes at the crust-mantle boundary where olivine crystallizes out and is deposited in ultramafic cumulates. In this model, only evolved basaltic melts are erupted (Larsen and Pedersen, 2000). It is not clear why this does not prevent picrite cumulates from being erupted at the surface, however.

Several arguments have been proposed to explain the absence of the subsidence predicted by thermal models. It has been suggested that long-term, post-eruption magmatic underplating causes continuous crustal growth and bathymetric support (Clift, 2005). It is not clear how this accords with the Plume hypothesis, which predicts that initially erupted plume heads are transported by plate motion away from the long-term melt supply provided by the plume tail. The Ontong Java Plateau, for example, is predicted to have drifted, since its formation, some ~5000 km from the only possible candidate for a plume tail (Fig. 2.5).

Support from buoyant, depleted mantle residuum remaining after melt extraction has also been suggested to explain the present high stand of the Ontong Java Plateau (Robinson, 1988). A residuum model was proposed to explain the bathymetry of the Hawaiian swell (Section 2.5.3) (Phipps-Morgan et al., 1995). In the case of Hawaii, however, the residuum is proposed to be viscous and to flow and disperse with time, allowing progressive subsidence. Viscous dispersal for the residuum in the case of Hawaii but not for other melting anomalies is inconsistent. This also raises the question why residuum support did not preclude rapid subsidence in the North Atlantic, a problem that has been addressed by the suggestion that vigorous convection removed the residuum there. It is unclear how the rapid subsidence of the North Atlantic accords with the proposal that the persistent high stand of parts of the margin, for example, Britain, results from underplating. Such explanations tend to lack independent support, be proposed on an *a posteriori*, *ad hoc* basis, and their absence may be required elsewhere. The lack of high heat flow from the ocean floor where expected has been explained by hydrothermal circulation, a proposal that is not invoked where measurements are closer to what is predicted.

The absence of evidence for high temperatures has also been explained by postulating much lower temperatures for plumes than usually presumed. For example, temperature anomalies of <100 °C have been suggested for much if not all of the postulated Icelandic plume (e.g., Ribe et al., 1995). Such low temperatures may violate other model requirements, however. For example, a plume tail with a temperature anomaly of only 100 °C would imply a plume head cooler than this, which could not produce the volumes of melt observed in the North Atlantic Igneous Province (Cordery et al., 1997).

Downward-adjusting predicted temperatures also usher in a more fundamental problem. Measuring source temperature then ceases to become a test of the Plume hypothesis because low temperature anomalies are indistinguishable from normal, regional variations in mantle T_P that result from processes, such as recent continental insulation (Anderson, 1982; Lenardic et al., 2005),[22] subduction, and mantle flow associated with plate motions. In some cases,

complex temperature fields have been suggested, for example, that temperature in the postulated Icelandic plume varies radially, with a hot, ~1500 °C core surrounded by a cooler, ~1400 °C sheath (White et al., 1995). However, there is no example where this has been convincingly supported by observation. On the contrary, in the case of the postulated Icelandic plume, it has been claimed that the hottest lavas are found distally, in volcanic rocks attributed to the plume head (Larsen and Pedersen, 2000).

Finally, it has been suggested that plumes could rise from the core-mantle boundary as a result of compositional buoyancy alone. Anderson (1975) proposed that "chemical plumes" result from inhomogeneities that survive from when Earth originally accreted and was chemically stratified. This model is at odds with geochemical observations that suggest, on the contrary, relatively dense components such as eclogite in magmas at melting anomalies (Anderson, 2007a).

6.7 Discussion

The endeavor to use measurements of temperature to study the origins of melting anomalies is plagued by several fundamental problems. The first is our ignorance of the temperature distribution inside the Earth. A simple, classical view envisages rigid plates underlain by mantle that is well-stirred by convection and has an essentially constant T_P except where special features with extraordinary temperatures exist, such as subducting slabs and mantle plumes. Surface eruptives are envisaged to arise from this well-stirred mantle and thus to have the same T_P everywhere except when they arise from, or are influenced by, these special features.

This is, however, a conceptual model and there is no unique, reliable observational evidence for it. Mantle T_P cannot be constant, either laterally or with depth. The mantle is compositionally inhomogeneous, and thus the distribution of radioactive, heat-producing elements is

[22] http://www.mantleplumes.org/ GlobalMantleWarming.html

also inhomogeneous. The mantle also contains mobile, metasomatizing fluids, which advect heat around. The insulating effect of continents, their break-up and lateral drift, contributes tens of degrees Celsius to large-scale lateral variations in mantle T_P (Anderson, 1982; Coltice et al., 2007). Mantle convection, of course, requires T_P to vary.

Unfortunately, it is difficult to estimate the average T_P of the mantle or its variation, because there is no ready way of measuring it accurately and obtaining consistent results. The common assumption that the source temperature of mid-ocean ridge basalt is a measure of the T_P of the mantle as a whole underneath is unsupportable. Estimates of average T_P range from 1280–1400 °C, and estimates of the variations range from ±20 °C to ±200 °C (Anderson, 2000a; 2007b; Kaula, 1983; McKenzie and Bickle, 1988). As a result, even if the source T_P of a melting anomaly could be measured accurately, it cannot be uniquely interpreted because a reference value with meaningful accuracy to which it can be compared is lacking. These difficulties are exemplified by the history of heat flow measurements on the Hawaiian swell. Heat flow there was originally concluded to be anomalously high and consistent with the Plume hypothesis. This was subsequently found to be almost entirely an artifact of using an incorrect reference thermal model (Stein and Stein, 1992). Latterly it has been suggested that the lack of a heat flow anomaly is still consistent with a Hawaiian plume, because heat is dissipated by hydrothermal circulation (Harris et al., 2000).

The difficulty in choosing a reference T_P is compounded by the problem of usefully interpreting observed variations. The Plume hypothesis predicts occasional high-T_P anomalies embedded in a general background of mantle with T_P 200–300 °C lower. The Plate hypothesis expects the T_P of melts to vary according to their depth of extraction from within the surface conductive layer, maximally approaching the general local regional mantle T_P (Fig. 6.3). Thus, observed variations in melt-source T_P may

be intrinsically unable to distinguish between the Plate and Plume hypotheses. What may be required are estimates of the local geotherm, the thickness of the surface conductive layer, and the general, regional T_P of the mantle beneath it. For this, measurements of both the source T_P and depth of origin for suites of samples are required. In view of the predictions of the Plate hypothesis, it is puzzling that there is so little reliable evidence for sources with T_P higher than the mid-ocean-ridge-basalt source. This may be an indication that there is relatively little variation in the maximum temperatures and pressures (i.e., depth) at which surface-erupted basalts form.

In addition to philosophical problems, progress is hampered by the inadequacy of the tools available for estimating T_P. Seismology can in some cases contribute useful constraints, for example, by assessing attenuation or by distinguishing between thermal and chemical origins for thick crust. However, the widespread use of wave-speeds calculated using tomography, and topography on the transition-zone-bounding discontinuities, is not justified. There is fundamental ambiguity between composition, phase and temperature in the interpretation of these parameters, which can only rarely be resolved by carefully designed experiments. Seismology is not a thermometer, red is not equal to hot and blue is not equal to cold.

The same difficulty applies to petrological methods. The mere existence of picrite cumulates is not evidence for high T_P. The tendency to assume *a priori* that basalts at melting anomalies arise from hot sources may in some cases encourage over-confidence in a subset of interpretations that are consistent with this prejudice ("Hotspots are, by definition, hot." Courtillot et al. (2003), p. 299). The T_P anomalies predicted by even the hottest plume model are small compared with the repeatability between results from the methods available. This is well-illustrated by the results from olivine-control-line analysis. The estimates of T_P calculated by different investigators for mid-ocean-ridge basalts using that method vary by 195 °C. Estimates for T_P for the source of Hawaiian basalts vary by ~400 °C (Table 6.1). It will clearly be a helpful step forward when general agreement can be reached on the correct way to apply olivine-control-line analysis. Modeling vertical motions and heat flow are intrinsically unable to provide precise estimates of T_P at specific localities. These methods can only provide broad, imprecise indications of lateral variations in temperature. Methods that assume uniform source composition, for example, rare-earth-element inversion, are fundamentally impotent to test the Plate hypothesis.

Notwithstanding our limited ability to measure temperature variations in the Earth, it remains the case that there is little reliable evidence for unusually high T_P at melting anomalies. This is despite diverse methods being applied to localities that range from long-extinct flood basalts to currently active volcanic regions considered to be the strongest candidates for modern mantle plumes.

Only at the North Atlantic volcanic margins and Hawaii has significant evidence been found for elevated mantle T_P. In the North Atlantic, anomalously rapid subsidence of the oldest ocean floor suggests a mantle up to ~100 °C warmer than average shortly after continental break-up. Evidence for high T_P beneath Iceland has not been found, however, and the lack of anomalous subsidence on the Iceland-Faeroe ridge suggests long-term normal T_P at this latitude. At Hawaii the unique observation of picrite glass suggests elevated T_P. At both localities – the volcanic margins around the North Atlantic and Hawaii – the melt may be sourced deep within the conductive layer. The North Atlantic volcanic margins formed when continental lithosphere possibly as thick as ~200 km ruptured. Melt at Hawaii today rises from depths of ~150–200 km. Both these regions may thus owe their high T_P to their unusually deep origins within the conductive layer.

A question of critical importance to the Plate hypothesis is whether the heat required to melt the volumes of basalt observed at major near-ridge melting anomalies can be supplied by isentropic upwelling of a fusible source at a similar mantle T_P to the regional. Such a mechanism has been suggested to explain the large volumes of melt at the Ontong Java Plateau, Shatsky Rise, Iceland and other oceanic plateaus in the Atlantic and Indian oceans (Anderson, 2005b; Foulger and Anderson, 2005; Korenaga, 2005; Yaxley and Green, 1998). Approximate calculations suggest that this mechanism can account for the observed volumes. However, at present, the amount of energy required to melt the relevant minerals at the appropriate temperatures and pressures is not known sufficiently well to be able to answer this question precisely. In the case of intraplate Hawaii, the excess melt is presumed to pre-exist at depth. The petrology of present-day Hawaiian basalts suggests an unusually fertile source (Section 7.6.2). A through-lithosphere propagating fracture tapping melt from a low-velocity zone containing laterally variable amounts of retained melt could explain both the high source T_P there and the highly variable eruption rate along the Hawaiian chain (Fig. 3.4).

What, then, is the answer to the question posed at the beginning of this chapter? Are "hot spots" hot? There is no reliable evidence, from seismic, petrological, bathymetric or heat-flow data from the vast majority of melting anomalies that they are. The term "hot spot" is thus a misnomer. It is furthermore damaging because it is leading and pre-supposes an outcome of observations that is not borne out. Its general use should be discontinued in favor of a term that describes without prejudice the phenomenon under scrutiny.

6.8 Exercises for the student

1 Are "hot spots" hot?

2 How does potential temperature vary with depth in the Earth?

3 How can rare-earth-element modeling be used to test the Plate and Plume hypotheses?

4 Reconcile the wide variation in estimates of mantle potential temperature obtained using the olivine-control-line method.

5 Can seismology be used as a mantle thermometer?

6 How low may temperature estimates for plumes be, before they become implausible?

7 Are the temperature variations in the mantle predicted by the Plate and Plume hypotheses different?

8 Can the melt volumes produced at near-ridge melting anomalies be modeled by isentropic upwelling of a fusible source at normal mantle potential temperature?

9 Does Iceland have a thin, hot crust or a thick, cold one?

10 Is Hawaii the only place on Earth where the olivine control-line method of estimating mantle potential temperature may be safely applied?

11 Is Hawaiian melt extracted from unusually deep within the surface conductive layer, or is there truly an unusually hot region in the mantle beneath Hawaii?

7

Petrology and geochemistry

Nihil tam absurde dici potest quod non dicatur ab aliquo philosophorum.[1]

–Cicero (106–43 BC)

7.1 Introduction

Petrology and geochemistry are vast, specialist subjects, and it is only possible here to touch briefly on those aspects that are of most significance to the origin of melting anomalies. A contribution of fundamental importance to melting anomalies that petrology and geochemistry might make is revealing the original depth of origin of the parent rock that melted to form a surface lava. Did the parent rock originate in the deep mantle, possibly the core-mantle boundary, to be transported up in plumes, or did it reside long-term in the shallow mantle?

Unfortunately, petrology and geochemistry have little ability to reveal the ultimate region of the mantle where rocks parental to surface lavas originated. Virtually all magmas sampled at the near surface melted at relatively shallow depths. This is known because the temperature at which basalt is molten increases rapidly with pressure and reaches temperatures far higher than plausible mantle temperatures at depths of ~200 km. The fundamental problem is clear. At

the surface we are not only largely limited to sampling derivatives (partial melt) of parent source rocks, but we only have access to derivatives that formed when the parent rock resided in the shallow mantle. It is not at all clear how any earlier history of migration of the parent rock can be deduced.

A few simple conclusions can be drawn concerning relatively shallow depths of origin. Shallower than ~35–45 km, plagioclase is stable and will crystallize from a cooling melt. Deeper than this, and down to ~60 km, spinel will form and at depths >~60 km garnet is stable (Hall, 1996). Diamonds mostly form at pressures corresponding to depths of ~150–200 km and at relatively low temperatures, conditions that are met in the lithospheric mantle beneath cratons. A few diamond inclusions show evidence of former residence at the base of the transition zone near ~650 km depth (Brey et al., 2004). However, no petrological or geochemical observation requires any larger depth of origin.

Much petrology and geochemistry is, nonetheless, aimed at studying the internal composition and dynamics of the Earth. However, there is great ambiguity in interpreting petrological data and thus simple *a priori* models of the stratification and lateral heterogeneity of the crust and mantle are commonly assumed. Independent constraints, for example, from seismology and convection modeling, are also applied. Constraints from seismology have been critical, for

[1] No idea, however absurd, does not find a philosopher who will support it.

Plates vs. Plumes: A Geological Controversy, 1st edition. By Gillian R. Foulger.
Published 2010 by Blackwell Publishing Ltd.

example, in enabling the origins and processes that form back-arc volcanoes to be relatively well understood. The same success has, however, not been achieved for the rocks at melting anomalies.

The Earth is assumed to have originally formed by the accumulation of small bodies with "primordial" compositions, similar to chondritic meteorites (Table 7.1). As the Earth passed through its early hot, largely molten phase, its average composition was strongly altered by the loss to space of volatile elements. Chondritic meteorites are likely to be a good guide to the Earth's current inventory of refractory elements such as Al, Ca, Ti, and elements of the Lanthanide ("rare-earth") series, but not to volatiles such as Re, Os and particularly He, in which the Earth must be strongly depleted (Figs 7.1 and 7.2) (Anderson, 1989; 2007b).

Immediately following formation, the planet stratified strongly. Fe, Ni, and siderophile elements such as Os and Ir were concentrated in the core, and crustal elements such as Si, Ca, Al, K, Fe and large-ion lithophile elements were concentrated into melts that rose and formed

Table 7.1 Definitions of theoretical mantle compositions.

Term	Definition
Primordial mantle:	A theoretical rock with the composition of the silicate portion of the Earth after loss of core-forming elements but without loss of volatile elements by degassing, e.g., during the magma-ocean stage of Earth evolution, nor depletion in continental crust-forming elements.
Primitive mantle:	Also called "bulk silicate Earth", it is the silicate portion of Earth that remained after removal of the core but before division into the continental crust and the mantle. Its composition is obtained from the compositions of chondritic ("stony") meteorites.
Primary melt:	The initial liquid formed when a rock melts, before any differentiation has occurred.

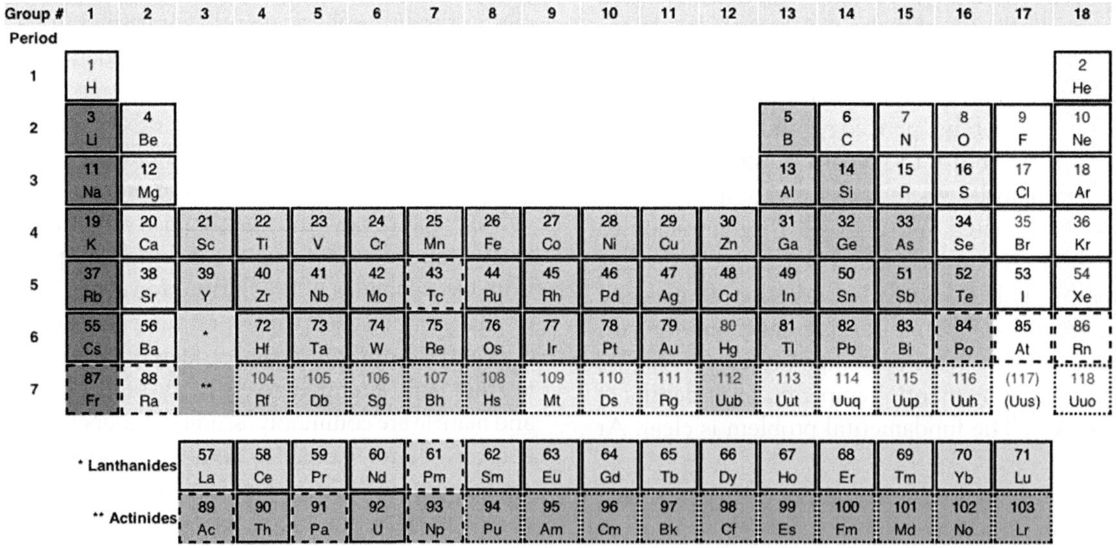

Figure 7.1 The Periodic Table of elements.

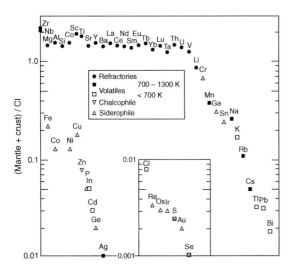

Figure 7.2 Abundances of elements in "primitive mantle" (Table 7.1) relative to chondritic meteorites (from Anderson, 1989).

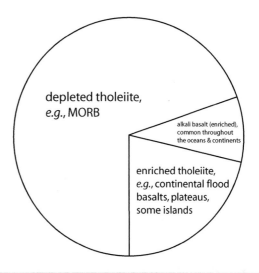

Figure 7.3 Basalt types found at melting anomalies.

the continental crust (Hamilton, 2003; 2007c). Up to 30–95% of the entire mantle must have been depleted in fusible constituents to form the continental crust. The exact percentage is poorly known, and it is by no means certain that only one part of the mantle was depleted, leaving other parts entirely undepleted and primitive. No petrology, geochemistry or model of planetary evolution requires the survival of primitive mantle.

The composition of the mantle can be studied using lavas, in particular basalts, that are sampled at the surface. In order to explain their variation, candidate global compositional stratifications in the mantle have been proposed. This exercise is problematic, however. Vertically stratified models proposed on the basis of petrology are speculative and non-unique. This non-uniqueness cannot be resolved using seismology, which is itself non-unique regarding geological interpretation (Section 5.1). Furthermore, the mantle is known to be maintained in a laterally heterogeneous state by processes associated with plate tectonics and volcanism, an inference that is supported by the diversity of mantle xenoliths.

Three lava types are most important to melting anomalies: depleted tholeiitic mid-ocean-ridge basalt (MORB), enriched tholeiite, and alkali basalt, plus candidate petrologies for their mantle sources (Fig. 7.3). At spreading ridges, MORB is thought to form where the mantle rises passively in response to plate divergence. The depth estimated for the base of the melting column varies from a few tens of kilometers to >100 km. Ridges are not stationary with respect to the mantle below, but migrate laterally, an inescapable consequence of the continual changes in size of plates. Ridges do not overlie long-lived, rising convection limbs, but rather sweep across the mantle like a vacuum cleaner over a carpet (Hamilton, 2007b). The mantle rises passively beneath in response to spreading. The mantle is not compositionally uniform, and the relative geochemical uniformity of MORB is partly a consequence of the huge productivity of ridges, and large-volume averaging in crustal magma chambers (Meibom and Anderson, 2004; Meibom et al., 2002).

Alkali basalt is non-tholeiitic basalt. It differs from MORB compositionally in its silica content, variability, fusible elements and radiogenic isotope ratios. The latter are strong evidence that the sources of MORB and alkali basalt are distinct. Most melting anomalies erupt only alkali basalts, and where tholeiite is also present it has an enriched character like alkali basalt (Fig. 7.3). The challenge is thus to understand if and why the sources of MORB and enriched basalts are different and, if possible, to discover how they formed and where they currently reside in the Earth.

7.2 Some basics

7.2.1 Tholeiite and alkali basalts

Most tholeiites differ from alkali basalts in containing sufficient silica that under equilibrium conditions the mineral quartz can exist (Hall, 1996). They are thought to be formed by high-degree melting of mantle rocks – 10–20% or more – depending on the source composition (Section 7.2.2). In contrast, alkali basalts are thought to derive ultimately from small-degree mantle melts, possibly <5%. As a result, those constituents in the source rock that melt most readily are strongly concentrated in alkali basalt, but diluted in tholeiite (Section 7.2.3).

Tholeiites are the most voluminous basalts. They can be divided into depleted and enriched types (Fig. 7.3). Most MORB is depleted tholeiite and most oceanic plateaus and continental flood basalts are dominantly enriched tholeiite. The annual production rate of MORB is ~20 km^3 a^{-1} It is relatively uniform in composition and has radiogenic isotope contents suggestive of a source depleted by an ancient melting event, probably formation of the continental crust (Section 7.2.4).

Alkali basalt tends to erupt in smaller quantities, and the annual production rate is only ~0.5 km^3 a^{-1}. It comprises the majority of oceanic islands, seamounts and small-volume continental volcanism. A remarkable exception to this is the Samoan archipelago, which is of the same order of magnitude as Hawaii in volume, and yet only alkali basalts have so far been found there (Kear and Wood, 1957; Natland, 1980). The radiogenic isotope contents of alkali basalts suggest that their parent rocks are the melt products of ancient melting events, not the depleted residuum, as is the case with MORB (Section 7.2.7).

7.2.2 Mantle composition and heterogeneity

Whereas basalts can be studied from samples collected near the surface of the Earth, the composition of the mantle can only be inferred from exhumed rocks, for example, in the Alps, xenoliths presumed to come from the mantle, for example, in kimberlite pipes, basalts and dikes, and indirect arguments. The main candidate rock types are peridotite, pyroxenite and eclogite (Table 7.2). Peridotite is the commonest xenolith brought up in kimberlite pipes. It is divided into lherzolite, which is fertile because it contains the basalt-forming minerals clinopyroxene and spinel or garnet, and harzburgite, which is infertile because it lacks them. Partial melting of lherzolite removes the clinopyroxene and spinel or garnet to produce basalt, leaving a harzburgite residuum. Pyroxenites are also found in kimberlite pipes. They have similar mineral assemblages to peridotites but are richer in pyroxene. Occasional eclogite xenoliths are also found. Eclogite is the high-pressure form of basalt and is thus composed almost entirely of the basalt-forming minerals.

The average composition of the mantle must match seismic wave-speeds and this is achieved for mixtures of harzburgite and basalt in the approximate proportions of 3:1. Pyrolite (Ringwood, 1975) and more-fertile piclogite (Anderson and Bass, 1984) are hypothetical compositions that have been suggested to satisfy this requirement.

Most of the continental and oceanic crust is underlain by lithospheric mantle, which is relatively stable and thought to move as part of the

Table 7.2 Mineralogy of common rock types.

Depth range, km	Rock type	olivine	ortho-pyroxene	clino-pyroxen	plagioclase	spinel	garnet	quartz
0–30	lherzolite, pyroxenite	√	√	√	√			
	harzburgite	√	√					
	basalt	√	√	√	√			
30–60	lherzolite, pyroxenite	√	√	√		√		
>60	lherzolite, pyroxenite	√	√	√			√	
	eclogite			√			√	√

plates. The mantle lithosphere is long-lived, but probably undergoes continual metasomatism, i.e., invasion by mobile, small-degree mantle melts from below. It is thus expected to be both fertile and compositionally variable.

Beneath the lithosphere is the more mobile, convecting, asthenospheric mantle, which is thought to be the source of tholeiitic MORB. Many observations suggest that the asthenospheric mantle is also inhomogeneous, including the following.

■ Subtle variations in the composition of MORB along the mid-Atlantic ridge are well explained by source heterogeneity (Section 6.5.1; Fig. 6.14) (Presnall and Gudfinnsson, 2008; Schilling, 1986).

■ Regional variations in MORB isotopic content occur in the Pacific and the Atlantic – the DUPAL ("Dupré-Allègre"; Dupré and Allegre, 1983) and SOPITA (South Pacific Isotopic and Thermal Anomaly; Hart, 1984) anomalies (Fig. 7.4). The DUPAL geochemical signature is thought to derive from recycled delaminated lower continental crust and continental lithospheric mantle (Regelous et al., 2009). SOPITA is a misnomer because there is no evidence for a regional thermal anomaly there (Section 6.5).

■ The variety of mantle xenoliths brought up by kimberlite pipes (Nixon, 1987).

Most of the inhomogeneities can be explained by recycled, near-surface material (Hofmann and White, 1982). The SOPITA anomaly, for example, may simply reflect a mantle region that is unusually rich in recycled material. Plate tectonics maintains recycling in several ways.

■ Extraction of MORB melt at ridges.

■ Re-injection of sediments, oceanic crust and lithospheric mantle at subduction zones. The rate of re-injection of oceanic crust is ~7% of the mass of the upper mantle per Ga (Yasuda et al., 1994).

■ Removal of the continental mantle lithosphere and possibly lower crust by gravitational instability, delamination or erosion by asthenospheric flow.

The scale and form of mantle heterogeneity is debated. Analogies have been drawn with marble cake and plum pudding, which imply relatively uniform, small-scale, fertile heterogeneities distributed in a lherzolite matrix (Allègre and Turcotte, 1986; Batiza, 1984; Smith, 2005). However, the nature of the dehomogenizing processes, which involve features ranging from old, subducting slabs to metasomatizing fluids, require that a wide variety of scales must be present, ranging from small up to ~100 km or

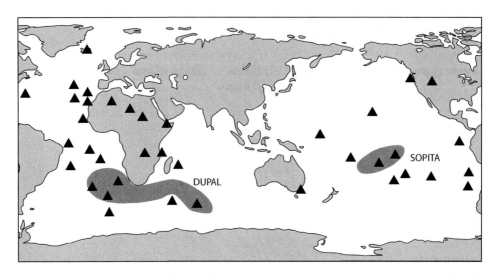

Figure 7.4 The DUPAL and SOPITA anomalies (dark gray) and melting anomalies (triangles) (adapted from White, 2005).

more (Meibom and Anderson, 2004; Meibom et al., 2002).[2] The heterogeneities must deform and alter, depending on how long they reside in the mantle. For example, the crust and mantle parts of subducting slabs may separate. The crustal part is expected to acquire a density ~5–10% higher than that of the underlying lithospheric mantle when it transforms to eclogite at ~60 km depth, gaining the heavy minerals clinopyroxene and garnet. It is not clear how or if this separation takes place, however, since the crustal part lies on top of the less-dense mantle lithosphere.

Convection works to re-homogenize the mantle, and the instantaneous form of the mantle must be a balance between this and the de-homogenizing action of plate tectonic processes. Numerical experiments show that the solid state convection associated with plate tectonics does not re-homogenize the mantle well. The mantle has a high viscosity, so flow is laminar and occurs only on large scales. The increase in viscosity with depth brought about

by rising pressure means that convection becomes even more sluggish with depth. Thus, analogies with familiar fluids such as water or the atmosphere, which convect rapidly and turbulently, are irrelevant to convection in the mantle. On the time-scale of the age of the Earth, heterogeneities ranging in size from small veins, through large blocks, to the scale of whole ocean basins must survive in the upper mantle (Davies, 1999; Davies and Richards, 1992).

What does this imply for basalt generation? When the mantle melts, the material with the lowest melting point melts first. The majority of near-surface material originated as partial melt from the mantle. Thus, when mantle containing recycled near-surface material melts, those constituents melt first. Basalt formed in this way will have subtly different compositions from basalt formed by partially melting peridotite. The diversity of near-surface materials is such that basalt geochemistry can be well-modeled by reasonable mixtures of materials known to exist in the mantle. There is no *prima facie* requirement for the involvement of remote "primordial" regions of the mantle surviving since early

[2] http://www.mantleplumes.org/He-Os.html

planetary formation. Furthermore, if fusible constituents are present in large quantities, then larger volumes of melt will be produced under the same pressure and temperature conditions than from peridotite alone (Section 3.3; Fig. 6.4).

7.2.3 From melt to source rocks

The relationship between the composition of a lava and that of its parent rock depends on several factors, including the following.

- The parent rock composition, including volatiles.

- The degree of melting.

- The pressure and temperature conditions under which the melt forms.

- Fractional crystallization during magma ascent to the surface.

- Possible assimilation of foreign material during ascent, for example, crustal material or magma conduit wall rocks.

In order to deduce the composition of the source of a lava, details of all these processes are required. This is virtually never possible, however. The nature of assimilated material is generally ambiguous. Also, no lava that contains crystals can confidently be assumed to represent an original melt because crystals may have been acquired or lost during magma ascent. Only glass can safely be assumed to represent a melt composition.

There are trade-offs between source composition and degree, pressure and temperature of melting. Also, lavas degas strongly on eruption, and it may be impossible to estimate the original volatile content reliably. Quantitative modeling of the major-element compositions of basalts usually assumes volatile-free rocks (e.g., Langmuir et al., 1992; McKenzie and Bickle, 1988). However, in addition to lowering the solidus of rocks by as much as hundreds of degrees Celsius, volatiles, in particular H_2O, CH_4 and CO_2, can affect melt compositions significantly.

At mid-ocean ridges, H_2O in the source lowers the solidus and increases the maximum depth of melting at a given T_P compared with a dry source. This not only increases the total amount of melt produced, but also counter-intuitively lowers the average degree of melting throughout the source region as a whole (Asimow and Langmuir, 2003). These effects increase both crustal thickness and incompatible-element concentrations in the melt, in agreement with observations from localities such as the Galapagos and the Azores (Asimow and Langmuir, 2003).

For a normal temperature gradient in the oceanic mantle, CO_2 exists as free gas down to ~50–60 km depth. At greater depths in the seismic low-velocity zone to depths of ~250 km, CO_2 exists as carbonate dissolved in a silicate melt. Over this depth range, compositions of melts vary widely and include carbonatites, various types of alkalic magmas, and tholeiitic magmas (Gudfinnsson and Presnall, 2005). When basalts erupt, most of the CO_2 degasses, whereas some of the H_2O is retained in minerals such as biotite and horneblende.

(a) Partition coefficients

Some ambiguities can be reduced by measuring the ratios of certain elements in melts, selected on the grounds of their partition coefficients (Section 7.2.4). When a source rock melts, elements concentrate in the liquid phase according to their partition coefficients. The partition coefficient of an element is the ratio of its concentration in the solid to that in the liquid phase. Elements with high partition coefficients are the so-called compatible elements, which tend to remain in the solid during partial melting. Examples of compatible elements are Mg, Fe, Ni and Cr. Incompatible elements are those with low partition coefficients. They do not fit readily into the lattices of common minerals, and tend to migrate preferentially into the liquid phase when a rock partially melts. Examples of incompatible elements and molecules are U, Th, Pb, Sr, Rb, K, Ti, H_2O and CO_2 (Hall, 1996).

7.2.4 Trace elements and rare-earth elements

Trace elements are those that comprise <1% of a rock. They vary widely in compatibility and their relative abundances show systematic behavior. This is often illustrated using "spider diagrams" – plots where the elements are ordered according to ascending compatibility and their abundances are normalized to those of chondritic meteorites, which are assumed to reflect primitive-mantle abundances, or to other estimates of primitive mantle (Table 7.1; Fig. 7.5).

Elements of the Lanthanum-Lutetium series (Fig. 7.1) are known as the rare-earth elements. They form a coherent group of trace elements with the same valence and external atomic orbital structure whose ionic radii progressively decrease with increasing atomic weight because of the increasing mass of the nucleus. As a con-

sequence, their compatibility in common minerals increases fairly regularly with atomic weight, because smaller ions fit more readily into crystal lattices (Fig. 7.5).

Rare-earth- and other trace-element concentrations, and ratios of concentration, can give insights into the degree of enrichment or depletion of a rock in incompatible elements. However, their interpretation is ambiguous because they are affected by a number of factors.

- *The degree of melting:* Because incompatible elements migrate most readily into the liquid phase, the earliest melt fractions have the highest concentrations of incompatible elements. Continued melting dilutes their concentrations, and thus low-degree melts are richer in incompatible elements than higher-degree melts.

- *Mixing of different sources:* This could occur, for example, as a consequence of heterogeneity in the

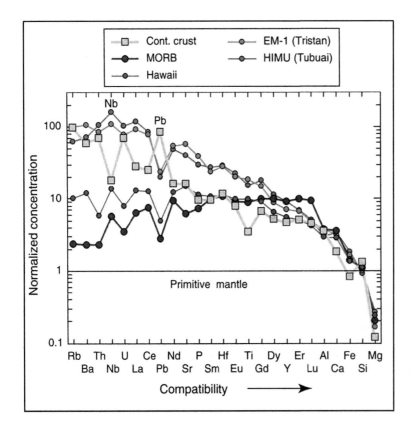

Figure 7.5 Trace element concentrations in terrestrial lavas and continental crust, normalized to primitive mantle and ordered according to increasing compatibility (from Hofmann, 1997).

original source or incorporation of additional materials during melt transport (contamination). For example, the assimilation of magma-chamber roof-rocks and sediments, that are enriched in incompatible elements, will raise concentrations.

■ *Crystallization during magma ascent:* Crystallization of refractory minerals such as olivine, which have low partition coefficients for incompatible elements, will concentrate incompatible trace elements in the remaining melt.

Despite these ambiguities, broad variations in trace element patterns can be used to subdivide basalts and draw certain conclusions. For example, the complementary trace-element patterns of continental crust and MORB are compatible with a model, whereby the former was produced by ~1% partial melting of primitive mantle and the residuum comprises the source of the latter (Hofmann, 1988).

7.2.5 Radiogenic isotope ratios

The concentration of an isotope produced by radioactive decay (a "radiogenic" isotope) increases with time and is dependent on the concentration of the parent isotope (Dickin, 1995). The concentration of a radiogenic isotope is usually expressed as its ratio with that of a stable ("primordial") isotope of the same element.

If the parent element has a partition coefficient very different from that of its daughter, then partial melting events in the past, or mixing with other materials, for example, seawater, may significantly change the growth history of the isotope ratio. Thus, materials with different melting histories may develop very different isotope ratios over time. These variations show that the sources must have experienced different histories of formation, and they can potentially be used for dating past melting events.

Different isotopes of the same element have the same partition coefficient and their ratios are thus less modified by melting and crystallization than the ratios of elements with contrasting partition coefficients. Such isotope ratios are

thus largely free from dependency on degree of partial melting and crystallization differentiation and the interpretive ambiguity that these introduce. They are often assumed to reflect the ratios of their source. This may not strictly be true for several reasons.

■ An isotope ratio may vary in the different minerals of a source rock, and the ratio in a partial melt may thus vary according to which minerals melted.

■ The lava may have undergone metamorphism.

■ It may include melt from more than one source region.

■ It may incorporate contributions from the rocks through which it rose ("contamination"). This is most commonly seen in continental flood basalts. For example, high initial $^{87}Sr/^{86}Sr$ seen in early lavas from both the Columbia River Basalts (0.703–0.714) and the Deccan traps (0.703–0.713) indicates the assimilation of up to 50% of crustal rock.

■ Hydrothermal alteration subsequent to eruption may alter the ratio.

■ Lavas on the sea-floor may have been contaminated by sea water. This would particularly affect $^{87}Sr/^{86}Sr$.

As for trace-elements, isotope ratios cannot reveal the location of the source, its composition, or its physical form, for example, whether it comprised a large block or small schlieren. In many cases, however, isotope ratios do indicate that lavas erupted in close proximity to one another from sources that have been isolated from one another for billions of years.

(a) Sr isotope ratios

^{87}Rb has a half-life of ~48.8 Ga and decays to form ^{87}Sr (Table 7.3). The concentration of ^{87}Sr is generally expressed as its ratio to its stable sister-isotope ^{86}Sr. Rb is more incompatible than Sr and thus melts are enriched in Rb/Sr relative to residuum and acquire higher $^{87}Sr/^{86}Sr$ over time.

The original $^{87}Sr/^{86}Sr$ of the Earth is thought to be ~0.699, which is the value measured for

Table 7.3 Radioactive parents and daughter products.

Radioactive parent	Daughter	Half-life (Ga)
^{238}U	^{206}Pb	4.5
^{235}U	^{207}Pb	0.7
^{232}Th	^{208}Pb	14.0
^{147}Sm	^{143}Nd	106.0
^{87}Rb	^{87}Sr	48.8
^{40}K	^{40}Ar, ^{40}Ca	1.3
^{190}Pt	^{186}Os	450
^{187}Re	^{187}Os	45.6
^{182}Hf	^{182}W	0.009

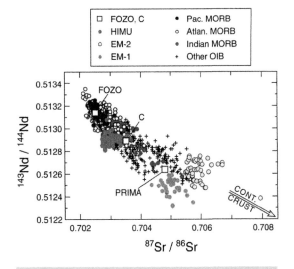

Figure 7.6 Plot of $^{87}Sr/^{86}Sr$ vs. $^{143}Nd/^{144}Nd$ for a variety of "end members" and volcanic provinces (from Hofmann, 1997).

meteorites that contain almost no Rb. If primitive mantle existed at the present day, its $^{87}Sr/^{86}Sr$ would be ~0.7045 (Fig. 7.6). Crustal rocks, for example, sediments have high values because much of the Earth's Rb was concentrated in the

crust when the latter formed early in Earth history. The crust has an average value of ~0.725, but is highly heterogeneous. Sea water has the high value of ~0.708, reflecting its time-integrated, steady-state value after interaction with continental and oceanic crust. It contaminates ocean floor basalts before subduction.

$^{87}Sr/^{86}Sr$ for MORB is ~0.703, but for alkali basalts it is 0.703–0.706 (Fig. 7.6). Some alkali basalts must thus come from a source or combination of sources with a higher concentration of Rb than primitive mantle, and MORB must come from a source with a lower concentration. Different melting anomalies have different $^{87}Sr/^{86}Sr$, and large variations in $^{87}Sr/^{86}Sr$ may even occur in successive lava flows at a single volcano, for example, Vesuvius, Italy (Di Renzo et al., 2007).

When a rock forms by crystallization of a melt, the Rb partitions differently into different minerals. $^{87}Sr/^{86}Sr$ thus grows more quickly in some minerals compared with others, and the time at which the rock crystallized can be calculated from measurements of these different $^{87}Sr/^{86}Sr$ values. Rough estimates suggest source-crystallization ages of 1–2 Ga.

(b) Nd isotope ratios

^{147}Sm decays to ^{143}Nd, which is expressed as its ratio to stable ^{144}Nd. The Sm-Nd isotope system is less sensitive than the Rb-Sr system, because the half-life of ^{147}Sm is longer than that of ^{87}Rb (Table 7.3), and Sm is less abundant than Rb. $^{143}Nd/^{144}Nd$ has thus only increased from ~0.507–0.513 during geologic time.

Both Sm and Nd are refractory and lithophile so their mean concentrations in the silicate Earth are assumed to be the same as in chondritic meteorites. The $^{143}Nd/^{144}Nd$ value of the silicate Earth is thus thought to be accurately known. Sm is less incompatible than Nd, and thus during partial melting it preferentially remains in the residuum. As a result, crustal $^{143}Nd/^{144}Nd$ is generally lower (~0.512) than in the mantle (~0.513).

(c) Pb isotope ratios

Two uranium isotopes (^{235}U and ^{238}U) decay to produce two different lead isotopes (^{207}Pb and ^{206}Pb) (Table 7.3). A stable Pb isotope (^{204}Pb) also exists and provides a reference. This is a more powerful isotope system than others because two parent- and two daughter isotopes enable the theoretical age of differentiation of the source to be inferred without the (usually unknown) original parent-daughter ratio. This is because the relative rates of production of ^{206}Pb and ^{207}Pb depend only on time.

On a plot of $^{206}Pb/^{204}Pb$ vs. $^{207}Pb/^{204}Pb$, MORB and basalts from different melting anomalies form distinct fields (Figs 7.7 and 7.8). Rocks with an average time-integrated U/Pb similar to the average silicate Earth are expected to plot along the 4.57-Ga geochron. Melts from depleted sources are expected to plot to the left of this,

and melts from enriched ones to the right. However, the fields for both MORB and enriched basalts from various melting anomalies all form linear arrays to the right of the geochron. This, curiously, indicates enrichments in time-averaged U/Pb for all their sources, in apparent disagreement with other isotope systems such as $^{87}Sr/^{86}Sr$ and $^{143}Nd/^{144}Nd$, which predict a depleted source for MORB.

Other puzzles are that the continental crust does not lie very far to the right of the geochron, despite its strong enrichment in U/Pb, and no complementary group of rocks is known with Pb isotopes indicative of time-averaged depletion in U/Pb. These unexpected observations are known as the "lead paradox". They probably result from depletion in Pb rather than long-term enrichment in U. Pb is extracted from basaltic oceanic crust by hydrothermal processes at spreading ridges and dehydration in

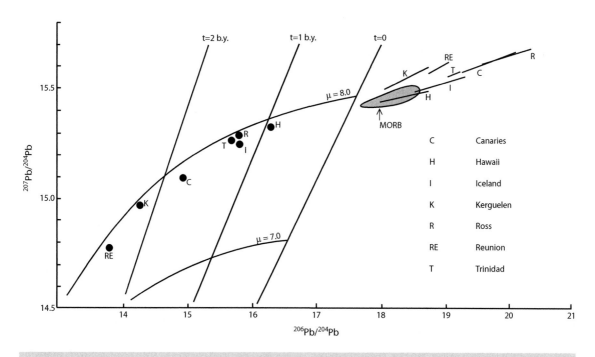

Figure 7.7 The lead isotope system, showing fields for MORB and various oceanic island groups. Lines indicate compositional ranges in each field and dots indicated the primary isotope ratios from which each is derived, assuming a two-stage model (after Chase, 1981).

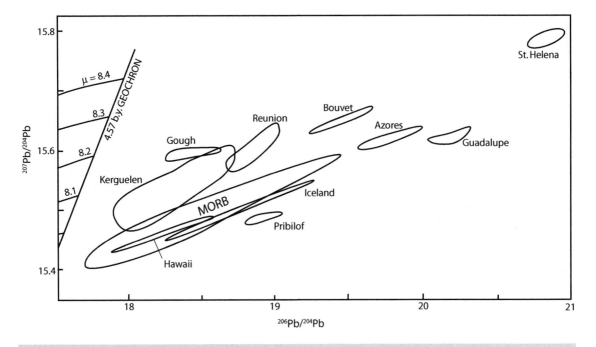

Figure 7.8 Lead isotope fields for MORB and various island groups (from Sun, 1980).

subduction zones. It is thus preferentially removed from the mantle and concentrated in the continental crust, raising the U/Pb of the former and lowering it in the latter.

In the simplest analysis, the linear arrays on the $^{206}Pb/^{204}Pb$ vs. $^{207}Pb/^{204}Pb$ plot can be extrapolated to the left and interpreted as source enrichments at various times in the past (dots, Fig. 7.7). These times lie in the range ~2.5–1.0 Ga, much later than formation of the continental crust (Chase, 1981). The events could have been partial melting or the addition of an enriched constituent, for example, subducted oceanic crust or recycled continental crust (Chase, 1981). If the enrichment event is assumed to be addition of continental crust, dates of 1–1.3 Ga are calculated. The observations could also be explained by many enrichment or melting events, or by continuous differentiation and remixing. Furthermore, magmas with different Pb isotope ratios could have been derived from melting minerals with

different parent isotope concentrations from an homogenous source. Thus even the powerful U/Pb system is highly non-unique. However, it does suggest first-order modification of upper mantle composition by plate tectonic or other processes.

(d) Os isotope ratios

^{187}Os forms by the decay of ^{187}Re, which has a similar half-life to ^{87}Rb (Table 7.3). It is usually expressed as a ratio with its stable sister isotope ^{188}Os. Both Re and Os are highly refractory and thus the Re/Os of primitive mantle is expected to be the same as in chondrites. Re is highly incompatible, concentrated in basalts by a factor of ~100 compared with the mantle source, and thus $^{187}Os/^{188}Os$ grows rapidly in melts. Xenoliths from the continental mantle lithosphere have low $^{187}Os/^{188}Os$, so this isotope system can also be used to detect recycled material of that kind.

(e) O isotope ratios

^{18}O comprises ~0.2% of the oxygen in nature and is ~13% heavier than the ^{16}O that comprises 99.76% (Taylor, 1968). It is significantly fractionated by meteorological processes, and variations are expressed as $\delta^{18}O$, the deviation in parts per thousand from an internationally agreed standard mean ocean water (SMOW). Water vapor is enriched in molecules containing the lighter isotope, ^{16}O, and thus all fresh water has a negative $\delta^{18}O$.

Normal values of $\delta^{18}O$ in basalts lie in the range ~5–7 but these can be drastically changed by hydrothermal alteration or contamination by sediments. $\delta^{18}O$ of sediments tends to be high and incorporation of such material in the source will raise the $\delta^{18}O$ above 7. Oxygen isotopes may thus be used as an indicator of surface materials in the source. Hydrothermal alteration can raise or lower $\delta^{18}O$.

7.2.6 The mantle zoo

The compositions of basalts often form elongated distributions on plots of isotope ratios. This has led to a schema nicknamed "the mantle zoo", whereby ranges of basalts are considered to arise from the mixing, in various proportions, of so-called "end members", which reside in "reservoirs". "End members" are defined solely by isotope ratios, they are based on empirical observations, and their definitions may be frequently changed (Stracke et al., 2005).[3] In some cases they may be postulated to be actual geological materials, but they are fundamentally based simply on isotope plots.

- EM-1 ("enriched mantle-1"): Basalts with Type EM-1 characteristics are high in $^{87}Sr/^{86}Sr$, low in $^{143}Nd/^{144}Nd$, and enriched in incompatible elements compared with primitive mantle (Fig. 7.6). $\delta^{18}O$ is high compared with MORB. These charac-

[3] http://www.mantleplumes.org/LowerCrust.html

teristics are thought to arise from the inclusion in the source of pelagic sediment from the deep sea floor far from continents. Examples of oceanic islands with Type EM-1-basalts include Pitcairn, Tristan, Kerguelen and Samoa.

- EM-2 ("enriched mantle-2"): Basalts of Type EM-2 have higher $^{87}Sr/^{86}Sr$ than Type EM-1 (Fig. 7.6). The EM-2 "end member" is thought to be mature, i.e., "granitic" terrigenous sediment eroded from continents and deposited on the sea floor. Both EM-1 and EM-2 can be explained by small amounts of sediment being subducted along with the plates on which they lie, and recycled in the mantle.

- DMM ("depleted MORB mantle"): This is the acronym given to a hypothetical mantle composition that is primitive, but depleted in elements that formed the continental crust, and presumed to approximate to the source of MORB.

- HIMU (high-μ, i.e., high-$^{238}U/^{204}Pb$): HIMU basalts are high in $^{206}Pb/^{204}Pb$, indicating high integrated $^{238}U/^{204}Pb$ in the source (Fig. 7.9). HIMU basalts are found in oceanic islands from widely separated parts of the world, including St Helena and the Cook-Austral Islands. A possible source for HIMU is old, subducted, dehydrated oceanic crust, recycled in the mantle (Weaver, 1991; Zindler and Hart, 1986).

- FOZO ("focal zone"): This is a mantle composition originally defined as being depleted in Sr and Nd

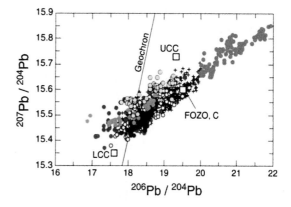

Figure 7.9 $^{206}Pb/^{204}Pb$ vs. $^{207}Pb/^{204}Pb$ using the same samples and symbols as for Fig. 7.6. LCC – Lower continental crust; UCC – Upper continental crust (from Hofmann, 1997).

but high in radiogenic Pb (Hart et al., 1992). It is considered to be an "end member" common to most basalts. Several other ubiquitous "end members" have been suggested, including C ("common component"), PREMA ("prevalent mantle") and PHEM ("primitive helium mantle"), the latter having high $^3He/^4He$ (Farley et al., 1992). It has been suggested that PHEM is similar to primitive mantle.

A large number of "end members" are generally needed to explain a single basalt. For example, at least four "end members" are required to explain the range of compositions in Icelandic basalts (Kempton et al., 2000), possibly with an additional component introduced solely to explain the He contents, an issue considered to be of critical relevance to the origin of melting anomalies (Section 7.5.1). The simplest explanation for natural basalt compositions in terms of isotopic end members is thus far from simple.[4] The ultimate envisaged origin of many of these "end members" is unclear. They amount to theoretical chemical compositions rather than known rock types. Unless their sources in terms of geological reality are known, reducing rocks to assemblages of "end members" does not contribute to understanding their origins (Armienti and Gasperini, 2007).

7.2.7 MORB, OIB, tholeiites and alkali basalt

Basalt terminology is confused because the terms MORB – mid-ocean-ridge basalts, and OIB – ocean island basalts, which are commonly used to signify composition, are geographical designations (Fig. 7.3). They do not take into account the fact that essentially identical rocks are found in other settings. MORB-like basalt is also found on continents (e.g., Greenland) and oceanic islands (e.g., Iceland). OIB-like rocks occur not only on islands but also on mid-ocean ridges, where they are called "enriched" MORB (E-MORB), to distinguish them from "normal"

[4] http://www.mantleplumes.org/LowerCrust.html

MORB (N-MORB), and on oceanic plateaus and continental flood basalts where they are called enriched tholeiite. Less siliceous OIB-like rocks – alkali basalts – are found throughout the oceans and continents.

MORB is usually understood to signify relatively depleted, large-volume tholeiitic basalt, the majority of which is erupted at spreading ridges. It makes up the bulk of the oceanic crust. It is thought to be formed largely from partial melt of a fertile mantle peridotite source – lherzolite. Some may be formed by high-degree melting of subducted oceanic crust, which itself has a basaltic composition.

The basalt generally thought of as MORB is relatively uniform in aspects of its composition compared with other basalts, with narrow ranges of isotope ratios and trace-element patterns (Figs 7.5 and 7.6). Although it is more enriched in incompatible trace elements than primitive mantle, it must come from a more depleted source because the enrichment is too low to derive from a source with primitive mantle composition. This is supported by its radiogenic isotope ratio contents, which indicate long-term depletion in ^{87}Rb and enrichment in ^{147}Sm (Fig. 7.6). MORB source could be largely mantle residuum left after formation of the continental crust.

The term OIB is commonly understood to signify basalts enriched in incompatible elements and radiogenic isotope ratios. However, the majority of rocks with the relevant compositions do not occur on oceanic islands and this term is thus inappropriate for them. Because of this, the terms enriched tholeiite and alkali basalt will be used here. Enriched tholeiite differs from MORB in having higher and more varied concentrations of incompatible elements and corresponding radiogenic isotope characteristics. These isotopes indicate source enrichment relatively late in geological time, in the period ~1–2 Ga (Figs 7.5 and 7.6). Enriched tholeiite occurs at some large-volume melting anomalies, for example, Hawaii, Iceland, the Galapagos and Yellowstone (Table 1.5), and commonly makes up the bulk of oceanic plateaus and continental flood basalts.

Alkali basalt has isotopic and trace-element contents similar to enriched tholeiite, but undersaturated in silica. It usually erupts in relatively small quantities, but larger accumulations do occur. The largest of these is probably the Savai'i/Upolu volcano pair in Samoa, which has a volume of tens of thousands of cubic kilometers and is comparable to the size of the largely-tholeiitic Big Island of Hawaii (Kear and Wood, 1957; Natland, 1980).

The settings in which alkali basalt erupts are diverse. They include tens of thousands of scattered seamounts and linear, time-progressive and non-time-progressive chains throughout the oceans. In the continents it occurs in both rift and non-rift settings, for example, East Africa, the Basin and Range province, East Asia and the North Sea (Barry et al., 2003). It may erupt only briefly, or be produced persistently for tens of millions of years. Most currently active melting anomalies erupt essentially solely alkali basalt.

A discriminator for depleted- and enriched basalts has been proposed based on a plot of Nb/Y vs. Zr/Y (Fig. 7.10) (Fitton et al., 1997; Meschede, 1986). ΔNb is defined such that N-MORB has ΔNb < 0 and enriched basalt has ΔNb > 0. Elevated ΔNb has been attributed to recycled crust and has been found to provide a useful discriminator at many localities on continents and the mid-ocean ridge system (Fitton, 2007). Nevertheless, although the contrasts between MORB and enriched basalts are typically emphasized, in reality these basalts are very similar. They are almost identical in the major elements that make up over 99% of the rocks, and only vary in their minor- and trace-elements. It is the apparent different ages of formation of their sources that is perhaps the most significant difference between them.

(a) Origin of enriched basalt

Because enriched basalt is common on oceanic islands, and distinct from the MORB that is most common on spreading ridges, it potentially holds important information about the origin of melting anomalies. Its petrogenesis is still disputed, however. Low-degree (3–7%) melting of peridotite, such as probably comprises much of the upper mantle, produces alkali basalt and concentrates incompatible elements in the melt. However, extracting both MORB and enriched

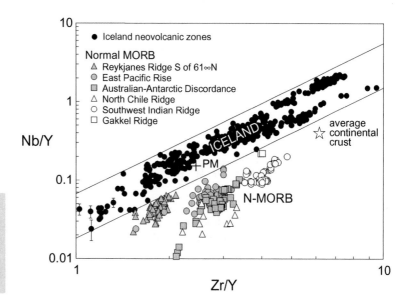

Figure 7.10 Nb/Y vs. Zr/Y for Icelandic basalts with MgO >5 wt.%, and N-MORB. +PM indicates estimated primitive mantle composition (from Fitton, 2007).

basalt from the same parent is at odds with the apparent different ages of their sources.

Explanations have been offered involving the mixing of different "end members". A simple scenario is that alkali basalts result from mixing, in various proportions, of melts from MORB source and an enriched end-member that might be primordial mantle. However, there is no convincing evidence that any primordial mantle has survived in the mantle and furthermore, two "end members" are insufficient to explain the compositions of alkali basalts – several are needed.

Basalt enrichment almost certainly arises in general from incorporation of recycled near-surface materials such as sediments, oceanic crust, oceanic and continental lithospheric mantle, and possibly continental lower crust. Partial melting to various degrees of subducted oceanic crust can explain the rare-earth-element patterns of tholeiitic and transitional basalts – those that approach alkalic composition (Fig. 7.11) (Hofmann and White, 1982). The inven-

tories of incompatible elements in the continental crust and the source of MORB fall short of the expected total, and if the balance resides in subducted oceanic crust, then this must account for several percent of the mantle.

Recycled oceanic crust cannot explain all the minor- and trace-element characteristics of enriched basalts, however, nor their isotopic compositions (Niu and O'Hara, 2003; Pilet et al., 2004; 2008).[5] During subduction, oceanic crust is dehydrated and elements mobile in the hydrous phase, including Rb, U, K and Pb, are removed. Nevertheless, even taking this into account, experiments involving melting oceanic crust still fall short of entirely reproducing the compositions of enriched basalts. Recycled, metasomatized oceanic or continental lithosphere provides a better match. Metasomatism – the infiltration of fluids – may be hydrous, for example, above a subducting slab, or involve melt rising from the asthenosphere and intruding the lithosphere. Experiments with samples of metasomatic veins from the French Pyrenees, partially melted both in isolation and enclosed in peridotite, match well the range of major-, minor- and trace-element compositions of both oceanic and continental alkali basalts (Pilet et al., 2008).[6]

Metasomatic veins are thought to be produced by hydrous, low-degree mantle melts that partially crystallize as they ascend from the asthenosphere, and form a compositional continuum. In this way, they inherit the trace-element and isotope characteristics of the original low-degree peridotite melts, for example, high incompatible-element contents. They have a lower melting point than the host rock in which they crystallize, and thus they re-melt first and acquire compositional diversity by reacting with rock through which they subsequently rise. Thus, instead of being a low-degree melt of the volumetrically large peridotic mantle, alkali

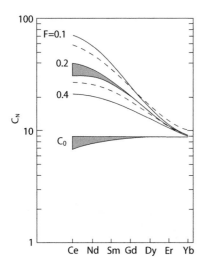

Figure 7.11 Normalized rare-earth-element concentrations in a melt with melt fraction F, calculated for garnet pyroxenite source. Dashed lines show rare-earth-element patterns for Hawaiian basalts (from Hofmann and White, 1982).

[5] http://www.mantleplumes.org/NotFromCrust.html
[6] http://www.mantleplumes.org/MetasomaticOIB.html

basalt may be a high-degree melt of a volumetrically small mantle constituent – metasomatic veins.

What does this explain that low-degree melt from mantle peridotite cannot? It provides a site where the source can age after separation from the ambient mantle, prior to re-melting and eruption. This provides a candidate physical location for the alkali basalt source. In order to develop different isotope ratios, that source must be isolated from the MORB source and not remelted and re-absorbed into the asthenosphere during convection. The large range in isotope ratios in alkali basalt is also explained, and the variations in composition with time observed at individual locations, because metasomatic veins form continually after creation of the mantle lithosphere host. A metasomatic source can also account for the uniformity in the ratios of trace-elements with similar partition coefficients in both MORB and alkali basalt, despite very different element concentrations, which suggests a common origin. The lithosphere tapped could be either *in situ*, or recycled in the convecting mantle.

A source in lithospheric metasomatic veins can explain why alkali basalt produced often remains uniform regardless of the volume erupted, the longevity of a site, or its lithospheric and tectonic setting. For example, the 2000-km-long Cameroon line has erupted compositionally uniform alkali basalt continually from ~65–0 Ma, even though part of the line is over continental- and part over oceanic lithosphere (Fitton, 1987; 2007). A source in metasomatic veins that are essentially the same in both kinds of lithosphere would explain the uniformity in composition. In the Scottish Midland Valley, compositionally uniform alkali basalt erupted continually throughout the 70-Ma time period from ~350–280 Ma, while the region drifted through 1600 km of latitude (Fig. 7.12), suggesting a source embedded in, and traveling with, the plate. Similar observations are reported for the Karoo volcanic province, Southern Africa, where magmas with similar compositions

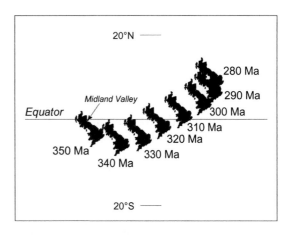

Figure 7.12 Migration of the Scottish Midland Valley ~1600 km in the ~70 Ma period during which alkalic basalt was erupted. Only migration in latitude is constrained (from Fitton, 2007).

erupted in close proximity, separated in time by 900 Ma (Jourdan et al., 2007; 2009).

Neither metasomatism nor low-degree melting of peridotite can explain large-volume enriched tholeiite such as is found at Iceland, Hawaii and the Galapagos islands. The melting of large, fusible inhomogeneities such as subducted slabs is required to account for both enriched geochemistry and large melt volumes. Flood basalts might also arise from melting hydrated subcontinental lithospheric mantle (Gallagher and Hawkesworth, 1992).

(b) Origin of MORB

MORB is generally considered to result from isentropic upwelling of the mantle beneath spreading ridges. However, the depth range throughout which melts form, and their compositions, are poorly known. Depths varying from as shallow as 0.4 GPa (~12 km) to as deep as 4 GPa (~120 km) have been suggested, both large and small depth ranges, and both picritic and non-picritic initial-melt compositions. Variations in MORB composition have been attributed to variations in temperature and source composition.

The melting processes that produce MORB have been modeled using their minor- and trace element contents. However, these elements make up only a tiny percentage of the rocks and they are mostly concentrated in highly mobile, low-degree melts that may not follow the bulk melt. The major elements have been studied in laboratory experiments that melt chemical assemblages that approximate the major-element composition of real lherzolites. The six-component system CaO-MgO-Al_2O_3-SiO_2-Na_2O-FeO (CMASNF) approaches the composition of natural rocks, accounting for 98% of MORB and 99% of lherzolite. It thus provides a robust basis for understanding the melting behavior of the mantle.

The results show that the production of MORB is well modeled by melting around the plagioclase-spinel transition in the relatively small depth range of ~35–45 km (~1.2–1.5 GPa). In this interval, the temperature of onset of melting drops by up to ~25°C and solid mantle rising isentropically beneath ridges is most likely to melt (Presnall and Gudfinnsson, 2008; Presnall et al., 2002). The mineral phase relationships cannot constrain the absolute proportions of minerals in the source, but they require the presence of critical minerals. For MORB generation, these are olivine, orthopyroxene, clinopyroxene, spinel and, at lower pressures, plagioclase. For Hawaiian basalts, olivine, orthopyroxene, clinopyroxene and garnet are required. Relatively uniform MORB is produced from a considerable range of source compositions because the required minerals need only coexist – similar melts are produced regardless of their relative abundances. Source heterogeneity is required to explain minor variations in MORB composition, however, for example, in the Fe_8 and Na_8 contents, which were formerly attributed to temperature variations (Section 6.2.2).

Abyssal peridotites, dredged from the sea floor, are thought to be the residue left over after MORB extraction. These rocks contain no original plagioclase, suggesting that melting near the plagioclase-spinel lherzolite solidus continues to the point of plagioclase exhaustion. Spinel is exhausted almost simultaneously, leaving only olivine, orthopyroxene and clinopyroxene, an assemblage that has a higher solidus temperature. Thus, melting ceases immediately. The fractional crystallization behavior of the liquids produced in experiments is similar to that of MORB, producing olivine and plagioclase at low pressures and pyroxene and plagioclase at higher pressure. Natural MORB glasses all conform to olivine-plagioclase fractionation, with no tendency to fractionate olivine alone.

Worldwide, olivine-controlled fractionation is only observed in glasses from Hawaii (Section 6.5.3). This might be explained by a second region of tholeiite generation at depths >100 km, in the seismic low-velocity zone. The phase relations of CO_2-bearing lherzolite – the system CaO-MgO-Al_2O_3-SiO_2-Na_2O-FeO-CO_2 (CMASNF-CO_2) – show an abrupt drop of ~280°C in the lherzolite solidus at ~60 km depth (Fig. 6.6). The temperature of the solidus then falls below reasonable geotherms at this depth. This probably explains the seismic low-velocity layer, which almost certainly contains low-degree partial melt. The phase diagrams predict that melts formed there range in composition from carbonatites, at low degrees, to alkali basalts and tholeiites, depending on the depth of formation. Hawaiian tholeiite may thus come from the low-velocity zone (Presnall and Gudfinnsson, 2008).

Near-complete melting of eclogite will also produce tholeiite. The liquidus temperature of eclogite is close to the solidus temperature of lherzolite, and the temperature interval over which eclogite melts completely is small – no more than ~200°C compared with several hundred degrees Celsius for lherzolite. Thus, under the same temperature conditions, if eclogite exists beneath ridges a larger volume of melt will be produced than from lherzolite (Natland, 2007).

7.3 Predictions of the Plume hypothesis

The original Plume hypothesis proposed that relatively primordial material rises adiabatically from the deep mantle up to the asthenosphere, where it partially melts and erupts to form melting anomalies. A primordial composition was invoked for plume material to explain the geochemical differences between basalts on oceanic islands and MORB, including the K and rare-earth-element patterns. The aged isotope characteristics were interpreted as reflecting the time for which the primordial material had been isolated in the lower mantle (Morgan, 1971). A key element of the hypothesis was that the distinct geochemistry of enriched basalts arose from a source deeper than the source of MORB.

Initial support for this model came from the Iceland region, where trace- and rare-earth element patterns and radiogenic Sr and Pb isotopes were found to contrast with those of MORB, and to apparently grade from Iceland southerly along the Reykjanes ridge. This was interpreted in terms of a two-reservoir model involving the influx of compositionally distinct, primordial plume material beneath Iceland, and mixing to varying degrees with MORB source along the Reykjanes ridge (Schilling, 1973).

This view led to the widespread concept of a simple, chemically stratified mantle. In its most basic form, the mantle above the 650-km discontinuity is viewed as depleted early in Earth history by formation of the continental crust, and subsequently maintained in an homogenous state by vigorous convection associated with plate tectonics. The lower mantle is viewed as a separate "reservoir" with a more primordial, relatively enriched composition.

Advances in petrological and geochemical knowledge exposed the following problems with this simple model.

■ Basalt geochemistry cannot be explained by the mixing of only two "end members" – as many as five may be required, for example, at Iceland (Section 7.2.6).

■ Different melting anomalies have different compositions and require different suites of "end members".

■ The enrichment in basalts from melting anomalies almost certainly arises from near-surface, recycled materials.

■ The 650-km discontinuity is a mineralogical phase change and there is no evidence for a compositional change there.

■ The volumes of melt produced in flood basalts cannot be simulated by melting in plume heads with the composition of primordial mantle peridotite (Cordery et al., 1997).

As a result, the Plume hypothesis was revised. In the new model, subducting slabs are postulated to sink to the base of the lower mantle and accumulate there in a "slab graveyard" (Kellogg et al., 1999). High-pressure mineral transformations and gravitational segregation are postulated to physically separate different parts of the slabs, providing "reservoirs" of material with different geochemical characteristics. The material is stored for long enough to develop aged isotopic signatures, and it is then incorporated into plumes that nucleate at the core-mantle boundary (Fig. 7.13). This model provides a candidate

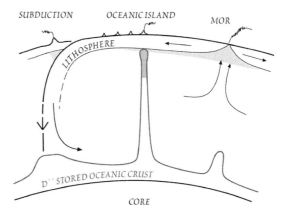

Figure 7.13 The model for recycling of oceanic crust from the surface, down to the core-mantle boundary, and back up to the surface again in plumes (Hofmann and White, 1982).

explanation for the presence of near-surface materials in the source of enriched basalts, and it provides a source for the fusible material needed to produce large magma volumes in numerical melting simulations.

7.4 Predictions of the Plate hypothesis

The Plate hypothesis envisages that the lavas at melting anomalies formed from passive melting in the shallow mantle. Direct involvement of the lower mantle is not expected. The melt is predicted to arise from materials more fertile than the MORB source,[7] with larger magma volumes produced as a consequence. These materials originate at the near-surface and are re-introduced into the mantle by plate-tectonic processes and recycled by mantle convection.

Plate tectonics is envisaged to maintain mantle inhomogeneity by extracting melt at some places, for example, spreading ridges and arcs, and re-injecting fusible materials at others, for example, by subduction or gravitational detachment of lithospheric mantle and lower crust. Subducted slabs are not expected to sink deep into the lower mantle or to the core-mantle boundary. They are predicted to be largely recycled in the upper mantle, with little or no material sinking below ~1000 km depth.

A contribution from recycled, near-surface, fusible material to lavas at melting anomalies is thus a primary prediction of the Plate hypothesis. This is a contrast with the modern Plume hypothesis, which considers such a contribution to exist only by chance and not to be required. The recycled materials derive largely from melts, formed at earlier times. They are thus expected to be enriched in incompatible elements and

---

[7] http://www.mantleplumes.org/Eclogite.html;
http://www.mantleplumes.org/LowerCrust.html;
http://www.mantleplumes.org/WTurkey.html;
http://www.mantleplumes.org/ChinaOIB.html;
http://www.mantleplumes.org/TopPages/
RecyclingTop.html

with various isotopic ages, depending on how long they have been stored in the upper mantle. Some inhomogeneities survive largely intact in the upper mantle for long periods. The time needed to process the entire upper mantle once at spreading ridges is ~1 Ga, but the entire upper mantle need not be reprocessed so frequently. For example, some material may have been stored in the lithospheric mantle beneath cratons since its creation early in Earth history. Because mantle heterogeneities are produced by diverse processes, they are expected to vary in geochemical character. The different geochemical isotopic "end members" are associated with distributed materials that arise from particular generic processes, for example, mantle metasomatism. They are not envisaged to arise from spatially distinct containment regions in the mantle.

The composition of the lower mantle is thought to be grossly similar to that of the upper mantle, but it is expected to be more homogeneous and to convect increasingly more sluggishly with depth. The deep mantle is not expected to have any direct involvement in surface lavas. The composition of the lower mantle thus cannot be estimated from the petrology of lavas erupted at the surface.

7.5 Proposed deep-mantle- and core-mantle-boundary tracers

7.5.1 Helium (He) isotope ratios

(a) The helium system

The He isotope system is essentially unique in being commonly assumed to be an unambiguous geochemical indicator of lower mantle material, and thus a plume diagnostic. This concept did not comprise part of the original Plume hypothesis. It arose as a result of later observations made in Hawaii, which was assumed to be underlain by a plume.

Table 7.4 Noble gas isotopes and their sources.

Isotope	Source
^{3}He	stable
^{4}He*	decay of U + Th. $^{238}\text{U} = 8(^4\text{He}) + {}^{206}\text{Pb},$ $^{235}\text{U} = 7(^4\text{He}) + {}^{207}\text{Pb},$ $^{232}\text{Th} = 6(^4\text{He}) + {}^{208}\text{Pb}$
^{20}Ne	stable
^{21}Ne*	nucleogenic decay chains of ^{24}Mg, ^{18}O when irradiated by U + Th decay products. Thus produced at the same rate as ^{4}He
^{22}Ne	some is primordial and some produced in a similar way to ^{21}Ne but only in very small amounts

In practice, observations of ^{3}He/^{4}He higher than ~8 times the current atmospheric value (Ra), which is 1.38×10^{-6}, tend to be interpreted as proof of a plume from the lower mantle (Anderson, 2000b, c; 2001b). This interpretation is, nevertheless, not required. In common with other geochemical signatures in basalts, high ^{3}He/^{4}He can arise from near-surface materials recycled in the shallow mantle.[8]

^{3}He is a primordial isotope, incorporated into the Earth during planetary accretion (Table 7.4) (Craig and Lupton, 1976). The primordial ^{3}He/^{4}He ratio is thought to have been ~200 Ra. A small amount of ^{3}He is constantly being added to the Earth's surface by interplanetary dust particles (Anderson, 1993), and "cosmogenic" ^{3}He is formed by the impingement of cosmic rays on rocks at high altitudes. However, ^{3}He is not produced in significant quantities by any radiogenic decay process.

^{4}He is constantly being created, however. The ^{4}He nucleus is an alpha particle and ^{4}He is thus

[8] http://www.mantleplumes.org/HeliumFundamentals.html

produced by the decay of ^{238}U, ^{235}U and ^{232}Th. As a result, the ^{3}He/^{4}He ratio in the Earth is constantly diminishing – it is not thought to be increasing significantly in any part of the Earth. ^{4}He accumulates rapidly in rocks that are rich in U and Th but slowly in rocks poor in those elements. The integrated lifetime U and Th content of mantle rocks and recycled material varies by 3 or 4 orders of magnitude. Thus, large variations in ^{3}He/^{4}He can develop, depending on how much Earth's original ratio is reduced by newly created ^{4}He.

The amount by which ^{3}He/^{4}He decreases in a given rock is also dependent on the absolute abundance of He. For a given production rate of ^{4}He, ^{3}He/^{4}He reduces only slowly in a rock rich in He. On the other hand, if a rock is poor in He, its ^{3}He/^{4}He ratio will decrease rapidly. The lowering of ^{3}He/^{4}He over time is thus dependent on the history of degassing and re-gassing by metasomatization of the rock. Once He enters the atmosphere it rapidly escapes to space because of its low atomic weight. The approximate lifetime of He in the atmosphere is ~1–2 Ma.

Continental rocks are generally high in U and Th and often have ^{3}He/^{4}He $\ll$ 1 Ra. MORB typically has relatively uniform values in the range ~6–12 Ra (Anderson, 2000b, c). In contrast, like other geochemical species, ^{3}He/^{4}He is much more variable in lavas from melting anomalies, ranging from <1 Ra to ~50 Ra (Stuart et al., 2003).

(b) The Plume model

Values of ^{3}He/^{4}He > 20, much higher than those typical of MORB, were first observed in basalts from Hawaii (Kaneoka and Takaoka, 1978). Since Hawaii had earlier been postulated to be underlain by a plume from the lower mantle, it was suggested that the lower mantle is characterized by high ^{3}He/^{4}He – perhaps as high as 25 or 30 Ra – and that this ratio can be used as a lower mantle tracer. The value of ^{3}He/^{4}He postulated for the lower mantle has had to be continually adjusted upward as progressively higher ^{3}He/^{4}He values have been discovered at other localities, for example, at Iceland and Baffin Island.

An explanation for this proposed spatial variation in ^{3}He/^{4}He was presented in a model in which the upper mantle is largely degassed and He-poor. In this model, progressive addition of ^{4}He over time reduced ^{3}He/^{4}He to a uniform value of ~8 Ra. The lower mantle was postulated to be entirely or in part little processed, little degassed, and with a near-primordial composition. It would thus be rich in He and its ^{3}He/^{4}He would have reduced much less over geological time. This relatively high-^{3}He/^{4}He material is postulated to be transported up to the surface by plumes.

There are a number of difficulties with this model.

1 Mass balance calculations show that unreasonably high concentrations of ^{3}He are required in the lower mantle, relative to other volatiles (Anderson, 1989). This is because, over the lifetime of the Earth, radioactive decay of U and Th in the lower mantle has generated a large amount of ^{4}He. In order for the value of ^{3}He/^{4}He in the lower mantle to have remained high, the concentration of He there would also have to be high – about an order of magnitude less than that in chondritic meteorites (Kellogg and Wasserburg, 1990). This is inconsistent with an Earth that was strongly degassed in volatiles during planetary formation (Section 7.1; Fig. 7.2). Even primitive mantle is expected to have many orders of magnitude less He than chondrites.

The problem of impossibly high predicted concentrations of ^{3}He in the lower mantle has become worse as the maximum ^{3}He/^{4}He observed at the surface has progressively increased. The value of ^{3}He/^{4}He of ~50 Ra reported from Baffin Island (Stuart et al., 2003a), is much higher than the value of ~30 Ra assumed by Kellogg and Wasserburg (1990) for the lower mantle. As a result, the proportion of the lower mantle suggested to be rich in He has been adjusted downward, including the suggestion that it is limited to a thin layer at the core-mantle boundary layer.

2 If high-^{3}He/^{4}He ratios in basalts from melting anomalies arise from a He-rich lower mantle, then lavas at melting anomalies would be expected to

be richer in He than MORB. However, the opposite is observed – lavas from melting anomalies are 2–3 orders of magnitude poorer in He. This has been attributed to stronger degassing a result of the shallow eruption depths on and near oceanic islands. This postulate has been tested using the concentrations of He relative to heavier noble gases.

Ne and Ar have a greater tendency to degas upon eruption than He, and thus He/Ne and He/Ar are expected to be higher for more strongly degassed magmas. These ratios should thus be higher in enriched basalts than in MORB. However, the opposite is observed, suggesting that basalts at melting anomalies are less degassed than MORB, not more (Figs 7.14 and 7.15) (Moreira and Sarda, 2000; Ozima and Igarashi, 2000).[9]

3 The geochemistry of enriched basalts does not suggest a primordial parent. At individual melting anomalies, ^{3}He/^{4}He does not correlate systematically with Sr, Nd and Pb isotope ratios, and globally, basalts with the highest-^{3}He/^{4}He tend to resemble MORB in these isotope ratios and in trace- and incompatible elements (Class and Goldstein, 2005). The highest ^{3}He/^{4}He values found anywhere on Earth occur in basalts otherwise indistinguishable isotopically from MORB (Dale et al., 2009; F.M. Stuart et al., 2003b). This is not what is expected

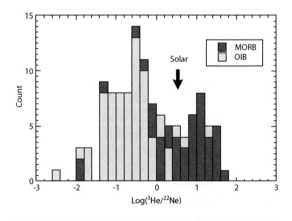

Figure 7.14 Histogram of ^{3}He/^{22}Ne ratio in MORB and alkali basalts (OIB) (from Ozima and Igarashi, 2000).

[9] http://www.mantleplumes.org/Ne.html

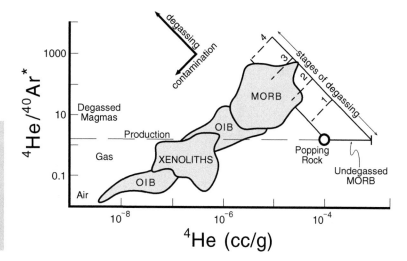

Figure 7.15 He and Ar isotopes in MORB, alkali basalts (OIB) and xenoliths. Arrows show the effects of degassing (i.e., acquisition of air) and contamination (i.e., acquisition of the degassed fluid) (from Anderson, 2000b).

for derivation from a primordial, little-processed source.

4 High $^3He/^4He$ is not found at several major proposed plume localities, for example, the Canary islands, St Helena and Tristan (Table 1.4; Courtillot et al., 2003). Also, there is no correlation between $^3He/^4He$ and the depths of seismic anomalies beneath melting anomalies. The seismic anomaly beneath Iceland is robustly shown to be confined to the upper mantle (Section 5.5.1), but $^3He/^4He$ there is among the highest found anywhere in the world.

(c) The Plate model

The Plate hypothesis postulates that, in common with other geochemical features of lavas from melting anomalies, high-$^3He/^4He$ arises from near-surface materials recycled in the shallow mantle. In contrast to the Plume hypothesis that postulates that old, high-$^3He/^4He$ is preserved in a high-^3He environment, the Plate model postulates that it is preserved in a low-^4He environment, i.e., a host with low time-integrated U and Th (Albarede, 2008; Anderson, 1998b, c; Class and Goldstein, 2005). Candidate host materials of this kind include the residuum left after basalt is extracted from mantle peridotite, compacted, olivine-rich cumulates in the lowermost oceanic crust, for example, dunite, and almost completely depleted Archean sub-

continental lithospheric mantle (Castro et al., 2009; Meibom et al., 2005).

When peridotite melts, garnet and clinopyroxene, in which the bulk of the U and Th resides, melt first and thus U and Th are preferentially lost early. As melting progresses, the residuum becomes increasingly dominated by orthopyroxene and olivine, which retain very little U and Th (Brooker et al., 2003; Parman et al., 2005). As a result, He stored for long periods in residuum could preserve its old, high-$^3He/^4He$ little changed. Such residuum is thought to form the shallowest oceanic mantle lithosphere – the part that lies directly beneath the crust.

He may either be stored in U- and Th-poor olivine or orthopyroxene crystal lattices, or, at depths shallower than ~60 km, in gas bubbles in olivine crystals (Natland, 2003). Bubbles in olivine encapsulate He as they grow, and are shielded from the addition of ^4He by the surrounding U and Th-poor crystal matrix. He diffuses very slowly, so little ^4He is likely to be added in that way. Furthermore, diffusion is not driven simply by concentration gradients, but by differences in chemical potential, and these work to retain He in the bubbles. He is highly soluble in CO_2-rich bubbles but essentially insoluble in the olivine crystal matrix itself.

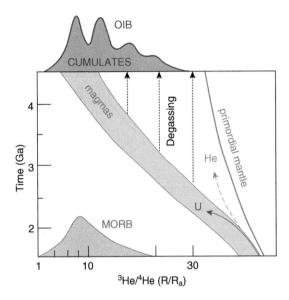

Figure 7.16 Models for the evolution of ^{3}He/^{4}He. "Primordial mantle" indicates how ^{3}He/^{4}He would evolve in a hypothetical primordial reservoir, "U" indicates how U and Th-enriched melts would evolve and "He" indicates how U and Th-poor melts would evolve. "Degassing" indicates how the ^{3}He/^{4}He ratio is preserved in gas that is removed from U and Th-bearing rocks. The MORB and OIB histograms are plotted at arbitrary positions (from Anderson, 2007b).

Olivine crystals may thus be time capsules for He, preserving old, high-^{3}He/^{4}He little changed for long periods (Fig. 7.16).

In addition to accounting for the general geochemistry of melting anomalies, recycled oceanic lithosphere thus also provides a source for variable ^{3}He/^{4}He, including the unusually high values. The ^{3}He/^{4}He from such a source will depend on its age and abundance, the quantity of trapped He, and the U and Th content of the surrounding rock, factors that are likely to vary throughout the material. Small-volume, stochastic sampling of such a source will produce Gaussian distributions of ^{3}He/^{4}He similar to those expected for MORB but with a greater range and no global correlation with other geochemical

species. The large volume-averaging that produces MORB averages variable ^{3}He/^{4}He by suppressing extreme values and reducing the range. Such a distribution is precisely what is observed (Anderson, 2000c; Meibom et al., 2003).[10]

This model for ^{3}He/^{4}He does not suffer from the problems of the plume model:

1 Very high concentrations of He are not required in any part of the mantle.

2 The observed low concentrations of He in high-^{3}He/^{4}He lavas require no special explanation – they simply reflect low concentrations in the source, in agreement with the low values of He/Ne and He/Ar observed.

3 A depleted, not enriched, source is predicted, in keeping with the observed geochemical affinities.

How can the shallow model for He be tested? He is light, mobile, does not react with other elements, and the complex degassing, metasomatism and time-integrated U and Th history of a source rock can never be completely known down to small spatial scales. As a result, its long-term evolution in all possible hosts cannot be calculated. Observational evidence is needed.

High ^{3}He/^{4}He is observed in Samoan xenoliths known to be of upper mantle origin (Poreda and Farley, 1992). It is also observed in fumaroles at Yellowstone, where 90% of the lavas are rhyolite and must result from remelting the lower continental crust (Christiansen, 2001).[11] High ^{3}He/^{4}He has also been measured in diamonds from kimberlite pipes, where they have been shielded from the addition of cosmogenic ^{3}He experienced by detrital diamonds that have lain on the surface for long periods. The He/CO_2 systematics at Hawaii also provide evidence. CO_2 is a carrier of He, and the ^{4}He/CO_2 and ^{3}He/CO_2 ratios are consistent with high-^{4}He and not low-^{3}He (Anderson, 1998b, c). Ideally, high ^{3}He/^{4}He will eventually be found in ancient, recycled oceanic lithosphere.

[10] http://www.mantleplumes.org/Statistics.html;
http://www.mantleplumes.org/HowMany.html
[11] http://www.mantleplumes.org/SLIPs.html

7.5.2 Neon (Ne) isotope ratios

The solid Earth is continually degassing Ne, which is heavier than He and does not escape to space. It is thus accumulating in the atmosphere, where it is more abundant than He. Three isotopes of Ne exist, of which only ^{21}Ne is created in significant quantities by radiogenic processes. It is produced as a result of U and Th decay, and it thus accumulates in a similar way to ^{4}He (Tables 7.3 and 7.4).

Ne isotope ratios are usually displayed on a "3-isotope plot" of ^{20}Ne/^{22}Ne vs. ^{21}Ne/^{22}Ne (Fig. 7.17) (McDougall and Honda, 1998). The ^{20}Ne/^{22}Ne contents of most rocks from the mantle lie between the atmospheric and solar values. ^{21}Ne/^{22}Ne extends to values higher than atmospheric, reflecting ingrowth of radiogenic ^{21}Ne. It is not understood why ^{20}Ne/^{22}Ne in the atmosphere should be different from that of the mantle, but it suggests that the atmosphere was not derived simply from degassing the mantle.

The Ne isotope characteristics of Hawaiian basalts and MORBs form distinctive fields. Ne in Hawaiian basalts is close to atmospheric, whereas MORB values extend to higher ^{20}Ne/^{22}Ne and ^{21}Ne/^{22}Ne. The latter indicates higher time-integrated U and Th. These distributions have been interpreted as mixing lines between the atmosphere and hypothetical "components" – "OIB source" and "MORB source" – that correspond to solar wind values plus various amounts of radiogenic ^{21}Ne.

These solar-like "components" are purely theoretical – there is no evidence that "reservoirs" of them exist. Furthermore, distributions of Ne isotopes, and Ne and He isotope ratios combined, that also match mixing of crustal and solar components, are found unassociated with any past or present melting anomaly, in groundwater from the Michigan Basin, USA, which overlies Archean basement (Castro et al., 2009). This has led to the suggestion that the "solar component" – material with high ^{20}Ne/^{22}Ne – is stored in ancient continental lithosphere.

7.5.3 The Rhenium-Osmium (Re-Os), Platinum-Osmium (Pt-Os) and Hafnium-Tungsten (Hf-W) isotope systems

Detection of a geochemical signature in basalts at melting anomalies that is unambiguously sourced from the core would provide support for

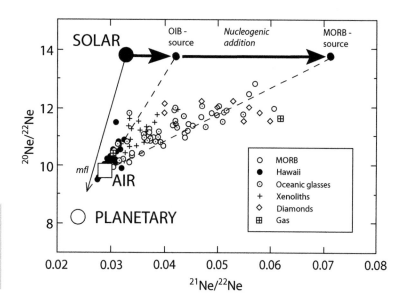

Figure 7.17 "Three-neon" plot of ^{20}Ne/^{22}Ne vs. ^{21}Ne/^{22}Ne for basaltic rocks (adapted from McDougall and Honda, 1998).

the Plume hypothesis. Isotope systems involving highly siderophile elements, which are thought to be strongly partitioned into the core, can potentially provide such tracers. The Re-Os, Pt-Os and Hf-W systems are candidates.

^{186}Os and ^{187}Os are formed by the decay of ^{190}Pt and ^{187}Re, respectively, and are usually expressed as ratios with ^{188}Os, a stable isotope (Table 7.3). Os, Pt and Re are highly siderophile and mostly migrated to the core during early planetary differentiation. The core is expected to have higher ^{186}Os/^{188}Os and ^{187}Os/^{188}Os than chondrites, whereas the mantle is expected to have sub-chondritic ratios. Supra-chondritic ratios have been observed in Hawaiian basalts and komatiites from Gorgona Island, and interpreted as a contribution from the core brought up in deep mantle plumes (Brandon et al., 1998; 2000; 2003).

This conclusion conflicts with other work, however. When the solid inner core crystallized, Os, Re and Pt partitioned between it and the liquid outer core. Because ^{190}Pt has an extremely long half-life, the data from Hawaii require that the inner core crystallized early and quickly, to allow enough time for the high-^{190}Pt liquid outer core to grow the observed ^{186}Os/^{188}Os. However, early, quick formation of the inner core is at odds with geophysical models of core evolution.

A core contribution to Hawaiian basalts also lacks the predicted support from the Hf-W isotope system. ^{182}W is produced from the decay of the short-lived ^{182}Hf isotope (Table 7.3), and its abundance is expressed relative to stable ^{184}W. W is highly siderophile, while Hf is lithophile. Thus, if a core contribution exists in Hawaiian basalts, a sub-chondritic ^{182}W/^{184}W ratio is predicted, along with a relatively high absolute abundance of W. However, lavas from Hawaii and South African kimberlites do not display these anomalies (Scherstén et al., 2004).[12]

The Os isotope ratios observed can, on the other hand, be simply explained by petrological

heterogeneities known to exist in the upper mantle (Meibom and Anderson, 2004; Meibom et al., 2002). The high Pt/Os and Re/Os ratios required to generate supra-chondritic ^{186}Os/^{188}Os and ^{187}Os/^{188}Os exist in pyroxenites and metasomatic sulfides derived from pyroxenite or peridotite melts (Luguet et al., 2008; Smith, 2003).[13] The ^{186}Os/^{188}Os vs. ^{187}Os/^{188}Os signatures found in basalts from Hawaii and komatiites could be produced in as little as ~150 Ma from peridotite containing 30–60% pyroxenite (Fig. 7.18). If isotopic evolution took place over storage times of the order of the isolation age of enriched basalts (1–2 Ga), only 5–10% pyroxenite is required.

7.6 A few highlights from melting anomalies

7.6.1 Iceland and Greenland

Iceland is a uniquely vast exposure of active oceanic volcanism, and the rocks there are uniquely variable for a mid-ocean-ridge setting. They include large amounts of enriched tholeiite, smaller-volume alkali basalts, and in addition ~10% of the lavas are andesite and rhyolite (Walker, 1963) – one of very few examples of modern calc-alkaline rocks that do not erupt in an arc or back-arc setting (Ulmer, 2001). Similar lavas are found elsewhere on the global spreading ridge, but it is the large proportion of acid and intermediate rocks and the unparalleled breadth and volume exposed above sea level that suggest a combination of circumstances at Iceland found nowhere else on Earth.

Although the main focus is usually on the mantle when considering the source of lavas at melting anomalies, the thick, complex Icelandic crust doubtless has a significant impact on the lavas erupted. The 500-km separation of the ~15-Ma isochrons requires the capture of a

[12] http://www.mantleplumes.org/Os-W.html

[13] http://www.mantleplumes.org/Pt-Os.html

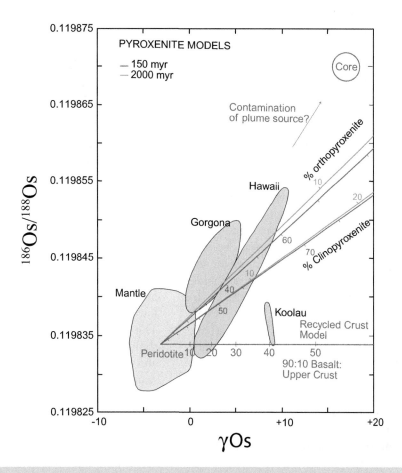

Figure 7.18 Mixing models for the generation of $^{186}Os/^{188}Os$-$^{187}Os/^{188}Os$ signatures in intraplate volcanic rocks by mixing mantle, recycled crust, pyroxene and outer core materials (Brandon et al., 2003; Smith, 2003).

microplate at least 210 km wide, which is now submerged beneath younger lavas (Section 4.5.1; Fig. 3.12; Foulger, 2006). This microplate is probably mostly oceanic crust older than ~15 Ma, but some may also be continental, possibly a southerly extension of the submerged Jan Mayen microcontinent. Evidence for this is found in elevated $^{87}Sr/^{86}Sr$ and Pb isotope ratios found in both basalts and rhyolites from the Öraefajokull volcano in south Iceland (Prestvik et al., 2001). The abundant andesites and rhyolites in the Icelandic crust may result largely from remelting the thick, subsiding crust itself,

as new mantle melts rise through it (Oskarsson et al., 1982).

There is no correlation between crustal thickness and petrology. For example, the Snaefellsnes volcanic zone, which erupts alkali olivine basalt, is underlain by crust 20–24 km thick, whereas the tholeiitic Western Volcanic Zone and the southern part of the tholeiitic Northern Volcanic Zone are underlain by crust 30–40 km thick (Fig. 3.11). There is thus no evidence that Icelandic alkali olivine basalts are produced at greater depths than tholeiites as a result of thicker overlying crust. This is consistent with their being derived from lithosphere (Section 7.2.7) (Pilet et al., 2008).

Systematic variations in rare-earth- and trace-element concentrations along the Reykjanes Ridge and northward across Iceland were among the first geochemical observations used to support a plume model (Section 7.3) (Schilling, 1973). The model initially proposed involved simple binary mixing of mantle peridotite and enriched plume material. However, it had to be progressively elaborated as new observations required greater complexity. Increasingly larger numbers of different "components" had to be introduced, including North Atlantic depleted mantle, fertile peridotite, both enriched and depleted "plume components", and an additional, special component to host the high-^{3}He/^{4}He observed (e.g., Stracke et al., 2003). Some models may involve as many as five "components" (Kempton et al., 2000; Stuart et al., 2003).

The radial geochemical symmetry, or bilateral symmetry about a north-south axis, that is predicted for a plume (Condomines et al., 1983; Schilling et al., 1983) does not exist. Instead, the geochemistry of lavas in north Iceland differs from those in the south. Major geochemical discontinuities occur across relatively minor tectonic structures presumed to be of shallow provenance, such as the 120-km-long Tjörnes Fracture Zone north of Iceland. The magnitude of geochemical anomalies in the region are small in both geographical extent and amplitude compared with those of other, smaller, proposed ridge-centered plumes, for example, the Azores (Schilling et al., 1983).

Icelandic basalts are broadly similar to those of other melting anomalies, and they can thus also be explained by inclusion in the source of recycled, subducted lithosphere. The concentrations of trace- and rare-earth elements and radiogenic isotopes, and the calculated compositions of parental melts, can be modeled using a mixture of remelted oceanic crust of Caledonian age and peridotite mantle (Breddam, 2002; Chauvel and Hemond, 2000; Foulger et al.,

2005b; Korenaga and Kelemen, 2000; McKenzie et al., 2004). The larger the crustal contribution, the more fusible the source and the greater the melt volume produced at a given temperature. The major-element systematics show that the source of Icelandic lavas lies in the depth range ~35–45 km (~1.2–1.5 GPa), which, beneath much of Iceland, is within the shallow layer that has crust-like seismic wave-speeds (Sections 5.5.1 and 7.2.7). This curious observation is as yet unexplained.

The extreme end-member possibility, that Icelandic lavas arise from a source made entirely of subducted crust, was studied in detail by Foulger et al. (2005b). Subducted slabs are made up of a variety of lithologies including sediments, altered basaltic upper crust, gabbroic lower crust, and depleted lithospheric mantle. The crustal part includes depleted and enriched tholeiite, alkali basalt and a variety of differentiates such as are found on spreading ridges and seamounts today. Primitive Icelandic tholeiite has a composition similar to gabbroic oceanic crust (Breddam, 2002; Chauvel and Hemond, 2000). The composition of Icelandic lavas was modeled using a long section of gabbroic crust cored at ODP Hole 735B on the Southwest Indian Ridge (Natland and Dick, 2001). The gabbroic section there is diverse, with rocks ranging in composition from granitic to troctolitic, and their average composition corresponds to a magnesian but non-picritic basalt. The concentrations of species such as TiO_2, Zr, Y and Nb are low, a consequence of the rocks being cumulates and thus depleted in species that partition preferentially into the liquid during crystallization (Foulger et al., 2005b).

This average resembles the composition of Icelandic lavas. The concentrations calculated for species such as Y, Zr and TiO_2 are lower than typical MORB but similar to primitive Icelandic tholeiite. The rare-earth-element patterns, positive Sr- and negative Pb anomalies, low-$\delta^{18}O$ indicative of assimilation of hydrothermally altered rocks, and elevated $^{87}Sr/^{86}Sr$ indicative of seawater alteration, match well. Mismatches, for

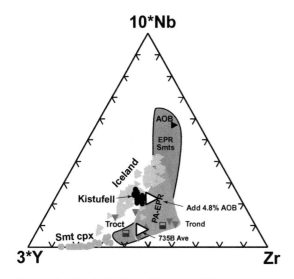

Figure 7.19 Ternary diagram showing proportions of Zr, 10*Nb and 3*Y in various rocks. Primitive tholeiites from Kistufell in central Iceland are well modeled by a melt corresponding to the average composition of gabbroic crust cored at Hole 735B, plus 4.8% of enriched material, for example, alkali olivine basalt (from Foulger et al., 2005b).

example, in Zr/Y and Nb, can be accounted for by the inclusion of <5% of enriched tholeiite, alkali basalt or silicic material, for example, rhyolite such as is also present in small quantities in oceanic crust (Fig. 7.19) (Natland, 1989; 2007; Niu et al., 2002). Isotopic ratios suggest an age of several hundred million years for the source of Icelandic basalts, consistent with oceanic crust that was young when Eurasia and Greenland collided, closing the Iapetus Ocean at ~440 Ma (Fig. 3.13) (Korenaga and Kelemen, 2000; McKenzie et al., 2004). The enriched signal could also be provided by mixing with, or assimilation of, the rhyolite that is plentiful in Icelandic crust (Carmichael and McDonald, 1961; Walker, 1963).

Such a model is consistent with the major-element characteristics of Icelandic basalts. Basalt and lherzolite have essentially the same mineralogy at moderate pressures – olivine,

orthopyroxene, clinopyroxene, and either plagioclase, spinel or garnet, depending on pressure. Only the mineral proportions and compositions are different. Thus, there will be only minor variations in the compositions of their partial melts. The major-element compositions of Icelandic basalts differ only slightly from MORB, and the relatively small changes in composition between the Reykjanes Ridge and Iceland are consistent with large increases in the amount of recycled basalt in the source.

What degree of melting would be required of a source made largely of recycled crust? At depths >~60 km, such crust would exist as the high-pressure mineral assemblage eclogite. Initial, small-degree melts of eclogite are similar to andesite (Yasuda et al., 1994; Yoder and Tilley, 1962). As the degree of melting increases, compositions progressively approach those of the parent rock. At 10–30% partial melting, liquids are ferrobasaltic with low SiO_2, and at 80% they are olivine tholeiite similar to those of Iceland (Ito and Kennedy, 1974). Thus, 60–80% melting of the original gabbro is probably required. Given the lower solidus of eclogite, such a degree of melt could be attained at approximately the same temperature as is required to produce a 20% partial melt in peridotite (Foulger and Anderson, 2005; Yaxley, 2000). The difference between 20% partial melting of eclogite, producing a ferrobasaltic liquid, and 80%, producing olivine tholeiite, may only correspond to a few tens of degrees of temperature difference at a given pressure.

It may also be that eclogite melt leaks upward and homogenizes with surrounding peridotite, and that melts from this fertilized peridotite rise to the surface. Such fertilized peridotite might have a composition similar to piclogite (Section 7.2.2) (Anderson, 1989) or pyroxenite (Smith, 2009). The latter scenario has been used to estimate a contribution to the source of only ~10% of recycled crust (Sobolev et al., 2007).

High $^3He/^4He$ isotope ratios are found throughout Iceland, including the highest non-cosmogenic ratios found at any currently active melting anomaly – up to ~42 Ra (Breddam and Kurz, 2001; Hilton et al., 1999). This material could arise from olivines stored in the recycled Caledonian lithosphere, either in dunites in the basal crustal section or in the low-U, low-Th lithospheric mantle. The highest non-cosmogenic ^{3}He/^{4}He ratios in the world, up to ~50 Ra, are found in basalts in Baffin Island (Stuart et al., 2003). These occur both in enriched tholeiites and depleted tholeiites otherwise indistinguishable from MORB. This is inconsistent with high-^{3}He/^{4}He being an intrinsic characteristic of enriched plume material.

Basalts on the Reykjanes and Kolbeinsey Ridges differ in composition from those in Iceland. They apparently arise from sources with a different composition – mainly depleted, refractory peridotite. This contrasts with the more fertile peridotite that apparently supplies many other ridges. The nearest such fertile mantle is south of the Charlie-Gibbs Fracture Zone (Fig. 2.14) (Kempton et al., 2000). Still further south, the Azores platform has basalts with yet different, typically alkalic compositions. The mantle underlying the Atlantic Ocean thus appears to be compositionally variable on a regional scale along much of its length.

7.6.2 The Emperor and Hawaiian chains

The geochemistry and style of volcanism on the Emperor and Hawaiian chains have both similarities and differences with Iceland. Comparatively little is known about the geochemistry of the seamounts because of the difficulty in sampling them. A few shallow drill holes have retrieved fresh basalts, but many samples were obtained by dredging, they only represent the tops of the volcanoes, and they are typically altered.

In common with Iceland and the Galapagos, the Emperor and Hawaiian chains contain large volumes of tholeiite. This may not always have been the case, however, because at times the magmatic rate was much lower than at present

(Figs 2.12, 3.2 and 3.4). Many lower-volume Pacific islands and seamounts are entirely alkalic.

The oldest Emperor seamounts have basalts similar to Pacific MORB, with depleted, and relatively unradiogenic compositions. Along the chain, radiogenic isotope contents and trace-element enrichment is higher in both tholeiitic and alkalic lavas (Fig. 7.20) (Regelous et al., 2003). This has led to the suggestion that the Emperor chain started at a spreading ridge that was recently subducted into the Aleutian trench (Norton, 2007).[14] Whether or not the volcanic chain started with an oceanic plateau is of interest. No plateau exists at present and there is no evidence that one has been obducted onto Kamchatka or contributed to volcanics there.[15]

The growth of recent Hawaiian volcanoes is thought to begin with the eruption of alkalic basalts, followed by a large-volume tholeiitic stage lasting several million years. That stage builds large shield volcanoes. As this dwindles, volcanism enters the post-shield stage, characterized by the formation of calderas and a return to alkalic volcanism.

There typically follows a hiatus in volcanism lasting a few million years, followed by a final stage of alkalic volcanism. This is known as the "post-shield rejuvenescent" stage because alkalic mafic lavas typically erupt onto eroded surfaces and from fissures that are oblique to the older rift systems of the shield volcanoes. Such an alkalic-tholeiitic-alkalic sequence is rare elsewhere and has only been reported from Gran Canaria, one of the Canary Islands (Hoernle and Schmincke, 1993; Walker, 1990).[16] It is unknown in Iceland, where alkalic and tholeiitic lavas tend to occur in different rift zones. At Hawaii, eruption rates during the tholeiitic stage may be several hundred times greater than during the alkalic stages.

In common with heterogeneity elsewhere, the alkalic-tholeiitic-alkalic sequence at Hawaii has been interpreted in terms of a compositionally

[14] http://www.mantleplumes.org/Hawaii2.html

[15] http://www.mantleplumes.org/Kamchatka2.html

[16] http://www.mantleplumes.org/Canary.html

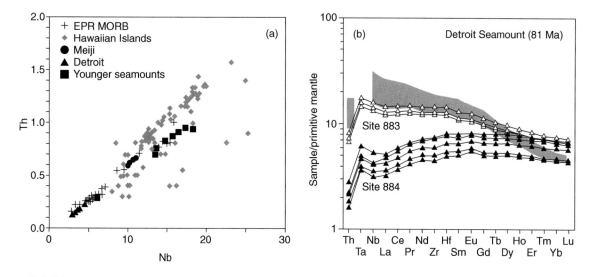

Figure 7.20 Left: Variation in Th vs. Nb for Emperor, Hawaiian and East Pacific Rise basalts. Right: Incompatible trace-element concentrations, normalized to primitive mantle composition, for tholeiitic lavas from Kilauea (gray) and Detroit seamount (lines: Sites 883 and 884) (from Regelous et al., 2003).

zoned plume, mostly enriched in its central core and surrounded by a sheath of cooler, less-enriched material. In this model, volcanism is viewed as sampling in the sequence sheath–core–sheath. The Plate hypothesis suggests that the chain is built by permissive volcanism over a lithosphere-traversing, propagating crack that extends into the partially molten seismic low-velocity zone. Phase relations predict that alkalic lavas are extractable from both shallower and deeper levels in the low-velocity zone, and tholeiitic basalt from intermediate depths (Fig. 6.6) (Presnall et al., 2010). In this model, the alkalic-tholeiitic-alkalic sequence would result from melt being drawn from progressively greater depths in the low-velocity zone until the diamond stability field was reached at ~6 GPa (~200 km). This is supported by the existence of nanodiamonds in mantle xenoliths in post-erosional alkali basalts at Salt Lake Crater, Koolau (Fig. 3.18) (Wirth and Rocholl, 2003). Volcanism ceases when the low-velocity zone is locally depleted of melt, or when the stress conditions that permit melt to escape to the surface cease to

exist. The extraordinarily high magma production rate that has existed for the last ~2 Ma at Hawaii is without precedent (Fig. 3.4) and requires an unusually large amount of pre-existing melt in the source that is currently being tapped.

Regarding the source of Hawaiian basalts, geochemical arguments have been used to argue for pyroxenite, eclogite, refertilized peridotite, harzburgite or a combination of these (e.g., Kogiso et al., 2003; Sobolev et al., 2007; Takahashi et al., 1998). There is general agreement, however, that recycled slab material is involved (Hofmann et al., 2000; Kogiso et al., 2003). One suggestion is mantle peridotite, possibly refractory harzburgitic Cretaceous lithosphere, refertilized by rising melts from recycled, subducted oceanic crust (Falloon et al., 2007a; b; Green and Falloon, 2005; Green et al., 2001; Yaxley and Green, 1998). Melts from partially melted eclogite react with peridotite to form pyroxenite, which has a similar mineralogy to peridotite but contains less olivine and is more fertile (Sobolev et al., 2007).

Several studies suggest that the inferred recycled slab retains its original structure. Specific isotopic "end-members" deduced for Hawaiian lavas thus could represent different slab materials originally present in layers, including pelagic sediments, basalts and abyssal gabbros. Because the geochemistry of slabs is altered when they pass through subduction zones, the signature of recycled, subducted oceanic lithosphere can be distinguished from the Cretaceous lithosphere on which the Hawaiian islands are built. For example, the trace-element patterns of olivine-hosted melt inclusions from Mauna Loa resemble those of layered gabbros found in ophiolites, which have high-Sr plagioclase. The major-element compositions indicate that these melts could not have been scavenged from the Cretaceous oceanic crust by through-passing magmas, but must have originated in gabbro transformed into high-pressure eclogite by passage down a subduction zone (Hofmann et al., 2000).

It has been argued that the high Ni and Si contents and low Mn, Ca and Mg of most parental Hawaiian magmas are inconsistent with a deep olivine-bearing source because this mineral, together with pyroxene, buffers both Ni and Si at lower levels. Such buffering has been attributed to rising partial melt from recycled oceanic crust reacting with the olivine in mantle peridotite to form pyroxenite. Up to half of Hawaiian magmas formed during the past 1 Ma may have been derived from such a source. The proportion of recycled oceanic crust involved is estimated to be 20–30%, compared with 10% at Iceland and up to 60% for some continental flood basalts (Sobolev et al., 2007).

An ongoing curiosity of Hawaiian geochemistry is that the products of volcanoes separated by only a few tens of kilometers, for example, Mauna Loa and Kilauea, are geochemically distinct (Ihinger, 1995). These differences are manifest in radiogenic isotopes and trace- and major elements, and are thought to indicate spatial variations in the composition of the source. This distinctive geochemistry is thought to characterize two lines of volcanoes that form distinct, curvilinear, parallel trends known as the Loa and Kea trends (Fig. 3.18).

Concentric compositional zoning about the centre of a cylindrical plume conduit has been suggested to account for these variations (e.g., DePaolo et al., 2001). However, detailed analysis tends to undermine models of a coherent pattern. Kea- and Loa-like major- and trace-element compositions are distinct on the scale of single lava flows, but both are present in olivine-hosted melt inclusions from both "trends" (Ren et al., 2005). The geochemical structure of the postulated Hawaiian plume has thus evolved with time from one of concentric zoning, through a bilaterally zoned bundle of filaments, to a streaky model (Fig. 7.21).

7.6.3 Flood basalts

Flood basalts occur in diverse tectonic settings, and vary by orders of magnitude in surface area, volume and eruption rate (Table 3.1). They are also petrologically and geochemically variable, both throughout a single province and between provinces. They have been attributed to plume heads, but their variability often requires fundamental modifications of this hypothesis in order to match the observations.

Many flood basalts are dominated volumetrically by enriched tholeiite, with subordinate volumes of depleted tholeiite and evolved lavas, for example, rhyolites. The Columbia River Basalts are of this kind. The enriched character of the main lavas has been attributed to assimilation of continental mantle lithosphere or lower crust into asthenospheric melt on its journey to the surface (Hooper et al., 2007).[17] A similar model has been proposed for the Deccan Traps (Sheth, 2005a) and the Scottish Tertiary Basalts, whose enriched character has been attributed to underlying Lewisian gneisses. Quantification of

---

[17] http://www.mantleplumes.org/LowerCrust.html

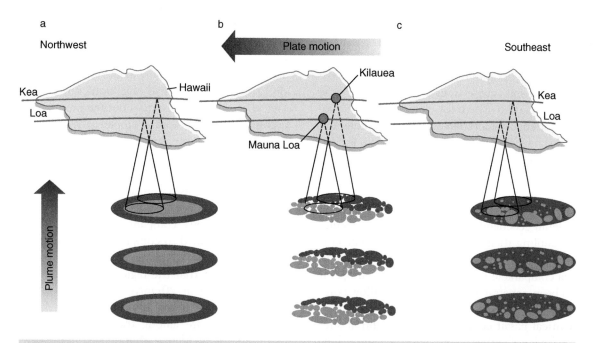

Figure 7.21 The disintegration of the Hawaiian plume – models of the distribution of Kea and Loa source materials beneath the Big Island. (a) An early, concentrically zoned model; (b) a later bilaterally zoned "bundle of filaments" model; and (c) a recent, streaky model (Ren et al., 2005). Light gray – Loa source, Dark gray – Kea source (from Herzberg, 2005).

the latter model has been attempted but was unsuccessful because very different amounts of contamination are required to explain the major elements, trace elements, and Sr-, Nd- and Pb isotopes (Thompson et al., 1982).

The Paraná Basalts have high $^{87}Sr/^{86}Sr$ and low $^{143}Nd/^{144}Nd$ and these characteristics, along with the trace-element patterns, defy modeling as asthenospheric melts altogether, even if contamination by sub-continental lithospheric mantle or crust is allowed. Furthermore, the geochemistry of the Paraná basalts is unlike any associated with the Rio Grand Rise, the Walvis Ridge, or the Tristan melting anomaly. As a result, Paraná geochemistry is concluded to be intrinsic to the source, and it is inescapable that they derived largely or wholly from remelting the continental mantle lithosphere (Comin-Chiaramonti et al., 1997; Peate, 1997; Piccirillo,

et al. 1989; Turner et al., 1996).[18] The presence of a DUPAL signature further suggests that lower continental crust, delaminated when the South Atlantic opened, may also have contributed to both the Paraná and the Etendeka basalts (Regelous et al., 2009). The geochemistry of the Karoo volcanic province can also be explained entirely as remelting of the sub-continental mantle lithosphere (Jourdan et al., 2007).[19]

The Siberian Traps are unusual, if not unique among continental flood basalts, in having enriched trace-element patterns but almost exclusively MORB-like Nb-Y-Zr systematics (Fitton, 2007). Some lavas have DUPAL isotopic characteristics (Section 7.2.2), which

[18] http://www.mantleplumes.org/Parana.html; http://www.mantleplumes.org/SAtlantic.html
[19] http://www.mantleplumes.org/Karoo2.html

suggests involvement of eclogite in the source, and these alternate with non-DUPAL basalts in the lava pile (Ivanov, 2007). Rare meimechites are found – porphyritic lavas with MgO >18 wt% that may come from a hot source or one rich in volatiles. The trace-element signatures of the enriched basalts have features in common with subduction-related basalts, and it has been suggested that they derived from subducted slabs that accumulated in the transition zone (Ivanov, 2007).

Very few oceanic plateaus have been studied in detail geochemically because of the difficulty of sampling them. The Ontong Java Plateau is one of the best studied, but although it has an area comparable with that of western Europe, it has been penetrated by only eight shallow boreholes (Fitton et al., 2004a, b). Remarkably, near-identical basalt compositions are found at sites hundreds of kilometers apart. Ontong Java basalts are typically only slightly enriched. Their compositions can be fit by a high (~30%) degree of partial melt in a hot, standard peridotite mantle source (Fitton and Godard, 2004) or by near-complete melting of an eclogite-rich source at a spreading ridge (Korenaga, 2005).

7.6.4 Back-arc regions

Crustal extension and volcanism occur in back-arc regions in response to slab roll-back (Doglioni et al., 1999; Hamilton, 2007b). Extension rates may be as much as 15 cm/year, requiring a large melt-extraction rate. Seismic images show some of the most strongly depressed wave-speeds found anywhere in the upper mantle (Nakanishi and Anderson, 1984). Nevertheless, volcanism in some of these localities has been attributed to deep mantle plumes largely, if not entirely, on the basis of geochemistry. Examples include Japan (Nakamura et al., 1985), the Manus back-arc basin (Macpherson et al., 1998) and Italy (Bell et al., 2004; Gasperini et al., 2002; Vollmer, 1989).

It is difficult to imagine a less plausible environment for a mantle plume, which would have

to puncture completely through a cold, downgoing slab, possibly as thick as ~100 km. For rapidly subducting slabs, such a puncture would also have to be continually maintained. In addition to these mechanical objections, the petrological evidence at best does not require plumes and may preclude them. The petrochemistry of basalts from Kyushu, Japan, are consistent with relatively low-temperature, low-pressure extraction (Mashima, 2005).[20] The basalts of the Manus back-arc basin were attributed to a plume solely on the basis of $^3He/^4He$ ratios of 12.2 Ra, hardly anomalous compared with mid-ocean ridges, but assumed to comprise unquestionable evidence for a deep-mantle plume. This disregarded the fact that the region is blocked by a slab subducting at 20 cm/year, and supporting seismic data are absent. Indeed, these helium measurements are among the strongest direct observational evidence that exists that high $^3He/^4He$ ratios can derive from the shallow upper mantle (Anderson, 2000b, c).

The attribution of Plio-Quaternary volcanism in Italy to one or more deep mantle plumes is at best not required and in reality is implausible (Avanzinelli et al., 2009; Peccerillo and Lustrino, 2005). Arguments in support of a plume explanation are based on the assumption that geochemical "end-members" such as HIMU and FOZO derive from the lower mantle (e.g., Cadoux et al., 2007). Such an assumption is untenable. Geochemical "end member" signatures reflect contributions from subducted, near-surface materials such as oceanic crust, oceanic mantle lithosphere, and delaminated lower crust and continental mantle lithosphere. There is no requirement that they derive from the lower mantle and it is more likely that they derive their geochemistry from materials in the shallow mantle (e.g., Stracke et al., 2005). Furthermore, volcanics similar to those found in Italy are widespread in regions that are untenable as plume localities. Carbonatitic lithologies, sometimes assumed to arise from plumes, can be

[20] http://www.mantleplumes.org/Japan1.html

explained by remelting of, or contamination by, carbonate sediments (D'Orazio et al., 2007).

Plio-Quaternary Italian volcanism forms discrete, compositionally distinct magmatic provinces that are separated by major structural boundaries (Lustrino and Wilson, 2007; Peccerillo, 1999; Peccerillo and Lustrino, 2005).[21] This is readily explained by the tapping of melt from a heterogeneous shallow mantle. Such heterogeneity is inevitable in this setting. The upper mantle beneath the Italian region has experienced a ~300-Ma history of complex geodynamic events including multiple ocean-basin openings alternating with subduction. These include consumption of the proto-Tethys Ocean and opening and closure of a new Tethys Ocean. Compression between Africa and Eurasia likely involved the subduction of continental crust, and complex deformations included the rotation of large blocks and formation of extensional oceanic sub-basins in response to southeastward retreat of the subduction hinge. Some volcanoes, including Etna, lie on major tectonic structures such as extensional faults, which clearly control the locus of volcanism. Shallow mantle heterogeneity is detected by seismic observations (Panza et al., 2007). The near-unparalleled diversity in composition of Italian Plio-Quaternary volcanics can be explained by recyling of Oligocene to present subducted African lithosphere beneath the Italian peninsula, with Sardinia requiring a somewhat older source. There is no necessity to invoke input from the deep mantle (Peccerillo, 2005).

7.7 Plume variants

The modern Plume hypothesis envisages plumes as raising up deep mantle material that is enriched and more nearly primordial than the upper mantle, and which contains high-^{3}He/^{4}He and Ne with "solar" characteristics, i.e., high ^{20}Ne/^{22}Ne. Other materials found in lavas from melting anomalies are also assumed to reside in the deep mantle, for example, subducted slabs. Upper-mantle material is expected to be entrained en route, and diverse lavas to result from the heterogeneity brought about by these many contributions. Nevertheless, despite the flexibility of this versatile model, observations have still been made that require additional model complexity to account for them. A few examples are now given.

(a) Iceland and Greenland

The depleted nature of much Icelandic basalt conflicts with the expectation that a lower-mantle plume source is enriched. To account for this, it has been proposed that the Icelandic plume tail is surrounded by a sheath of depleted upper-mantle material (Fitton et al., 1997; Kempton et al., 2000; Saunders et al., 1997). The termination in the transition zone of the seismic anomaly beneath Iceland is at odds with derivation of high-^{3}He/^{4}He from the lower mantle, and it has been suggested that the postulated plume rises from the base of the upper mantle and sucks up some lower mantle material from beneath. This, in turn, conflicts with the postulated restriction of the high-^{3}He/^{4}He reservoir to thin layer at the core-mantle boundary, necessitated by mass balance calculations (Section 7.5.1).

The later discovery of ^{3}He/^{4}He up to ~50 Ra in basalts, otherwise indistinguishable from MORB in west Greenland and Baffin Island, required an explanation for why it apparently does not reside in enriched plume material. It was thus suggested that the lower-mantle part of the envisaged plume is compositionally distinct from the upper mantle solely in He, which arises from the deep lower mantle or even the core, and is decoupled from everything else (Fitton et al., 2008).

[21] http://www.mantleplumes.org/Italy.html; http://www.mantleplumes.org/Italy2.html

Voluminous cumulate picrites in West Greenland have been explained by several different models, each adapted to the observations current at the time. When it was thought that the postulated Icelandic plume impacted beneath East Greenland, a separate, short-lived "Davis Strait plume" was proposed (McKenzie and Bickle, 1988). This model was abandoned when it was shown that if the postulated plume were fixed with respect to others in the Indo-Atlantic ocean, then it would have lain closer to the West Greenland picrites at the time they erupted (Fig. 3.10). Nevertheless, the chronological development of volcanism in the North Atlantic does not follow the pattern predicted by the Plume hypothesis and a pulsing plume has been suggested to account for this (Ito, 2001; Saunders et al., 1997). There is still speculation regarding the time of arrival of the postulated Icelandic plume head, even with the extreme proposal that it is represented by the Siberian Traps (Saunders et al., 2005), despite the fact that no time-progressive volcanic trail connects Siberia to the North Atlantic region.

(b) Hawaii

The distinct geochemistry of the Loa and Kea trends in Hawaii have inspired a series of models to explain them (Fig. 7.21) (Herzberg, 2005). Originally, a vertically continuous, concentric plume was proposed (DePaolo et al., 2001). Subsequent observations required instead a plume with geochemical heterogeneities ordered like the strands in a bunch of spaghetti (Abouchami et al., 2005). Still later observations necessitated quasi-random vertical and lateral heterogeneities (Ren et al., 2005).

(c) The Canary Islands

Cyclic variations in the degree of SiO_2 saturation and melt productivity in volcanoes in the Canary Islands has been explained by the rising of discrete blobs of plume material. In this model, the upper margin of the rising blob is sampled first, followed by the core, and lastly the lower margin. Small-volume, undersaturated melts are derived from the cooler margins, and larger-degree tholeiitic melts from the hotter blob centers. Volcanic hiatuses between and within cycles represent periods when residual blob material or cooler entrained shallow mantle fill the melting zone (Hoernle and Schmincke, 1993).

(d) The Paraná Basalt

The geochemistry of the Paraná basalts cannot be explained either by asthenospheric melt or plume material similar to any at the Rio Grand Rise, the Walvis Ridge or the Tristan melting anomaly. The basalts have thus been attributed to the continental lithospheric mantle, postulated to have been heated and melted by a "passive", "stalled" or "incubating" hot plume head that did not spawn off "baby plumes" or contribute melt itself (Ernst and Buchan, 2003; Peate, 1997; Turner et al., 1996).

(e) The ubiquity of enriched basalts

Enriched basalts are commonly attributed to plumes, and yet such lavas are widespread throughout the oceans and continents where there is no other indication or expectation of plumes. These basalts have been attributed to lateral flow from distant plumes, for example, enriched basalts on the East Pacific Rise have been attributed to lateral flow of ~5000 km from a plume postulated to underlie Hawaii (Niu et al., 2002). A generalization of this concept suggests that there is a ubiquitous layer of plume material in the shallow mantle that is transiently sampled by migrating ridges and other volcanic regions, which can thus erupt plume material without needing to overlie one (Davies, 1999; Yamamoto et al., 2007a; b).

7.8 Discussion

Where and what is the source of enriched basalts? The leap from basalt geochemistry to source composition is usually highly non-unique. Further decoupled from the observations is the nature of the geological materials that combined to make that source, and still more difficult to identify is the physical location in the mantle. Traditional models assume that the ultimate sources of MORB and enriched plume basalts reside in separate, compositionally distinct and relatively homogenous layers – the upper and lower mantle – which are little mixed by convection. These layers would have to be tapped in order to extract the lavas erupted, which immediately presents a conceptual difficulty. The preservation of long-term, compositionally distinct regions in the Earth's mantle by layered convection is at odds with the continuous exchange of material via subducting slabs and rising plumes.

Such a model also sits awkwardly with geological observations. Thousands of small seamounts that litter the ocean floor, so-called "hot lines", for example, Cameroon, and widespread small-volume continental volcanism cannot reasonably be attributed to plumes and must derive from the upper mantle (Aldanmaz et al., 2006; Fitton, 2007; Hofmann, 1997). Yet their chemistry is essentially the same as provinces such as Hawaii, and localities where plumes have been proposed solely on the basis of geochemistry, for example, St Helena. Lavas at the "petit spots" and in the East African Rift valley are similar to those in Pacific island chains and to enriched basalts found on spreading ridges (Natland, 2007).

Enriched basalts cannot be considered to be a diagnostic feature of plumes, but this assumption is nevertheless widespread along with the lesser, but equally invalid, assumption that enriched geochemistry strengthens the case for a plume. Assumption-driven nomenclature encourages such views to persist and hinders the evolution and improvement of models. Enriched basalts are commonly labeled "ocean-island basalt" (OIB), or "OIB-like", a term that implies a fundamental association with oceanic islands that they do not have. The fact that essentially identical rocks are found on mid-ocean ridges is obscured by labeling those rocks "E-MORB".

The perceived chemical uniformity of MORB is partly an artifact of eliminating all diverse lavas from the MORB set. This perceived uniformity may, in turn, encourage the common assumption that the upper mantle itself is homogenous (Meibom and Anderson, 2004; Meibom et al., 2002). Other lava compositions are then viewed as anomalies requiring anomalous sources. Nevertheless, the phase relationships for common mantle minerals are such that relatively homogeneous magmas are produced by source rocks with extreme variations in mineral proportions. The relevant phases need only be present – they need not have any particular relative abundances. The study of basalt petrogenesis tends to rely heavily on minor and trace elements and isotope ratios. However, these elements make up only a minute percentage of the mantle, they are relatively mobile, and over-emphasis of them in interpretations may camouflage more robust results provided by the major elements that make up a large majority of the rocks.

It is well established that enriched material is distributed throughout the upper mantle. The bulk of the upper mantle may comprise relatively depleted material that has the most influence on the major-element chemical signature of MORB. However, it probably also contains abundant enriched material as streaks or blobs of various sizes ranging from the very small to the scale of a subducted slab or delaminate, i.e., tens of kilometers (Meibom and Anderson, 2004; Meibom et al., 2002). Regional variations in composition within and between the Pacific and Atlantic oceans testify to source variations on even larger scales.

Upper mantle heterogeneity is maintained by the convection in which plate tectonics is an integral part. At ridges, basaltic melt is extracted to form oceanic crust, leaving behind depleted mantle. At subduction zones, this basaltic crust, along with the underlying oceanic mantle lithosphere, is re-injected. The basaltic part is not uniform MORB but a stack of layers that includes sediments, lava flows, dikes, differentiates associated with processes in shallow magma chambers, and gabbro. These have diverse petrologies and geochemistry. The gabbroic part comprises cumulates in which the strongly incompatible elements that are abundant in basalt have low concentrations and different ratios. Recycled oceanic crust is thus a diverse source. Eclogites and depleted harzburgites in the mantle likely arise from subducted basaltic crust and melt-stripped residuum within ancient and deeply emplaced subducted slabs.

Mantle wedge material is fluxed with volatiles from the down-going slab, adding to diversity. Subduction injects $\sim 20 \, km^3 \, a^{-1}$ of oceanic crust into the upper mantle, an amount that far exceeds the total $\sim 1 \, km^3 \, a^{-1}$ of magma produced at melting anomalies (Fig. 3.2). The eclogite to which subducted crust transforms at $\sim 60 \, km$ depth is commonly assumed to be denser than ambient mantle and to consequently sink unless strongly reheated. However, passive mantle convection upwellings, for example, beneath ridges, rise sufficiently quickly that eclogite blocks on a scale of the oceanic crust (a few kilometers) can entrain and rise with them (Korenaga, 2005).

Sub-continental lithospheric mantle may be very old – as old as the crust above it. The South African craton, for example, is 3–3.5 Ga. Depleted and enriched source material could co-exist for long periods, each developing its own distinctive isotopic signature, without being widely separated in space. It takes $\sim 1 \, Ga$ for the entire surface of the Earth to be swept by the migrating spreading ridge system, so even shallow heterogeneities can survive for long periods. The diversity of recycled near-surface materials is such that they are sufficient to explain essentially all enriched-basalt geochemistry.

What is the proportion of recycled material in enriched basalts? Knowing this is important to calculations of whether magma volumes can be accounted for simply by increased source fusibility, without the need for elevated temperature. Estimates of 10–60% have been made for MORB and enriched oceanic basalts on the basis of petrological models of melting and resorption of eclogite with peridotite (Sobolev et al., 2007). Qualitative models aimed at explaining large magma volumes have suggested sources containing up to 100% of eclogite (Cordery et al., 1997; Foulger et al., 2005b; Korenaga, 2005). Enriched material such as eclogite is the most fusible and will begin to melt first and be more completely melted than peridotite at any temperature below the peridotite liquidus. It is commonly assumed that a fusible source constituent does not affect melt volumes, and that variations in volume are dependent on source temperature only (e.g., Fitton et al., 1997). However, this clearly cannot be the case.

Seismology is appealed to for information on the destination and present location of subducted slabs. In the upper mantle, slabs have high wave-speeds, which are conventionally color-coded blue in tomographic maps and cross-sections. Low-wave-speed regions are conventionally color-coded red. Although the interpretation of seismic wave-speeds is ambiguous, a "blueium and redium" interpretive approach has evolved (Section 5.1.2). Blue is assumed to indicate cold, subducted lithosphere, and red is assumed to correspond to hot, rising, and even He-rich material. This approach is insupportable and flies in the face of well-known facts that rule it out. For example, subducted slabs thermally re-equilibrate over a similar period to the time it takes them to sink to the base of the upper mantle (Toksöz et al., 1971) and thus high-wave-speed anomalies at the core-mantle boundary cannot be slabs that have subducted from the surface and are still cold. Modest compositional variations can affect seismic wave-speeds as much as large temperature variations (Chen et al., 1996).[22] Despite this, seismic tomography has been used to support the notion that slabs in general sink to the base of the lower mantle and accumulate there, ready to be transported back up in plumes.

Nothing in the geochemistry of lavas from melting anomalies requires a source that rose from great depth. Both the Plate and Plume hypotheses attribute enriched geochemical signatures to recycled, near-surface materials in the source. The question thus boils down to whether this material sank to the core-mantle boundary, whence it was swept back up to the surface in rising plumes, or whether it was simply mined from the upper few hundred kilometers of the mantle by passive melt extraction. Geochemical evidence from both Iceland and Hawaii suggests direct links between observed petrologies and particular layers in subducted slabs, for example, the gabbroic lower crust (Chauvel and Hemond, 2000; Foulger et al., 2005b). In the Plume hypothesis, it is thus required that slabs are subducted 3000 km down to the core-mantle boundary, and transported back up to the surface in hot plumes, a round trip of 6000 km, with their structure remaining intact.

The only geochemical tracer that is argued to be an unambiguous lower-mantle tracer is high-^{3}He/^{4}He isotope ratios. This is seriously contested. Interestingly, an extraordinary diversity of origins has been proposed for high-^{3}He/^{4}He, including meteorites, cosmogenic ^{3}He creation, continental lithospheric mantle, recycled slabs in the upper mantle, the lower mantle, the core-mantle boundary and even the core itself. He is highly mobile, and ^{3}He/^{4}He correlates globally with little else, limiting possible research approaches. High-^{3}He/^{4}He has not been found at all melting anomalies, but is most common at those whose melts have the greatest similarities to MORB. Estimates of the mass flux from the lower mantle predicted by the assumption that

high-^{3}He/^{4}He comes from the lower mantle are extremely small. Of all aspects of the geochemistry of melting anomalies, the origin of high-^{3}He/^{4}He is perhaps the most difficult to address and the most urgently in need of significant new data.

7.9 Exercises for the student

1 Are there any unequivocal geochemical tracers of lower-mantle material?

2 What is the definition of "ocean-island basalt"?

3 Where and what is the source of enriched basalts?

4 Do alkali basalts arise from metasomatic veins?

5 Can the quantities of recycled material in the sources of melting anomalies be determined?

6 Can different types of recycled material in the upper mantle be identified and their distributions mapped?

7 How do melts react with their surroundings as they rise through the mantle?

8 How does source composition affect magma volumes?

9 What is the scale of source regions that are compositionally distinct?

10 How can the controversy regarding the origin of high-^{3}He/^{4}He be solved?

11 Isotopic fractionation is most extreme where there is a large difference in atomic weight, and this is most extreme in light elements. Are there any geological processes that could fractionate ^{3}He and ^{4}He?

12 Are any biological isotope-fractionating processes of significance to basalt geochemistry?

13 How can the eruptive sequence alkalic-tholeiitic-alkalic at Hawaii be reconciled with the Plate model?

14 What is the origin of the nanodiamonds in Hawaii?

15 Are the Loa and Kea trends at Hawaii mythology?

[22] http://www.mantleplumes.org/ InterpretingSeismicVelocity.html

16 What is the petrology of the Icelandic lower crust and upper mantle?

17 Mineralogical phase diagrams require that Icelandic basalts are extracted from a depth of ~35–45 km, and yet beneath much of Iceland this lies within the layer that has seismic wave-speeds characteristic of crust. How can these two observations be reconciled?

18 Why is alkali basalt erupted through thinner crust than tholeiite at Iceland?

19 Why is volcanism at Samoa almost exclusively alkalic?

20 What is the relationship between structure and volcanism in East Africa?

8

Synthesis

En hvatki er missagt er í fræðum þessum, þá er skylt að hafa það heldur, er sannara reynist.[1]

–Ari Thorgilsson (1067–1148)

8.1 Introduction

8.1.1 In the beginning

At the beginning of this book, many questions were asked. Why was the Plume hypothesis proposed? What is a plume, and how is one defined? How many are there thought to be, and where are they? What predictions does the hypothesis make, how may they be tested, and is the hypothesis in truth unfalsifiable as a practical matter? Can volcanic provinces as different as Iceland and Hawaii be rooted in the same cause? What is the alternative "Plate" hypothesis, and can the diversity of processes it proposes for volcanism be justified, when compared to the simple Plume hypothesis? What are the predictions of the Plate hypothesis and how, in turn, may these be tested? Does the Plate hypothesis explain observations better than the Plume hypothesis, and what is the way forward?

It is not the objective of this book to answer all these questions for everyone. A book that

aspires to answer all questions discourages independent thought. Rather, it seeks to assemble, in a critical manner, a compendium of observations to assist each reader to think things through independently and develop a personal view. It is a matter of style whether a scientist interprets observations to support a favored hypothesis, or to highlight misfits and follow a path of criticism and challenge to current ideas. However, the facts themselves are not a matter of choice. They are things that everyone can all potentially agree on. Answers to some of the questions posed above, and others, are left as an exercise for the reader.

Terminology that pre-supposes a cause or an outcome is a hindrance to free thought. For this reason, the term "hot spot" is not used in this book. Whether or not a volcanic province is supplied by an anomalously hot source is a matter for investigation, not presumption. Whether a volcanic province is well-described as a "spot" is also a valid question. Perhaps the only truly unquestionable observation from (most of) the regions of interest is the presence of volcanism. How much is also a question. Only a few (although not zero) "hot spots" have been proposed where there is no volcanism.

The term "melting anomaly" has been used in this book. Even this term is unsatisfactory because it implies a distinction between the volcanic provinces whose ultimate origin is disputed, for example, the Azores, and other

---

[1] Wherever there is a mistake in this book, take instead that which is more correct.

Plates vs. Plumes: A Geological Controversy, 1st edition. By Gillian R. Foulger.
Published 2010 by Blackwell Publishing Ltd.

volanic provinces. Most of the Earth's volcanism occurs along spreading ridges and subduction zones, and although details of processes there are still debated, the fundamental cause of that volcanism is not. If the volcanic provinces referred to herein as melting anomalies are merely part of a continuum of the Earth's volcanism, then the term "anomaly" is also misleading. How voluminous must volcanism on a spreading ridge be, before it is an "anomaly"? Is Iceland an anomaly, or merely a statistical end member? When is a phenomenon an "anomaly" and not an end member? A natural break in a continuum, for example, of volume, is needed before distinct and separate categories can reasonably be proposed. It is not clear that these exist among the Earth's volcanic provinces.

8.1.2 The rise and fall of plumes

A primary prediction of the Plume hypothesis is kilometer-scale, domal precursory uplift a few million years prior to the onset of flood volcanism (Campbell, 2007; Farnetani and Richards, 1994). It has been suggested that this is the single most important diagnostic that can unequivocally identify a plume source. The Plate hypothesis, in contrast, does not require such uplift, and nor does it require the same pattern of vertical motion to accompany all flood basalt eruptions. Instead, vertical motions are expected to depend on the cause of flood volcanism. In the case of continental break-up, many kilometers of shoulder uplift along the new ocean basin margins are expected. Loss of thickened subcontinental lithospheric mantle by gravitational instability or delamination will produce precursory subsidence, not precursory uplift,[2] and the melt extraction process itself will produce large uplifts (Fig. 2.4).

Early attempts to assess the vertical motions at some localities initially supported the predictions of the Plume hypothesis, for example,

around the North Atlantic, and at the Paraná Basalts, the Deccan Traps and the Emeishan flood basalt (Cox, 1989; He et al., 2006). However, more critical scrutiny eroded and even eliminated these initial positive results rather than reinforced them. The pattern and rate of uplift in and around the North Atlantic was found to be inconsistent with the timing, shape and rates predicted, and radial drainage patterns reported from the Paraná Basalts and Deccan Traps were found to have developed long after the flood volcanism (Anell et al., 2009; Hegarty et al., 1995; Sheth, 2007a),[3] The uplift claimed to have occurred prior to eruption of the Emeishan basalts was found to have been based on mis-identification of basal marine hydromagmatic deposits as continental alluvial fan conglomerate formed in response to pre-volcanic domal uplift and erosion (Ukstins Peate and Bryan, 2008),[4] Why did these domal uplifts all collapse?

Particularly difficult for the Plume hypothesis was the discovery that at the world's largest flood basalts – the Ontong Java Plateau and the Siberian Traps – uplift was either much smaller than predicted, or subsidence preceded volcanism (Czamanske et al., 1998).[5] On the other hand, an apparently plume-like uplift has been reported from the back-arc volcanic area of Anatolia, where a plume cannot have risen because of the blocking down-going slab (Keskin, 2007).[6] Is the Columbia River Basalt Group a form of back-arc volcanism? The plume-like uplift postulated to have occurred there has also been challenged (Hales et al., 2005)?[7]

To confuse the issue still further are the many major vertical motions for which there is no obvious cause. The Bermuda Rise, East Greenland, the Columbia Plateau, the Brazilian

[2] http://www.mantleplumes.org/LithGravInstab.html

[3] http://www.mantleplumes.org/DeccanUplift.html
[4] http://www.mantleplumes.org/Emeishan.html
[5] http://www.mantleplumes.org/Siberia.html;
http://www.mantleplumes.org/OJ_Impact.html
[6] http://www.mantleplumes.org/Anatolia.html
[7] http://www.mantleplumes.org/CRBDelam.html

coastal plain and southern Africa have all experienced large regional uplifts at times when no major, local volcanic or tectonic events apparently occurred.[8] The "egg box" or "piano key" deformation patterns observed around Britain currently have no ready explanation (Anell et al., 2009; Stoker et al., 2005). Such short range uplifts and subsidences may be a more common phenomenon than hitherto thought. A wide range of processes can cause uplift, but uplifts are also observed that seem to defy our current powers of explanation.

To what accuracy can uplift be estimated? In continental regions where sequences immediately below the earliest lavas are unconformable, incomplete or inaccessible, there may be little evidence to work on. Markers such as basal conglomerates may indicate local conditions only. River incision and conglomerates do not necessarily require uplift – they are most likely to simply reflect climate change (Bull, 1991). Easier to study are oceanic plateaus and rifted volcanic margins. These are less prone to the effects of erosion and tectonics that can obfuscate the vertical motions of land areas. Subsidence of the sea floor is expected to follow a relatively simple, smooth pattern. The temperature variations expected to be associated with plumes cause deviations from this, which can be measured. However, study of numerous plateaus has confirmed that anomalous subsidence following eruption is rare – of some 14 plateaus studied, this has only been detected for one, and then only to a modest degree (Clift, 2005).[9] As the subject stands today, at no oceanic plateau or continental flood basalt anywhere on Earth do the observations require regional, kilometer-scale, domal precursory uplift.

8.1.3 The ins and outs: Volume

The existence of volcanic rocks at a melting anomaly is the primary observation that requires

[8] http://www.mantleplumes.org/Bermuda.html
[9] http://www.mantleplumes.org/SedTemp.html

explanation. Everything else is postulated, secondary and potentially disputable. The second most significant observation is arguably quantity. The volumes of melting anomalies range over about eight orders of magnitude (Fig. 3.2).

Are different processes required to explain volcanic fields of different sizes? In particular, is a special process needed to explain the largest? If so, one or more natural breaks might be expected in the size distribution, and possibly also other parameters such as rate and duration of eruption. No such breaks have been shown to occur, but how carefully has anyone looked? The Plate hypothesis predicts a wide range of volumes, areas, time-distributions and rates. Not all flood basalts are expected to be followed by the building of a time-progressive volcanic chain, and the variation in volcanic rate with time is not predicted to decrease monotonically. The standard space-time eruption pattern predicted by the Plume hypothesis is not expected.

Plumes cannot explain large-volume flood basalts erupted over thick lithosphere, for example, at the Siberian Traps, the Paraná Basalts and southern Africa (Cordery et al., 1997). A hot diapir, rising and stalling beneath thick lithosphere cannot decompress sufficiently to produce the volumes observed on the same time-scale as the eruptions. It has been claimed that only plumes can explain the huge volumes and exceptional rates of eruption of flood basalts. However, it may be that these high volumes and rates are, ironically, the very observations that invalidate plumes and demand, instead, melt accumulation and storage on a longer time-scale than eruption (Silver et al., 2006). If so, then the mechanism that releases this melt need not be directly related to the melt-accumulation process. It could instead result from the reversal of a stress field inclined to contain melt to one inclined to release it.

Further evidence that melt reservoirs must exist comes from the seismic low-velocity zone, which requires partial melt, from the need for metasomatization of the mantle lithophere to explain the geochemistry of continental basalts

(Pilet et al., 2008),[10] and from small volcanic fields such as the "petit spots" that form on the up-warp of the sea floor as it approaches the Japan Trench (Hirano et al., 2006). The intriguing observation that larger seamounts are built on thicker oceanic lithosphere, including the volcanic edifices of the Hawaiian archipelago, is the opposite of what is predicted by the Plume hypothesis (Hillier, 2007). This cannot merely be because thicker, stronger lithosphere is capable of supporting larger edifices, because that does not explain why more melt is erupted. This pattern is instead consistent with deeper cracks penetrating further into the surface conductive layer, tapping broader melt-storage regions, and releasing hotter pre-existing partial melt.

Many major melting anomalies occur where ridge tectonics are unusually complex. Iceland is a diffuse spreading plate boundary with multiple parallel spreading ridges, propagators and microplates (Fig. 3.12).[11] The Galapagos islands lie between two families of propagating ridges migrating in opposite directions, in a region where major variations in the direction of subduction induces extension in the plate (Fig. 3.16). The Easter chain emanates from the southern edge of a rotating oceanic microplate (Fig. 2.10). In the continents volcanism is clearly associated with re-activated lithospheric structures (Bailey and Woolley, 2005; Comin-Chiaramonti et al., 1999; Liégeois et al., 2005).[12] Some volcanism in regions of clear tectonic complexity has been attributed to lateral flow from distant plumes. An extreme view is that most of the asthenosphere comprises material supplied by plumes throughout geological time. In this view, it is "normal" mantle that is fundamentally anomalous (Yamamoto et al.,

2007a, b). In the Plate hypothesis anomalously hot mantle is not piped into productive parts of the spreading ridge – the excess magmatism results from local, fertile source material. If the largest flood basalts do not result from plumes, then some other process or processes must be capable of generating them, and if so, such a process or processes can generate them all. The Earth does not produce things she does not need, and the question then reduces to whether the Earth needs plumes.

8.1.4 The when and where

The Plume hypothesis predicts that when flood volcanism wanes, small-volume "plume tail" volcanism takes over and builds a time-progressive volcanic chain on the plate moving overhead. It is thus curious to note how rarely an association between a flood basalt and a time-progressive volcanic chain is observed (Tables 1.4 and 8.1). Only 3 such associations out of 49 melting anomalies are considered certain by Courtillot et al. (2003), and only an additional 11 are considered to be possible.

Much volcanism fits neither the "plume head" nor the "plume tail" ideal. This is well-illustrated by the broad swathe of volcanic edifices and products that carpet the Pacific Ocean from the

Table 8.1 Numbers of currently active melting anomalies listed by Courtillot et al. (2003) as having flood basalts and/or time-progressive chains, or neither.

System	# certain	# questionable
Flood basalt and chain	3	11
Chain only	10	12
Flood basalt only	1	0
Neither	8	4

[10] http://www.mantleplumes.org/MetasomaticOIB.html
[11] http://www.mantleplumes.org/Iceland1.html
[12] http://www.mantleplumes.org/Africa.html; http://www.mantleplumes.org/Hoggar.html; http://www.mantleplumes.org/Parana.html

East Pacific Rise west to the Mariana Trench (Fig. 1.10). Attempts to identify time-progressive chains have failed, and the volcanoes clearly erupt from numerous, widespread fissures. The Plume hypothesis offers no explanation for these first-order observations.

A primary prediction of the Plume hypothesis is that melting anomalies are fixed relative to one another. The Plate hypothesis permits relative fixity between melt-extraction loci on the same plate, where the stress field is relatively stable. It does not predict relative fixity between melting anomalies on different plates, however.

The error budget in calculations of fixity is dismayingly large. A great deal of flexibility is permitted by our imperfect knowledge of global plate motions, insufficient ages for the volcanoes in chains, and variations in the positions of the Earth's palaeomagnetic and spin axes. Factors such as the finite time it takes to build a volcano, diffuse volcanism within chains, and the lack of longitude information in palaeomagnetic positioning weaken constraints still further.

Nevertheless, despite the great flexibility offered by this large suite of error sources, and the possibility of selecting small subsets of the most compliant chains, the degree of misfit of the observations to the "fixed hot spot" hypothesis is profound. The greatest irony is that the melting anomalies that fit the worst are the Emperor-Hawaii system, the type plume example, and the massive volcanic province of Iceland (Sager, 2007).[13] Relative fixity can be ruled out with confidence for the Emperor chain, which formed in the period ~80–50 Ma, and for the Iceland melting anomaly, which has been active subsequently. Even powerful methods that increase the number of degrees of freedom, for example, by permitting deflections by convection currents in the mantle, are unable to make the data fit the model (Steinberger et al., 2004). Only arguments such as allowing *ad hoc* lateral plume flow where required, for example, at

Iceland, but not elsewhere, for example, at Réunion, in the absence of quantitative physical models, can force a fit.

On the other hand, pre-existing lithospheric structures such as sutures, fault zones and the edges of cratons are clearly important in controlling the location of magmatism. Such structures are implicated at the Greenland-Iceland-Faeroes ridge, which formed over the frontal thrust of the Caledonian suture (Foulger and Anderson, 2005; Foulger et al., 2005a, b), and along the line of opening of the North and the South Atlantic Oceans, the majority of which formed along pre-existing sutures (Lundin and Doré, 2005). The Eastern Snake River Plain-Yellowstone volcano chain is developing along a major discontinuity between cratonic lithosphere to the north and the hot, thinned lithosphere of the Basin and Range province to the south (Christiansen et al., 2002). It is probably forming as a result of differential motion between these two blocks, a consequence of their different lithospheric strengths and spatially varying directions of stress. Volcanism in Paraguay, including the Paraná basalts, CAMP magmatism, and igneous activity in the Italian region is all related to major lithospheric discontinuities, faults and sutures (Beutel, 2009; Comin-Chiaramonti et al., 2007; Peccerillo and Lustrino, 2005). In the oceans, the volcanism that built the Walvis Ridge and the Rio Grande Rise has been attributed to leakage from reactivated, pre-existing wrench and shear faults (Fairhead and Wilson, 2005), and the Lakshadweep-Chagos and Ninetyeast ridges, claimed to represent plume tail volcanism, both formed along major reactivated oceanic transform faults.

Old sutures and major, trans-lithospheric fault zones are likely to contain fusible material such as trapped, late-subducting slabs, and they provide pathways for rising metasomatic fluids and melts. They are thus likely to comprise not only pathways preferentially exploited by rising melt, but also sources of fusible material. Such features are thus not only important in

[13] http://www.mantleplumes.org/HawaiiBend.html; http://www.mantleplumes.org/WhatTheHell.html

controlling the location of volcanism, but probably also in enhancing the volume produced.

8.1.5 Redium and blueium

Widespread trust is vested in seismology to provide clinching evidence, even clear images, of plumes extending down into the deep mantle. However, seismology is less powerful to do this than might be hoped. The transition zone cannot be used as a thermometer. The temperature sensitivity of the 650-km discontinuity is complicated by multiple phase transitions and the depths of the 410- and 650-km discontinuities are sensitive to composition and water content, in addition to temperature.[14] Seismic wave-speeds, which can be mapped in three dimensions using tomography, are ambiguous. They, and other seismic parameters, are dependent on petrology, composition, mineralogical phase, degree of partial melt, flow direction and temperature. Seismic wave-speeds are not a thermometer, and tomography images cannot be viewed as mapping "redium and blueium", corresponding to hot and cold.[15]

A particularly good example of this is provided by the mantle beneath the Ontong Java Plateau, where there is a downward-extensive low-wave-speed anomaly similar to that beneath Iceland (Foulger et al., 2001; Richardson et al., 2000). The Ontong Java seismic anomaly also has unusually low attenuation, and as a consequence requires interpretation as a viscous, compositional body, not a hot one (Klosko et al., 2001). The celebrated "Farallon slab", a high wave-speed body extending from the shallow upper mantle beneath North America down to the base of the lower mantle, is popularly interpreted as subducted material purely because it comprises a quasi-continuous zone of relatively high wave-speeds. This neglects considerations

of volume and the thermal re-equilibration time-scale of slabs, which is of the order of just tens of millions of years (Toksöz et al., 1971). The "superplumes", large low-wave-speed bodies in the lower mantle that are widely assumed to be hot – an assumption on which many plume models have been based – have been shown by normal-mode analysis to be chemical, not thermal bodies (Brodholt et al., 2007; Trampert et al., 2004).

Of the three currently active melting anomalies that have been targeted in the greatest detail possible using seismology – Iceland, Eifel and Yellowstone – the most striking outcome is how similar the results are, despite the fact that it would be difficult to think of three melting anomalies more dissimilar. In each case, a low wave-speed anomaly is detected extending from the surface downwards, petering out above or in the transition zone. The 410-km discontinuity is depressed somewhere in the neighborhood of all three anomalies but there is no correlating topography on the 650-km discontinuity. It is difficult to think of a process that could cause similar, hot risers to form spontaneously in the transition zone in such different tectonic environments and create such different surface phenomena. Another possibility is that the anomalies form from the top down, as a consequence of melt and volatile extraction at the surface, which in turn occurs in response to extension. In this model, the anomalies grew downward, just as the depletion zone in an exploited geothermal or hydrocarbon reservoir grows outward from the fluid extraction locus (Gunasekera et al., 2003).

As for all geological analysis methods, various problems hinder the mapping of the interior of the Earth using seismic rays. The repeatability of the results between independent experiments is often poor. This problem tends to be ignored in assumptions that the latest result must supercede all earlier ones. Different tomographic inversions commonly yield different depth extents for low-wave-speed bodies, and proposed plume tilt directions are particularly prone to multiple,

[14] http://www.mantleplumes.org/TransitionZone.html
[15] http://www.mantleplumes.org/
TomographyProblems.html

mutually exclusive assertions. Much of the Earth's mantle is poorly sampled by rays because of the restricted distribution of earthquakes along narrow plate boundaries, and the sparsity of seismic stations in the oceans. This is typically obscure in tomographic maps and cross-sections, which provide no information on the spatial variation of quality of the result. Resolution may be too poor to image the structures sought. Global seismology, for example, has resolution at the thousand-kilometer level (Ritsema, 2005). Because of resolution limits and, more fundamentally, interpretive ambiguity, seismology cannot answer all the questions popularly asked of it.

Seismology must be applied to suitable problems. It can potentially detect the ponded melt that is required to erupt a flood basalt through thick continental lithosphere. The discovery of seismic anomalies similar to that beneath Iceland under long-extinct volcanic regions such as the Ontong Java Plateau and Paraná hint that the thickness of lithosphere transported along with plates may be much greater than usually supposed, a finding that could have significant cross-disciplinary implications. The radical difference in seismic character between the upper and lower mantle may be the most powerful constraint currently available on the nature of material exchange between these two regions (Gu et al., 2001). Studying this in detail is more likely to yield significant advancements in understanding mantle convection than continuing to attempt direct observations of plumes, an approach that has, for nearly four decades, failed to yield the expected results.

8.1.6 Are "hot spots" hot?

A high-temperature mantle source is a fundamental prediction of the Plume hypothesis that is difficult to argue away. Nevertheless, the temperature anomalies predicted have been downward adjusted, and some estimates are as low as ~100°C. The Plate hypothesis does not require

high T_P, but temperature variations are nevertheless expected in the mantle as a result of local cooling by down-going slabs, melt migration and extraction, continental insulation and variable radioactive heat production. The first potential difficulty is clear – are the temperature differences predicted by the Plate and Plume hypotheses different?

The second problem concerns the precision of the methods available for measuring mantle T_P. There are many of these, including heat flow, diverse seismological, petrological and geochemical approaches, and modeling of ocean floor bathymetry and vertical motions. However, the uncertainties are often as large as the anomalies sought. Ocean-floor heat-flow may only have a resolution of 100–200°C. Seismological methods for probing mantle temperature are insensitive and ambiguous. Petrological and geochemical methods are hampered by lack of repeatability between different studies, and lack of clarity concerning the validity of certain approaches calls the results into question.[16] As a result, the uncertainties in estimates of T_P may be up to hundreds of degrees Celsius.

There is no unequivocal evidence for anomalously high T_P at the vast majority of melting anomalies. Heat flow, bathymetry and vertical motions in the oceans, where the lithosphere has the simplest behavior, yield no consistent evidence for high temperature anomalies over bathymetric swells or near melting anomalies. Seismology in general cannot constrain temperature. A rare exception to this is the demonstration, using normal modes, that the Pacific and Atlantic lower mantle "superplumes" are chemical anomalies and not hot risers (Brodholt et al., 2007; Trampert et al., 2004). Where seismological experiments have been designed to probe crustal temperatures, they find no evidence for unusually hot sources (Korenaga and Kelemen, 2000; Menke and Levin, 1994; Sallares

---

[16] http://www.mantleplumes.org/GreenlandHot.html; http://www.mantleplumes.org/GreenlandReply.html

et al., 2005).[17] Petrological methods vary in their findings, depending on the method used. Reports include uniform, low source T_P for all melts except at Hawaii (Presnall and Gudfinnsson, 2008), low source T_P for spreading ridges and high for many melting anomalies (Putirka, 2008), and relatively high, uniform source T_P for all melts, including at Hawaii (Falloon et al., 2007a).

Despite these problems, it is a simple, basic observation that only at Hawaii is picrite glass found. Picrite glass requires high T_P. Despite many thousands of analyses, none has been found anywhere along the global mid-ocean ridge system, at Iceland, or at any other melting anomaly or flood basalt. What this means is disputed. Are other volcanic regions equally hot, but the evidence merely not yet found? Is Hawaiian melt extracted from unusually deep within the surface conductive layer, or is there truly an unusually hot region in the mantle beneath Hawaii? Settling this dispute is one of the most important current issues regarding the origins of melting anomalies.

8.1.7 Chemical messengers

The original Plume hypothesis predicted characteristic geochemistry at melting anomalies, representing primordial material from the deep mantle. Ironically, this prediction initially seemed to be supported by observation. Subsequently, however, in a total reversal of the conclusions, the characteristic geochemical signatures were found to represent not unprocessed material from great depth, but processed material from the surface. This did not result in abandonment of the Plume hypothesis, however, but absorption of this exact inverse finding into a revised plume model. Near-surface materials were proposed to subduct to the core-mantle boundary and be transported from there back up again in plumes.

[17] http://www.mantleplumes.org/GalapagosIsostasy.html

Today, one of the most important issues for melting anomalies is whether the near-surface material found virtually universally in their sources was drawn from the deep mantle or simply directly from the upper mantle by shallow-sourced melt-extraction processes. Solving this problem is a difficult challenge because petrology and geochemistry have essentially no power to determine the ultimate depth of origin of melt sources from deeper than ~100 km. Efforts to find geochemical core tracers have failed.

Nevertheless, new evidence is emerging that is revealing the nature of recycling in the mantle. Some lavas at melting anomalies are traceable to recognizable layers in subducted slabs, suggesting that their structures are preserved (Chauvel and Hemond, 2000). This raises the serious question whether it is reasonable to expect that slab structure survives intact a 6000-km round trip to the bottom of the mantle and back up again. At the same time, spatial variations in composition at individual melting anomalies have necessitated elaborate models of plume conduit structure, for example, sheaths and bundles of parallel filaments, different for various proposed plumes (Herzberg, 2005). Lavas compositionally indistinguishable from those attributed to plumes occur almost universally throughout the oceans and continents, rendering the term "ocean-island basalt" (OIB) at best inappropriate and at worse obfuscating (Fitton, 2007). In detail, the geochemistry of melting anomalies varies greatly.

Geochemistry has evolved to such a high degree of technical and interpretive complexity as to present serious challenges to non-specialists, who nevertheless need to understand the issues involved. Acceptance of magisterial decree is not an option in good science (Glen, 2005). Interpreting lava compositions as mixtures of discrete isotopic "end members" can only help if the nature and origin of the geological materials represented is understood. The primary importance of explaining the major elements that make up 98% of the contents of lavas has been too de-emphasized in favor of modeling the minor- and trace elements that make up only a tiny fraction of the rocks.

Despite widespread acceptance that the mantle is inhomogeneous, both thermally and chemically, many models still involve simple "anomalies" embedded in "normal" mantle. Variations in fusibility are disregarded in considering melt volumes, which are traditionally linked solely to temperature. The enriched, aged source material tapped at melting anomalies cannot easily be explained as stored in large, deep, isolated regions. It is more naturally explained as comprising variable, distributed material that resides close at hand. This applies also to variations in ^{3}He/^{4}He. Old, high values have been preserved in the mantle for periods comparable to the isolation ages of alkali basalt sources. Distributed inhomogeneities are best described as continuums in size and composition (Meibom and Anderson, 2004). Drawing dividing lines in distributions such as trace-element contents and ^{3}He/^{4}He values is statistically insupportable and not useful (Anderson, 2000b, c).[18] Clear definitions of terms such as "MORB" and "OIB" do not exist. What is considered to be the normal variability of "non-plume" ^{3}He/^{4}He is directly controlled by the elimination from the data set of values assumed to represent "plume contamination" (Anderson, 2000b, c).[19]

The presence of recycled near-surface materials in lavas at melting anomalies is, on the other hand, in elegant agreement with the predictions of the Plate model. Geochemistry alone is unlikely ever to be able to resolve the depth of any melt source deeper than ~200 km. Ironically, however, if the Plate hypothesis is correct, melt sources are largely contained within this depth range. The sources of melting anomalies may be nothing more than simply the volumes within

which melt forms. There may then be little relevant happening that does not affect the composition of lavas and cannot be studied using their petrology and geochemistry.

What is the way forward? In common with other sub-disciplines, geochemistry is fundamentally impotent to answer some of the questions presently asked of it, which focus on assumed deep mantle sources. These are likely the wrong questions, but regardless of this it is more fruitful to ask questions that can potentially be answered. Can different types of recycled material in the upper mantle be identified and their distributions mapped? Can their quantities in the sources of melting anomalies be determined? How do melts react with their surroundings as they rise through the mantle, and how does source composition affect magma volumes? Do alkali basalts arise from metasomatic veins, and what is the scale of source regions that are compositionally distinct? Answers to these questions will contribute to unraveling the history of plate tectonics and subduction further back in time than can be achieved by mapping the sea floor.

8.2 Mantle convection

Mantle convection has been studied using both fluid-filled tanks in the laboratory and numerical fluid-dynamical computer models in order to explore the Plume hypothesis (e.g., Davies, 1999; Griffiths and Campbell, 1991b; Phillips and Bunge, 2005). Although influential because of their visual appeal, laboratory models are intrinsically unable to reproduce realistic and influential conditions in the mantle such as the pressure dependence of thermal expansivity.[20] Furthermore, they often rely on generating buoyancy contrasts by non-thermal means, for example, by injecting compositionally light material such as syrup into the base of the tank,

[18] http://www.mantleplumes.org/Statistics.html; http://www.mantleplumes.org/HowMany.html
[19] http://www.mantleplumes.org/Statistics.html; http://www.mantleplumes.org/HowMany.html

[20] http://www.mantleplumes.org/Convection.html

in order to get the desired results. Low-density risers produced in this way are not internally self-generated and are thus irrelevant to real mantle convection. There is a tendency to consider plumes to be a distinct, separate mode of convection that is decoupled from, and independent of, the general background convection in the mantle. Such behavior is not physically self-consistent and thus not possible. All upwellings must be an intrinsic part of the main convecting system (Tozer, 1973).

Other influential factors that cannot be simulated in laboratory tanks include spherical geometry, temperature and pressure dependence of rheology, internal heat generation, mineral phase transformations including the phase transformations at 410- and 650-km depths, and compositional layering at the base of the mantle. These factors can be incorporated into numerical models, however, which have become increasingly sophisticated with time as progressively more powerful computers have become available, continued progress has been made in theory and programming, and more information has accumulated on the properties of the mantle at great depth.

Radically different results can be obtained from numerical modeling, depending on the physical parameters included and the values adopted for them, for example, the balance between heat input into the mantle from bottom heating by the core, heat loss from the surface, and internal heat production by radioactive decay.[21] For mostly bottom heating, strong, hot plumes rising from the core-mantle boundary dominate convection. Surface cooling has the opposite effect, producing narrow, cold downwellings that drive convection, though pressure and internal heating destroy the symmetry between the upper and lower boundary layers (Lenardic and Kaula, 1994). For mostly internal heat production, convection is also dominated by narrow down-wellings and upwelling is weak

[21] http://www.mantleplumes.org/
GlobalMantleWarming.html

and diffuse (e.g., Anderson, 2001a; Phillips and Bunge, 2005; Tackley, 1998).

Because of pressure effects, viscosity and radiative heat transfer increase with depth and thermal expansivity reduces, so heating yields less thermal buoyancy. These factors work to reduce the vigor of convection in the mantle with depth, and to increase the scale of convection limbs and slow their rise, increasing convective overturn times in the lower mantle. Knowledge of the physical conditions in the mantle is still improving, and older studies in particular involve unrealistic assumptions. The commonly used Boussinesq approximation, for example, ignores the approximately four-fold reduction in thermal expansivity that occurs as pressure increases from the top to the bottom of the mantle. Other common assumptions that tend to over-estimate the vigor of convection in the deep mantle are that the mantle is incompressible and chemically homogeneous.

Knowledge of key input parameters is incomplete, and a wide range of different values and combinations of values are possible. As a result, numerical convection modeling can produce a wide range of results, including small numbers of strong plumes, numerous small plumes, tall, thin plumes, short, wide ones that do not penetrate the entire mantle, pulsing, tilting, forking, headless, tail-less or blob-like plumes, or no plumes at all that connect the core-mantle boundary to the surface.

Two fundamental issues are of primary concern. The first is whether the true conditions in the mantle have been sufficiently closely approached to yield a realistic picture of mantle convection, i.e., whether mantle convection is dominated by the Plume or the Plate mode (Fig. 1.2). The second is whether the predicted effects of plumes simulated in convection models match observed quantities. The present book is primarily concerned with the latter issue. Factors that have been predicted by plume convection modeling include the volume of melt produced, the amount of uplift expected when plume heads impact the base of the lithosphere, the amplitude of the geoid, topographic and heat-flow anomalies for steady state plumes, and the width of the seismic anomaly associated with the plume tail. Critically, no self-consistent convection model has reproduced plate tectonics itself (Bercovici and Ricard, 2003). Predictions must be matched by real observations for models to be candidate possibilities – it is not enough merely to simulate visually satisfying plume-like convective upwellings.

The volumes of melt observed in small flood basalts can be simulated by numerical plumes if temperature anomalies of up to 300°C, purely eclogite melting, and impact under thin lithosphere are assumed (Cordery et al., 1997). However, large flood basalts have erupted over thick lithosphere, where unequivocal evidence for high temperature anomalies is lacking (Chapter 3). Predicted precursory surface uplifts are at the ~1–4 km level (Campbell and Griffiths, 1990; Farnetani and Richards, 1994; Griffiths and Campbell, 1990). The geoid anomalies predicted are typically several tens of meters, regional heat flow is generally $>10\,mW\,m^{-2}$, and the topographic anomaly over steady-state plumes is typically 2–5 km (King and Redmond, 2007). Plume conduits are predicted to be ~1000 km wide in the deep mantle. These effects are rarely observed and there is no example where several of them are observed at a single melting anomaly. The geoid and topographic anomalies observed are typically small, often close to zero and they do not require an origin in the mantle. No melting anomaly is associated with a reliably measurable, regional heat-flow anomaly. A hot conduit ~1000 km wide would be imaged in the deep mantle by seismology, despite deteriorating resolution with depth, but none have been reliably detected.

Morgan (1971) suggested that material raised up by narrow plumes is replaced by diffuse return down-flow uniformly distributed throughout the mantle. It is now known that this is wrong – down-flow occurs in narrow, localized form at subduction zones. The mantle below must rise to replace this material. However, it may well be, then, that in full reversal of the original proposal, it is this return upflow that is distributed and diffuse.

A convective scenario consistent with the Plate hypothesis involves physical properties for the lower mantle that result in broad, mantle-scale convection limbs that flow sluggishly and have an overturn age of the order of a billion years or more (Fig. 1.2) (Anderson, 2005a). Such models tend toward lower heat flux from the core, more internal heating from radioactivity, larger radiative heat transport, reduction of thermal expansivity with depth, larger viscosity in the lower mantle and possibly chemical stratification. If the deep mantle is chemically denser than the overlying layers, warming will be insufficient to allow vigorous upward escape. That the mantle convects, and upwellings occur, is not in dispute. The question is simply whether or not the upwelling convective limbs are in the form of narrow, quasi-vertical, anomalously hot conduits that traverse the whole mantle, actively penetrate the lithosphere, and give rise to melting anomalies and volcanic chains at the surface.

8.3 An unfalsifiable hypothesis

An hypothesis must be falsifiable in order to be scientific, and for it to be meaningful to question and test it (Kuhn, 1962; Popper, 1959).[22] In this respect the modern Plume hypothesis is in disarray. The original hypothesis (Morgan, 1971) was clearly and unambiguously laid out, along with its envisaged relationship to Plate Tectonics and six predictions (Section 1.4). These have all been either falsified (e.g., that plumes drive plate tectonics), found to be unsupported by convincing observations (e.g., that plumes are rooted in the deep mantle), or found to be only rarely true (e.g., that time-progressive chains exist).

[22] http://www.mantleplumes.org/TopPages/GTofPT_Top.html

Nevertheless, the hypothesis has not yet been rejected by mainstream science. Instead it has exploded in complexity to a point where there is a bespoke variant for almost every melting anomaly. The modern Plume hypothesis currently amounts to a *pot pourri* of special cases, different for almost every melting anomaly, with many of them of an *ad hoc* nature, unsupported by any physical modeling. It has little power to describe volcanic regions in a form reduced below the observations themselves, it seeks to explain observations that have not been made, and it has little predictive power (Table 8.2).

Table 8.2 Observations that are problematic or unexplained by the Plume hypothesis, or have contrasting explanations in the Plate hypothesis.

Observation	Plume hypothesis	Plate hypothesis
Vertical motions		
lack of precursory domal uplift at many melting anomalies	downsizing proposed plumes; lateral flow from distant uplifted regions	precursory domal uplift not expected
lack of rapid post-emplacement subsidence of oceanic plateaus	not expected	post-emplacement subsidence not expected
subsidence prior to emplacement of Siberian Traps	Siberian Traps fed by lateral flow	Siberian Traps resulted from detachment of gravitational instability
insufficient precursory uplift at the Ontong Java Plateau	not expected	precursory uplift not expected
the athermal Hawaiian swell	hot source, but removal of the thermal signal by hydrothermal circulation	supported by buoyant residuum remaining after melt extraction
major vertical motions contemporaneous with the eruption of the North Atlantic Igneous Province	not expected	vertical motions associated with continental breakup
Domal uplift in the back-arc setting of Anatolia	plume penetrates down-going slab	asthenosphere upwelled after slab break-off
Volume		
continuum in size of volcanic provinces	no explanation	continuum expected
eruption of flood basalts over thick continental lithosphere	not expected	release of pre-accumulated, ponded melt
eruption of major part of flood basalts in 1–2 Ma	not expected	release of pre-accumulated, ponded melt
volcanic margins	continental breakup occurs coincidentally over plumes	volcanism results directly from continental breakup

Table 8.2 *Continued*

Observation	Plume hypothesis	Plate hypothesis
frequent location of melting anomalies on or near ridges and triple junctions	coincidence; "plume capture" by ridges	predicted, as ridges and triple junctions are extensional features
widespread scattered volcanism in the oceans and continents	no explanation	predicted to occur where the lithosphere is in extension
largest oceanic volcanoes erupted over thickest lithosphere	not expected	expected for volcanoes sourced in the asthenosphere
flood volcanism late in the volcanic sequence, *e.g.*, at Iceland	plume pulsing	any progression in volcanic rate permitted
chevron ridges about the Reykjanes Ridge	migration of pulses from a plume under Iceland	progression of ridge propagators
Time-progressions		
lack of relative fixity between melting anomalies	not expected; lateral flow	relative fixity not required
lack of time-progressive chains emanating from most flood basalts	not expected	time-progressive chains not required to follow flood volcanism
lack of time-progressions in linear volcanic ridges and chains in the Pacific and Atlantic Oceans	not expected, lateral flow	permissive volcanism through lithospheric cracks, small-scale sub-lithospheric convection
southerly migration of 1,500–2,000 km of the Emperor melting anomaly	part mantle wind, part unexplained	migration of the locus of extension that permits passive volcanism
130° bend in the Emperor-Hawaiian chain	not expected	change in plate-wide stress field
lack of a time-progressive track at Iceland	thick Greenland lithosphere; lateral flow; ridge capture	volcanism expected to remain co-located with Mid-Atlantic Ridge, so time-progressive track not expected
association of oceanic melting anomalies with micro-continents, *e.g.*, at Iceland, Lakshadweep Ridge and Kerguelen	coincidence	tectonic complexities allow permissive volcanism
Seismic		
lack of plume tails to the core-mantle boundary	tails invisible or too narrow to be detected	tails do not exist
lack of correlation between topography on the 410- and 650-km seismic discontinuities	not expected	upper-mantle and lower-mantle convection largely decoupled – convection limbs generally do not cross the 650-km discontinuity

Table 8.2 *Continued*

Observation	Plume hypothesis	Plate hypothesis
seismic anomalies beneath Iceland, Yellowstone and Eifel are upper-mantle-only features	deeper continuations invisible; too narrow to be detected; upper-mantle-only plumes	the source regions of melting anomalies are confined to the upper mantle
low-wave-speed anomalies beneath melting anomalies	high temperature	composition, mineralogical phase, partial melt and temperature anomalies
negative correlation between crustal thickness and lower-crustal seismic wave speed in the North Atlantic and the Galapagos region	not expected	thick crust results from melting of fertile source, not high-temperature source
a 40-km-thick crustal block beneath Iceland	plume pulse	submerged microplate
low-wave-speed mantle seismic anomalies beneath the Ontong Java Plateau, the Paraná Basalts, the Deccan Traps and Ireland	fossil plumes	chemical anomalies
Heat & temperature		
lack of high T_P at melting anomalies	contested; the expected observations have not yet been made	high T_P not required
lack of a heatflow anomaly on the Hawaiian swell	hydrothermal circulation	no heat flow anomaly expected
lack of picrite glass anywhere except at Hawaii	the expected observations have not yet been made except at Hawaii	picrite glass not expected except where melt extracted from deep in the low-velocity zone
picrite glass at Hawaii	a hot plume	melt extraction from deep in the low-velocity zone
lack of regional heat flow anomalies around swells	hydrothermal circulation	heat flow anomalies not expected
no evidence that volcano chains and the centers of flood basalts erupted over hotter sources than distal areas	the expected observations have not yet been made	evidence not expected
Geochemistry		
geochemical signature of recycled, near-surface materials in lavas at melting anomalies	near-surface materials subducted down to the core-mantle boundary and back up to the surface in plumes	fusible, near-surface materials recycled in the upper mantle and enhance magma volumes at melting anomalies

Table 8.2 *Continued*

Observation	Plume hypothesis	Plate hypothesis
the lack of any geochemical evidence that requires a deep source	melting occurs only in the shallow parts of plumes	lavas at melting anomalies do not arise from deep sources
enriched basalts similar to those at melting anomalies widely distributed throughout the continents and oceans	not expected; ubiquitous plume material	melting anomalies tap the same source as other volcanism–the shallow mantle
the requirement for up to five isotopic "end members" in enriched basalts	heterogeneous plumes	diverse recycled near-surface materials in the shallow mantle source
widespread mantle heterogeneity, *e.g.*, the DUPAL anomaly	no explanation	fusible, near-surface materials recycled in the upper mantle
core tracers not found	the expected observations have not yet been made	core tracers not expected
lack of high-^{3}He/^{4}He in the lavas at some melting anomalies	the expected observations have not yet been made; ^{4}He ingrowth	high-^{3}He/^{4}He permitted but not required
evidence that high-^{3}He/^{4}He is associated with a helium-poor, depleted, little-degassed source	not expected	high-^{3}He/^{4}He arises from a low-helium, low-U+Th, melt-depleted source
high-^{3}He/^{4}He observed at melting anomalies underlain by upper-mantle-only seismic anomalies	high-^{3}He/^{4}He diffuses up from the lower mantle or the core	high-^{3}He/^{4}He resides in the shallow-mantle melt source
identical geochemistry at oceanic and continental parts of the Cameroon line	a melt source in the asthenosphere	a melt source in lithospheric mantle metasomatic veins
"solar" Ne component observed in groundwater overlying Archean rocks	not expected	"solar" Ne component resides in the shallow mantle
the alkalic-tholeiitic-alkalic sequence of lava petrology at Hawaiian volcanoes	melt extraction from plume in the order sheath-core-sheath	melt extraction from the seismic low-velocity zone in the order shallow-deeper-deepest
non-asthenospheric geochemistry of the Paraná Basalts	heating and re-melting of the continental mantle lithosphere by a plume that does not contribute to surface eruptives	remelting of the continental mantle lithosphere by long-term incubation or processes associated with the opening of the South Atlantic

Although attempts have been made to define in scientific terms the modern Plume hypothesis (Campbell and Davies, 2005; Campbell and Kerr, 2007), there is, in practice, no consensus on **several critical basic points:**

■ *What is a plume – how is it defined?* Numerous different definitions of "plume" have been suggested.[23] Some are proposed for general use, and others solely to justify the usage of the term in a single publication. In most cases, however, the word is used vaguely, without definition, and it means different things to different scientists and in different situations. This problem, recognized early on by Tozer (1973), cannot be simply dismissed as a matter of puristic philosophy and semantics, unimportant to the technical detail of scientific practice. Understanding the causes of melting anomalies will not progress if every region is simply attributed to "the underlying plume", where in effect "plume" can mean whatever causes the melting anomaly. If the term "plume" signifies little more than simply the melt generation region beneath a volcanic area, then it is a truism that adds nothing to our understanding.

■ *How many plumes are proposed to exist, and where are they located?* Many lists exist (Tables 1.2–1.5), but all are different, and the numbers proposed range from a mere handful (Ritsema and Allen, 2003) to several thousand (Malamud and Turcotte, 1999). No two different sets of criteria, if rigorously applied, yield the same list, regardless of any cut-off in size or other parameter allowed. Of the five modern lists given in Table 1.5, only two localities appear on as many as 4 lists and 10 appear on only one. Of those 10, 7 are on the list of 16 originally proposed by Morgan (1971). All signs point to an attempt to draw a dividing line, between plume and non-plume origins, in what is, in reality, a continuum of volcanic phenomena. Even were such a line agreed upon, the question would still remain what causes the melting anomalies designated "non-plume", and whether such a cause might also explain those attributed to plumes.

■ *What predictions does the modern Plume hypothesis make?* The predictions of the modern Plume hypothesis have been clearly laid out (Campbell

[23] http://www.mantleplumes.org/ DefinitionOfAPlume.html

and Davies, 2005; Campbell and Kerr, 2007),[24] but in practice there is no general acceptance of any agreed criteria, nor adherence to the discipline of rejecting the hypothesis if a prediction is reliably shown to be violated. For example, although no uplift, or amounts far less than predicted, occurred prior to eruption of the Siberian Traps and the Ontong Java Plateau, these flood basalts are still widely attributed to mantle plumes, sometimes on the indefensible grounds that it is the best, or only explanation available (Allen et al., 2006; Fitton and Godard, 2004). At various proposed plume localities, all of the predictions, including precursory uplift, flood basalts, time-progressive volcanic chains, seismic observations of plume tails and high temperature are treated, in practice, as though they are optional. Not at any single proposed plume location are all of the five predictions listed by Campbell and Davies (2005) confirmed. These are not isolated, individual failures. It is so rare that the predictions survive detailed scrutiny that failure is the general rule rather than an exception. Despite this, the absence of evidence, sometimes in the face of extensive searching, is commonly not allowed to materially affect the default assumption that a plume exists. Missing uplift is attributed to lateral flow from the nearest neighboring uplifted area, absent flood basalts are eroded away or subducted, and invisible plume tails shrink to undetectable size. Convincing evidence for the absence of required observations is waived (Sheth, 2007a).

8.3.1 Building on sand

How do these factors manifest themselves in how we do science? Do we use data to ratify hypotheses or to attempt to refute them? Do we conduct studies with the objective of confirming our initial assumptions? Does an hypothesis serve us, or vice versa? All philosophers of science agree that observations should be used to challenge and attempt to disprove hypotheses – to test them to see if they can stand in the face of evidence. This, however, is not how the majority of present-day plume-related research is carried out. The Plume hypothesis is typically firmly assumed to be correct. New data are used

[24] http://www.mantleplumes.org/Plumes.html

only to iterate variant models, or they even have no influence at all. A wry illustration of this is the controversy surrounding the thickness and temperature of the Icelandic crust, where the most popular model oscillated from thick/cold to thin/hot and back to thick/cold again, without affecting the plume assumption at all (Section 5.5.1).

Terminology influences thinking, and even controls it. Leading terms such as "hot spot", "ocean-island basalt" and "large igneous province" hinder unbiased assessment of the observations. How can it not be awkward to discover that a "hot spot" is not hot, that a basalt from the interior of a continent is "ocean-island basalt", or that a flood basalt covers an area of 99,000 km^2 and thus is not a "large igneous province"? Most leading of all is the term "plume" itself, so often used for convenience that the difference between a label and an explanation has become lost. Leading terminology encourages both selective reporting and selective interpretation of data. High source temperature may be assumed in interpreting seismic results. Geochemical interpretations may build on the unsupported concept of large spatial separations between the sources of different basalts ("Since MORBs are rather clearly derived from the shallowest mantle…OIBs are by implication sampling deeper parts of the mantle" Davies (1999), p 367). The notion of a migrating, coherent, persistent volcanic supply system – a tangible object–may lead thinking, even if surface volcanism is suspended, perhaps for tens of millions of years (Vogt and Jung, 2007). Melt source compositions may be assumed to be those that require high temperatures to simulate observed geochemical data (Fitton et al., 2004a, b). Our terminology has become so impoverished that, as things stand today, discussions of many volcanic regions are almost impossible to conduct without using the words "hot spot" or "plume".

It will be clear from this why the objective of the present book is not to falsify the Plume hypothesis. It is because this is fundamentally impossible to do – the modern Plume hypothesis is invulnerable. Voluminous melt produced far from any proposed plume, for example, on the northern mid-Atlantic ridge at the time of Eurasian break-up, is explained as lateral flow, with no explanation as to why melt did not flow in other, equally permissive directions. The spectacular violation of the Emperor seamount chain of the prediction of relative fixity, obvious from the crudest data gathered by the very earliest surveyors (Karpoff, 1980; Kono, 1980), was first disregarded and later half-interpreted as "mantle wind" with the remaining misfit ignored (Steinberger et al., 2004). The total absence of an initial flood basalt at the end of the Louisville chain, and the absence of a volcanic chain emanating from the Siberian Traps, are ignored. The reliable, repeatable observation at Iceland, that the seismic anomaly does not extend down into the lower mantle has been disregarded by later workers or attributed to unsupported, *ad hoc*, sometimes fanciful models such as vanishingly thin lower-mantle plumes, or plumes that comprise vertical zones of heat and helium diffusion only. The almost complete absence of any evidence for high mantle temperatures at any melting anomaly, flood basalt or plateau except Hawaii is largely ignored.

In the field of geochemistry, the spectacular discovery that the geochemical signature in basalts from proposed plume localities is the exact opposite of a primordial component from the deep mantle – it is an evolved, shallow geochemical signature – was met with a model that transports surface material down to the core-mantle boundary and back up to the surface again, its structure preserved largely intact. This is not the hypothesis of least astonishment for these observations. The relentless accumulation of findings the opposite of those predicted for the ^{3}He/^{4}He isotope system – the failure of mass-balance calculations (Anderson, 1998b, c; Kellogg and Wasserburg, 1990), the discovery of partition coefficients the opposite of what is predicted (Parman et al., 2005), the correlation of ^{3}He/^{4}He with MORB-like geochemistry (Class and Goldstein, 2005), and the discovery of the solar noble gas "component" in groundwater over an Archean craton (Castro et al., 2009), are all either ignored or absorbed into modified plume models whose only foundation is a stubborn refusal to abandon a failed and no longer useful hypothesis that is a substitute for genuine understanding of the causes of melting anomalies.

Why does the Plume hypothesis so persistently cling on?[25] Is it because of the perfection of explaining surface eruptives as melt that was transported in from below – an irrefutable truism? Are plumes the modern-day equivalent of geosynclines – the linear, subsiding sediment troughs invoked to explain mountain ranges in the years before continental drift was accepted? The geosyncline concept also began as a useful working hypothesis but then, just as the Plume hypothesis, exploded in complexity as accumulating observations failed to accord with the original, simple model. The geosyncline hypothesis finally succumbed when a more successful explanation for mountains was ushered in by Plate Tectonics.

8.4 Diversity: a smoking gun

The great diversity of melting anomalies raises intrinsic doubts about whether it is realistic to expect that all their first-order characteristics can be explained by a single, one-size-fits-all causative model. There is no active melting anomaly anywhere on Earth similar to the vast basalt plateau of Iceland, where active volcanism is distributed between multiple rifts, some oriented normal to the spreading plate boundary, and extending up to 400 km laterally from it (Foulger and Anderson, 2005). Nowhere else is the same as Hawaii, with its huge tholeiite productivity and its remoteness from any plate boundary. Indeed, it would be difficult to think of two more dissimilar melting anomalies. Afar is a large flood basalt, erupted in the African continent in an area that later developed into

a triple junction, one arm comprising a continental rift. Samoa, the largest outpouring of alkali basalt on Earth, lies compactly in a region of extensional stress on the outside corner of a right-angled bend in the Tonga Trench (Natland, 1980; Stuart et al., 2007). The Easter melting anomaly is a persistent, small-volume volcanic chain whose active end lies close to a rotating microplate. Within continents, Yellowstone is unique.

Under the broad umbrella of shallow, plate-related processes, the details of postulated stress-related melt-release mechanisms at individual localities are as diverse as the melting anomalies themselves (Section 1.9). Does such complexity violate Occam's Razor – the requirement to adopt always the simplest explanation that can fit the data? Models should be as simple as possible, but no simpler. The Plate hypothesis has developed because the body of evidence that has accumulated in recent decades testifies that the Plume hypothesis is indeed simpler than possible. New information has tended to violate its predictions, while strengthening the case that shallow, stress-related factors are inevitably complicit in the volcanism. The frequent, common and varied relationships that melting anomalies have with plate boundaries can only be attributed to coincidence in the Plume hypothesis. In the Plate hypothesis they are integral elements of the fundamental cause of volcanism, and suggest at the same time both diversity and commonality of origin.

8.5 The need for joined-up science

There is an urgent necessity for joined-up work. It is not productive to base data interpretations and models on only one type of data, for example, geochemistry or seismology, when a plethora of other kinds are also available. Models must be able to account for all observations before they are viable candidates. The dangers inherent in interpreting only one kind of data at a time are well illustrated by the anecdote of the "Blind

---

[25] http://www.mantleplumes.org/Zombie.html

men and the elephant".[26] The problem is not alleviated merely by citing compliant interpretations of independent data types. Continued reiteration of common assumptions, beliefs, or conclusions based on older, less complete data, without frequent re-visiting, re-examination and re-confirmation, is a recipe for retarded progress.

Disregarding data and concepts across sub-discipline boundaries has generated discordant models, which immediately pose interesting problems for further study. Some examples follow.

- *Convection modeling and seismology:* Calculations of the bending and tilting required of postulated plumes to account for the absence of relative fixity of melting anomalies (Steinberger, 2000) are inconsistent with plume tilts deduced from seismology. This problem is extreme at Hawaii, Iceland and Yellowstone, where even different plume tilts proposed for the same locality are mutually inconsistent (Section 5.8).

- *Convection modeling and temperature:* Convection modeling of the deflections of postulated plume stems in the "mantle wind" assume that seismic wave-speed is a proxy for temperature. The low-wave-speed South Pacific and South Atlantic "super-plumes" are particularly influential in the results, but they have been shown to a high degree of confidence to be dense, chemical structures, not thermal and buoyant (Brodholt et al., 2007; Trampert et al., 2004).

- *Convection modeling and geology:* Convection modeling produces visually appealing images of plumes but the model predictions fall short of matching quantitative observations such as uplift, magmatic volume, swell height, geoid, heat flow and seismic observations (King and Redmond, 2007).

- *Upper-mantle plumes:* Upper-mantle plumes have been suggested to account for seismic observations,

for example, at Iceland and Yellowstone. No explanation is offered for what mechanism might then cause them to be fixed relative to one another, a factor that is also generally assumed. No explanation is offered either for how a small plume confined to the upper mantle could produce the extensive flood basalts attributed to their postulated plume heads, a volcanic province ~2400 km wide in the case of Iceland. It is not explained how such plumes could be generated in the transition zone, which is not a thermal boundary layer, nor how they could be continually replenished for tens of millions of years.

- *Volume and fusibility:* Numerical modeling of impacting plume heads must invoke a fusible source, for example, eclogite, to generate significant amounts of melt beneath moderately thick lithosphere (Cordery et al., 1997). However, the volume-enhancing effect of fusible material in the source at melting anomalies, for example, the Azores, is almost universally ignored as models seek to explain volume solely by elevated temperature (Georgen, 2008).

- *Seismic observations and $^3He/^4He$:* Iceland and Yellowstone, two of the melting anomalies where the most detailed seismic studies have been performed, both have seismic anomalies that extend no deeper than the transition zone and both have been suggested to represent "upper-mantle plumes". However, they both have high-$^3He/^4He$ anomalies, which is attributed to a lower-mantle- or even core-mantle-boundary source in the Plume hypothesis.

In addition to cross-disciplinary inconsistencies, work can also be inconsistent within a single sub-discipline. For example, plumes have been postulated solely on the basis of geochemistry, for example, St Helena (Schilling et al., 1985a), irrespective of the fact that basalt with essentially identical geochemistry is widespread throughout the oceans and continents where plumes are not expected (Fitton, 2007; Hofmann, 1997). In the field of convection, "upside-down drainage" is invoked to account for the lateral flow of plume material from beneath thick lithosphere to adjacent regions of thinner lithosphere. Nevertheless, drainage from thin to thick lithosphere is also invoked, for example, from the West Siberia basin to the Siberian Traps (Saunders et al., 2005).

[26] *The blind men and the elephant*, a poem by John Godfrey Saxe (1816–1887), describes how six blind men each feel a different part of an elephant. Each comes to a radically different and wrong conclusion regarding what kind of a beast it is, when pooling their observations and interpreting them jointly would have allowed them to eliminate all of their models.

Earth science is so diverse, and now involves so many highly technical sub-disciplines, that it is an almost insurmountable task for practitioners to be sufficiently proficient in all subjects to be able to judge the validity of conclusions and interpretations. Such is the present information overload that there is even more pressure to accept magisterial authority than there was in the past (Glen, 2005). Scientists may look to popular consensus to decide difficult scientific problems (Jordan, 2007). However, this is a dangerous path. All old, wrong, reigning paradigms were the most popular ones in their day.

It is a curious observation that specialists in almost every subdiscipline will readily admit that their subject does not require plumes. At the same time, it is commonly believed that other sub-disciplines, with which the practitioner is less familiar, do. Geochemists look to geophysics to provide the evidence needed (e.g., Meibom, 2008). However, geophysics may not be able to do this, especially if the size and strength of the proposed physical anomalies can be endlessly downward-adjusted. Seismologists commonly assume that geochemical observations require plumes from the lower mantle, but this is not the view of most geochemists (e.g., Hofmann, 1997). It is a widely held view that because convection modeling can simulate plume-like upwellings, then plumes must be responsible for volcanic provinces on the Earth's surface. However, it is first necessary to demonstrate that they can match the geological observations, and this has not yet been done. It is a primary goal of the present book to help to equip practitioners to judge for themselves the results of other sub-disciplines, and whether those subjects support most strongly deep, hot sources or comply more readily with non-thermal, shallow, plate-tectonic-related models.

8.6 The future

Although thermal convection has been extensively studied, stress-driven plate-based models

for melting anomalies are largely unexplored using numerical modeling. Such modeling requires shedding the perception of the sources of melting anomalies as proactive, coherent, persistent objects. Instead, they must be viewed simply as the loci of passive melt extraction, the result of local tectonic complexities. This offers a new perspective on volcanism, a schema in which nearby features such as propagating ridges and triple junctions are no longer coincidences, but potential causes of the volcanism. The tools needed for convection modeling already exist, but those for modeling inhomogeneous stress fields are less well developed. The effect of source fertility on magma volume also remains to be rigorously explored.

A peer-review system that functions to reject papers with which reviewers disagree is a relatively new system, introduced in the middle of the 20th century, and there has been much speculation regarding whether it hinders paradigm shifts by tending to favor mainstream ideas (Kuhn, 1962). It has been said that the peer review system filters out both the worst and the best ideas. Certainly, the experiences of the early continental drift advocates, in particular Lawrence W. Morley, are examples of this (Section 1.2). Peer-review approval bestows implied validation on papers that are published, working to increase confidence that published papers are right while simultaneously eroding the responsibility of the reader to make personal judgment. These effects work to support *status quo* ideas. With the advent of the World Wide Web, the problem of limited publishing capacity has disappeared, removing one significant rationale for rejecting papers with unusual viewpoints. The problem of how much time scientists have to spend on reading is a perennial problem, but it is not best solved by authority bodies filtering what is made available.

Many wrong theories were at one time mainstream thought. Major advances have not been achieved by re-iteration of current beliefs. Acceptance of an hypothesis because a better alternative is not perceived, or because "most people believe in it", are guaranteed to limit progress to small steps. The use of the best working hypothesis available is helpful and can be a springboard for much good work. Plume theory has inspired and justified over three decades of careful data collection and remarkable advances in field techniques, data analysis and modeling. However, a theory should be employed where it is useful, but not allowed to control thought. It should be discarded when no longer productive, and give way to better ones that can stimulate new ideas and new experimental approaches.

In Earth science, the precision of techniques and the amount of data is increasing hugely. Even the remotest of localities are now accessible to geologists. Rocks that were considered geochemically identical two decades ago are now thought to be different, and geologists have instant access to vast data bases of multi-component geochemical analyses. Digital, broadband earthquake recordings are now available in quasi-real time from world-wide networks and tools are available for full waveform modeling. These facilities and data sets were not dreamed of when the Plume hypothesis was first proposed, at which time scientists were largely limited to hand tools and paper – mass spectrometers, computers, digital data and the internet were highly limited. It would be surprising and disappointing if these radical technological advances did not bring with them radical advances in our understanding of the Earth. It would also be a failure of the Plume hypothesis – a world view that has worked so well to stimulate fascination, underpin science by providing testable predictions,

and inspire innovative experiments – if its legacy were not even newer and better advances.

The right path forward can be recognized when new data and discoveries tend to fall naturally into the existing hypothesis. If new discoveries tend to violate predictions, rather than confirm and extend them in a natural and consistent way, then the hypothesis is probably wrong. In the case of Plate Tectonics, new discoveries such as intraplate deformation fit naturally and compliantly into the original, simple hypothesis. Its predictions have been repeatedly re-confirmed over the years by many diverse investigative methods, culminating in the direct measurement of plate motions on a year-by-year basis using space-based methods such as Satellite Laser Ranging. The question now at hand is whether a second, independent theory is needed for the special class of volcanism termed melting anomalies in this book, or whether they do not also fit naturally into the Plate Tectonic hypothesis – whether Plate Tectonics will not serve, in effect, as a unifying theory of surface geological processes, more powerful than hitherto thought.

8.7 Exercises for the student

1 Where is the long-term storage region of subducted oceanic lithosphere?

2 How do plumes die?

3 Would plate tectonics be possible without CO_2?

4 How can the Plume and Plate hypotheses be falsified?

References

Abouchami, W., Hofmann, A.W., Galer, S.J.G. et al. (2005) Lead isotopes reveal bilateral asymmetry and vertical continuity in the Hawaiian mantle plume. *Nature*, **344**: 851–856.

Abratis, M., Mädler, J., Hautmann, S. et al. (2007) Two distinct Miocene age ranges of basaltic rocks from the Rhön and Heldburg areas (Germany) based on $^{40}Ar/^{39}Ar$ step heating data. *Chemie der Erde – Geochemistry*, **67**: 133–150.

Al-Lazki, A., Seber, D., Sandvol, E. et al. (2003) Tomographic Pn velocity and anisotropy structure beneath the Anatolian plateau (eastern Turkey) and the surrounding regions. *Geophysical Research Letters*, **30**: 8043.

Albarede, F. (2008) Rogue mantle helium and neon. *Science*, **319**: 943–945.

Aldanmaz, E., Koprubasi, N., Gurer, Ö.F. et al. (2006) Geochemical constraints on the Cenozoic, OIB-type alkaline volcanic rocks of NW Turkey: Implications for mantle sources and melting processes. *Lithos*, **86**: 50–76.

Allègre, C.J., and Turcotte, D.L. (1986) Implications of a two component marble-cake mantle. *Nature*, **323**: 123–127.

Allen, M.B., Anderson, L., Searle, R.C. et al. (2006) Oblique rift geometry of the West Siberian Basin: Tectonic setting for the Siberian flood basalts. *Journal of the Geological Society of London*, **163**: 901–904.

Anderson, D.L. (1967) Phase changes in the upper mantle. *Science*. **157**(3793): 1165–1173.

Anderson, D.L. (1975) Chemical plumes in the mantle. *Bulletin of the Seismological Society of America*, **86**: 1593–1600.

Anderson, D.L. (1982) Hotspots, polar wander, Mesozoic convection and the geoid. *Nature*, **297**: 391–393.

Anderson, D.L. (1989) *Theory of the Earth*, Blackwell Scientific Publications, Boston, p. 366.

Anderson, D.L. (1993) He-3 from the mantle – primordial signal or cosmic dust? *Science*, **261**: 170–176.

Anderson, D.L. (1998a) The edges of the mantle, in *The Core-Mantle Boundary Region* (eds M.E. Gurnis, E.K. Wysession and B.A. Buffett), AGU, Washington, DC, pp. 255–271.

Anderson, D.L. (1998b) The helium paradoxes. *Proceedings of the National Academy of Sciences*, **95**: 4822–4827.

Anderson, D.L. (1998c) A model to explain the various paradoxes associated with mantle noble gas geochemistry. *Proceedings of the National Academy of Sciences*, **95**(16): 9087–9092.

Anderson, D.L. (2000a) The thermal state of the upper mantle: No role for mantle plumes. *Geophysical Research Letters*, **27**, 3623–3626.

Anderson, D.L. (2000b) The statistics and distribution of helium in the mantle. *International Geology Reviews*, **42**(4): 289–311.

Anderson, D.L. (2000c) The statistics of helium isotopes along the global spreading ridge system and the Central Limit Theorem. *Geophysical Research Letters*, **27**: 2401–2404.

Anderson, D.L. (2001a) Top-down tectonics? *Science*, **293**: 2016–2018.

Anderson, D.L. (2001b) A statistical test of the two reservoir model for helium. *Earth and Planetary Science Letters*, **193**: 77–82.

Plates vs. Plumes: A Geological Controversy, 1st edition. By Gillian R. Foulger.
Published 2010 by Blackwell Publishing Ltd.

Anderson, D.L. (2002) Plate tectonics as a far-from-equilibrium self-organised system, in *Plate Boundary Zones* (eds S. Stein and J. Freymuller), American Geophysical Union, Washington, DC, pp. 411–425.

Anderson, D.L. (2005a) Scoring hotspots: The plume and plate paradigms, in *Plates, Plumes, and Paradigms* (eds G.R. Foulger, J.H. Natland, D.C. Presnall and D.L. Anderson), Geological Society of America, Boulder, CO, pp. 31–54.

Anderson, D.L. (2005b) Large igneous provinces, delamination, and fertile mantle. *Elements*, **1**: 271–275.

Anderson, D.L. (2007a) The eclogite engine: Chemical geodynamics as a Galileo thermometer, in *Plates, Plumes, and Planetary Processes* (eds G.R. Foulger and D.M. Jurdy), Geological Society of America, Boulder, CO, pp. 47–64.

Anderson, D.L. (2007b) *New Theory of the Earth*, Cambridge University Press, Cambridge, p. xv & p. 384.

Anderson, D.L. and Bass, J.D. (1984) Mineralogy and composition of the upper mantle. *Geophysical Research Letters*, **11**(3): 229–232.

Anderson, D.L. and Natland, J.H. (2005) A brief history of the plume hypothesis and its competitors: Concept and controversy, in *Plates, Plumes and Paradigms* (eds G.R. Foulger, J.H. Natland, D.C. Presnall and D.L. Anderson), Geological Society of America, Boulder, CO, pp. 119–145.

Anderson, D.L., Zhang, Y.-S. and Tanimoto, T. (1992) Plume heads, continental lithosphere, flood basalts and tomography, in *Magmatism and the Causes of Continental Break-up* (eds B.C. Storey, T. Alabaster and R J. Pankhurst), Geological Society of London Special Publication, London, pp. 99–124.

Anell, I., Thybo, H. and Artemieva, I.M. (2009) Cenozoic uplift and subsidence in the North Atlantic region: Geological evidence revisited. *Tectonophysics*, **474**: 78–105.

Armienti, P. and Gasperini, D. (2007) Do we really need mantle components to define mantle composition? *Journal of Petrology*, **48**: 693–709.

Arndt, N., and Nisbet, E.G. (1982) *Komatiites*, Allen & Unwin, London, p. 526.

Asimow, P.D. and Langmuir, C.H. (2003) The importance of water to oceanic mantle melting regimes. *Nature*, **421**, 815–820.

Avanzinelli, A., Lustrino, M., Mattei, M. et al. (2009) Potassic and ultrapotassic magmatism in the circum-Tyrrhenian region: Significance of carbonated pelitic vs. pelitic sediment recycling at destructive plate margins. *Lithos*, **113**: 213–227.

Bailey, D.K. (1992) Episodic alkaline igneous activity across Africa. *Magmatism and the Causes of Continental Breakup*, **68**: 91–98.

Bailey, D.K., and Woolley, A.R. (1995) Magnetic quiet periods and stable continental magmatism: can there be a plume dimension? Paper presented at Plume 2, Alfred-Wegener-Stiftung, Bonn.

Bailey, D.K. and Woolley, A.R. (1999) Episodic rift magmatism: the need for a new paradigm in global dynamics. *Geolines, Czech. Academy of Sciences*, **9**: 15–20.

Bailey, D.K. and Woolley, A.R. (2005), Repeated, synchronous magmatism within Africa: timing, magnetic reversals and global tectonics, in *Plates, Plumes and Paradigms* (eds G.R. Foulger, J.H. Natland, D.C. Presnall and D.L. Anderson), Geological Society of America, Boulder, CO, pp. 365–378.

Bailey, J.C. and Rasmussen, M.H. (1997) Petrochemistry of Jurassic and Cretaceous tholeiites from Kongs Karls Land, Svalbard, and their relation to Mesozoic magmatism in the Arctic. *Polar Research*, **16**: 37–62.

Baker, J., Snee, L. and Menzies, M. (1996) A brief Oligocene period of flood volcanism in Yemen: Implications for the duration oand rate of continental flood volcanism at the Afro-Arabian triple junction. *Earth and Planetary Science Letters*, **138**: 39–55.

Baksi, A.K. (1999) Reevaluation of plate motion models based on hotspot tracks in the Atlantic and Indian Oceans. *Journal of Geology*, **107**: 13–26.

Baksi, A.K. (2005) Evaluation of radiometric ages pertaining to rocks hypothesized to have been derived by hotspot activity, in and around the Atlantic, Indian and Pacific Oceans, in *Plates, Plumes, and Paradigms* (eds G.R. Foulger, D.L. Anderson, J.H. Natland and D.C. Presnall), Geological Society of America, Boulder, CO, pp. 55–70.

Baksi, A.K. (2007a) A quantitative tool for detecting alteration in undisturbed rocks and minerals–II: Application to argon ages related to hotspots, in *Plates, Plumes, and Planetary Processes* (eds G.R. Foulger and D.M. Jurdy), Geological Society of America, Boulder, CO, pp. 305–333.

Baksi, A.K. (2007b) A quantitative tool for detecting alteration in undisturbed rocks and minerals–I: Water, chemical weathering, and atmospheric argon, in *Plates, Plumes, and Planetary Processes* (eds G.R. Foulger and D.M. Jurdy), Geological Society of America, Boulder, CO, pp. 285–303.

Ballmer, M.D., van Hunen, J., Ito, G. et al. (2007) Non-hotspot volcano chains originating from small-scale sub-lithospheric convection. *Geophysical Research Letters*, **34**: L23310, doi:10.1029/2007GL031636,2007.

Ballmer, M.D., van Hunen, J., Ito, G. et al. (2009) Intraplate volcanism with complex age-distance patterns: A case for small-scale sublithospheric convection. *Geochemistry, Geophysics, Geosystems*, **10**(6): Q06015. doi:10.1029/2009GC002386.

Bargar, K.E., and Jackson, E.D. (1974) Calculated volumes of individual shield volcanoes along the Hawaiian-Emperor chain. *US Geological Survey Journal of Research*, **2**: 545–550.

Barry, T.L., Saunders, A.D., Kempton, P.D. et al. (2003) Petrogenesis of Cenozoic basalts from Mongolia: Evidence for the role of asthenospheric versus metasomatized lithospheric mantle sources. *Journal of Petrology*, **44**: 55–91.

Barton, J.M.J. and Pretorius, W. (1997) The lower unconformity-bounded sequence of the Soutpansberg Group and its correlatives – remnants of a Proterozoic large igneous province. *South African Journal of Geology*, **100**: 335–339.

Bath, M. (1960) Crustal structure of Iceland. *Journal of Geophysical Research*, **65**: 1793–1807.

Batiza, R. (1984) Inverse relationship between Sr isotope diversity and rate of oceanic volcanism has implications for mantle heterogeneity. *Nature*, **309**(5967): 440–441.

Beaman, M., Sager, W.W., Acton, G.D. et al. (2007) Improved Late Cretaceous and early Cenozoic Paleomagnetic apparent polar wander path for the Pacific plate. *Earth and Planetary Science Letters*, **262**: 1–20.

Beardsmore, G.R. and Cull, J.P. (2001) *Crustal Heat Flow: A Guide to Measurement and Modelling*, Cambridge University Press, Cambridge.

Beattie, P. (1993) Olivine-melt and orthopyroxene-melt equilibria. *Contributions to Mineralogy and Petrology*, **115**: 103–111.

Bell, K., Castorina, F., Lavecchia, G. et al. (2004) Is there a mantle plume below Italy? *EOS Trans, AGU*, **85**: 541–547.

Belousov, V.V. (1954) *Белоусов В.В. Основные вопросы геотектоники. Государственное научно-техническое издательство литературы по геологии и охране недр (The main questions of geotectonics)* State scientific-technical publishing house on geology and Earth interior protection, Moscow, p. 608.

Beloussov, V.V. (1962) *Basic problems in geotectonics (trans.)*, McGraw-Hill, New York.

Bercovici, D. and Ricard, Y. (2003) Energetics of a two phase model of lithospheric damage, shear localization and plate boundary formation. *Geophysics Journal International*, **152**: 581–596.

Best, W.J., Johnson, L.R. and McEvilly, T.V. (1975) ScS and the mantle beneath Hawaii. *EOS Trans. AGU*, **56**(12): 1147.

Beutel, E.K. (2005) Stress induced seamount formation at ridge-transform-intersections, in *Plates, Plumes and Paradigms* (eds G.R. Foulger, J.H. Natland, D.C. Presnall and D.L. Anderson), Geological Society of America, Boulder, CO, pp. 581–594.

Beutel, E.K. (2009) Magmatic rifting of Pangaea linked to onset of South American plate motion. *Tectonophysics*, **468**: 149–157.

Beutel, E.K. and Anderson, D.L. (2007) Ridge-crossing seamount chains: a non-thermal approach, in *Plates, Plumes, and Planetary Processes* (eds G.R. Foulger and D.M. Jurdy), Geological Society of America, Boulder, CO, pp. 375–386.

Bianco, T.A., Ito, G., Becker, J.M. et al. (2005) Secondary Hawaiian volcanism formed by flexural arch decompression. *Geochemistry, Geophysics, Geosystems*, **6**(Q08009).

Bijwaard, H. and W. Spakman, W. (1999) Tomographic evidence for a narrow whole mantle plume below Iceland. *Earth and Planetary Science Letters*, **166**: 121–126.

Bina, C. and Helffrich, G. (1994) Phase transition Clapyron slopes and transition zone seismic discontinuity topography. *Journal of Geophysical Research*, **99**: 15,853–815,860.

Bird, P. (1979) Continental delamination and the Colorado Plateau. *Journal of Geophysical Research*, **84**: 7561–7571.

Bjarnason, I.T., Menke, W., Flovenz, O.G. et al. (1993) Tomographic image of the mid-Atlantic plate boundary in south-western Iceland. *Journal of Geophysical Research*, **98**: 6607–6622.

Björnsson, A., Eysteinsson, H. and Beblo, M. (2005) Crustal formation and magma genesis beneath Iceland: magnetotelluric constraints, in *Plates, Plumes, and Paradigms* (eds G.R. Foulger, J.H. Natland, D.C. Presnall and D.L. Anderson), Geological Society of America, Boulder, CO, pp. 665–686.

Bohannon, R.G., Naeser, C.W., Schmidt, D.L. et al. (1989) The timing of uplift, volcanism, and rifting peripheral to the Red Sea: A case for passive rifting? *Journal of Geophysical Research*, **94**: 1683–1701.

Boschi, L., Becker, T.W., Soldati, G. et al. (2006) On the relevance of Born theory in global seismic tomography. *Geophysical Research Letters*, **33**(L06302).

Bott, M.H.P. (1985) Plate tectonic evolution of the Icelandic transverse ridge and adjacent regions. *Journal of Geophysical Research*, **90**: 9953–9960.

Bott, M.H.P. and Gunnarsson, K. (1980) Crustal structure of the Iceland-Faeroe ridge. *Journal of Geophysics*, **47**: 221–227.

Boutilier, R.R. and Keen, C.E. (1999) Small-scale convection and divergent plate boundaries. *Journal of Geophysical Research*, **104**: 7389–7403.

Brandon, A.D., Walker, R..J., Morgan, J.W. et al. (1998) Coupled ^{186}Os-^{187}Os evidence for core-mantle interaction. *Science*, **280**: 1570–1572.

Brandon, A.D., Norman, M.D., Walker, R.J. et al. (2000) ^{186}Os-^{187}Os systematics of Hawaiian picrites. *Earth and Planetary Science Letters*, **174**: 25–42.

Brandon, A.D., Walker, R.J., Puchtel, I.S. et al. (2003) ^{186}Os-^{187}Os systematics of Gorgona Island komatiites: Implications for growth of the inner core. *Earth Planetary Science Letters*, **206**: 411–426.

Breddam, K. (2002) Kistufell: Primitive melt from the Iceland mantle plume. *Journal of Petrology*, **43**: 345–373.

Breddam, K. and Kurz, M.D. (2001) Helium isotope signatures of Icelandic alkaline lavas. *EOS Trans. AGU*, **82**(47): Fall Meeting Supplement, Abstract V22B-1025, 2001.

Bréger, L. and Romanowicz, B. (1998) Three-dimensional structure at the base of the mantle beneath the central Pacific. *Science*, **282**: 718–720.

Brey, G.P., Bulatov, V., Girnis, A. et al. (2004) Ferropericlase – a lower mantle phase in the upper mantle. *Lithos*, **77**: 655–663.

Brodholt, J.P., Helffrich, G. and Trampert, J. (2007) Chemical versus thermal heterogeneity in the lower mantle: The most likely role of anelasticity. *Earth and Planetary Science Letters*, **262**: 429–437.

Brooker, R.A., Heber, V., Kelley, S.P. et al. (2003) Noble gas partitioning behaviour during mantle melting: A possible explanation for "The He Paradox"? *EOS Trans. AGU, AGU Fall Meet. Suppl.*, V31F-03.

Bryan, S.E. and Ernst, R.E. (2008) Revised definition of Large Igneous Province (LIP). *Earth-Science Reviews*, **86**: 175–202.

Bryan, S.E., Ewart, A., Stephens, C.J. et al. (2000) The Whitsunday Volcanic Province, central Queensland, Australia: Lithological and stratigraphic investigations of a silicic-dominated large igneous province. *Journal of Volcanology and Geothermal Research*, **99**: 55–78.

Bull, W.B. (1991) *GeomorphicResponse to Climatic Change*, Oxford University Press, New York, p. 326.

Bullen, K.E. (1963) *An Introduction to the Theory of Seismology*, Cambridge University Press, Cambridge, p. 381.

Burke, K. (1996) The African plate. *South African Journal of Geology*, **99**: 341–409.

Burke, K.C. and Wilson, J.T. (1976) Hot spots on the Earth's surface. *Scientific American*, **235**: 46–57.

Burov, E. and Guillou-Frottier, L. (2005a) The plume head-continental lithosphere interaction using a tectonically realistic formulation for the lithosphere. *Geophysical Journal International*, **161**: 469–490.

Cadoux, A., Blichert-Toft, J., Pinti, D.L. et al. (2007) A unique lower mantle source for Southern Italy volcanics. *Earth and Planetary Science Letter*, **259**: 227–238.

Calvès, G., Clift, P.D. and Inam, A. (2008) Anomalous subsidence on the rifted volcanic margin of Pakistan: No influence from Deccan plume. *Earth and Planetary Science Letters*, **272**: 231–239.

Campbell, I.H. (2006) Large Igneous Provinces and the mantle plume hypothesis. *Elements*, **1**: 265–269.

Campbell, I.H. (2007) Testing the plume theory. *Chemical Geology*, **241**: 153–176.

Campbell, I.H. and Griffiths, R.W. (1990) Implications of mantle plume structure for the evolution of flood basalts. *Earth and Planetary Science Letters*, **99**: 79–93.

Campbell, I.H. and Davies, G.F. (2005) Do mantle plumes exist? *Episodes*, **29**: 162–168.

Campbell, I.H. and Kerr, A.C. (2007) The Great Plume Debate: Testing the plume theory. *Chemical Geology*, **241**: 249–152.

Campbell, I.H., Griffiths, R.W. and Hill, R.I. (1989) Melting in an Archean mantle plume: heads it's basalts, tails it's komatiites. *Nature*, **339**(6227): 697–699.

Cannat, M. (1996) How thick is the magmatic crust at slow spreading oceanic ridges? *Journal of Geophysical Research*, **101**: 2847–2857.

Carmichael, I. and McDonald, A. (1961) The geochemistry of some natural acid glasses from the north Atlantic Tertiary volcanic province. *Geochimica et Cosmochimica Acta*, **25**(3): 189–222.

Castro, M.C., Ma, L. and Hall, C.M. (2009) A primordial, solar He-Ne signature in crustal fluids of a stable continental region. *Earth and Planetary Science Letters*, **279**: 174–184.

Cawthorn, R.G., Meyer, P.S. and Kruger, F.J. (1991) Major Addition of Magma at the Pyroxenite Marker in the Western Bushveld Complex, South-Africa. *Journal of Petrology*, **32**(4): 739–763.

Chalot-Prat, F. and Girbacea, R. (2000) Partial delamination of continental mantle lithosphere, uplift-related crust-mantle decoupling, volcanism and basin formation: a new model for the Pliocene-Quaternary evolution of the southern East-Carpathians, Romania. *Tectonophysics*, **327**: 83–107.

Chase, C.G. (1981) Oceanic island lead: two-stage histories and mantle evolution. *Earth and Planetary Science Letters*, **52**: 277–284.

Chaubey, A.K., Ajay, K.K., Krishna, K.S. et al. (2008) Structure and evolution of the southwestern continental margin of India. *EOS Trans. AGU*, **89**, Fall Meet. Suppl., Abstract T53G-07.

Chauvel, C. and Hemond, C. (2000) Melting of a complete section of recycled oceanic crust: Trace element and Pb isotopic evidence from Iceland. *Geochemistry, Geophysics, Geosystems*, **1**: 1999GC000002.

Chen, G., Spetzler, H.A., Gettinag, I.C. et al. (1996) Selected elastic moduli and their temperature deriviatives for olivine and garnet with different $Mg/(Mg+Fe)$ contents: Results from GHz ultrasonic interferometry. *Geophysical Research Letters*, **23**, 5–8.

Christiansen, R.L. (2001) *The Quaternary and Pliocene Yellowstone Plateau Volcanic Field of Wyoming, Idaho, and Montana*, US Geological Survey Professional Paper, US Government Printing Office, Washington DC, p. 144.

Christiansen, R.L., Foulger, G.R. and Evans, J.R. (2002) Upper mantle origin of the Yellowstone hotspot. *Bulletin of the Geological Society of America*, **114**: 1245–1256.

Clague, D.A. and Dalrymple, G.B. (1987) The Hawaiian-Emperor volcanic chain, in *Volcanism in Hawaii* (eds R.W. Decker, T.L. Wright and P.H. Stauffer), US Government Printing Office, Washington DC, pp. 5–54.

Clague, D.A., Weber, W.S. and Dixon, J.E. (1991) Picritic glasses from Hawaii. *Nature*, **353**(6344): 553–556.

Clague, D.A., Moore, G.J., Dixon, J.E. et al. (1995) Petrology of submarine lavas from Kilauea's Puna ridge, Hawaii. *Journal of Petrology*, **36**: 299–349.

Class, C. and Goldstein, S.L. (2005) Evolution of helium isotopes in the Earth's mantle. *Nature*, **436**: 1107–1112.

Clift, P.D. (1996) Plume tectonics as a cause of mass wasting on the southeast Greenland continental margin. *Marine and Petroleum Geology*, **13**(7): 771–780.

Clift, P.D. (1997) Temperature anomalies under the northeast Atlantic rifted volcanic margins. *Earth and Planetary Science Letters*, **146**: 195–211.

Clift, P.D. (2005) Sedimentary evidence for moderate mantle temperature anomalies associated with hotspot volcanism, in *Plates, Plumes, and Paradigms* (eds G.R. Foulger, J.H. Natland, D.C. Presnall and D.L. Anderson), Geological Society of America, Boulder, CO, pp. 279–288.

Clift, P.D. and Turner, J. (1995) Dynamic support by the Iceland Plume and its effect on the subsidence of the Northern Atlantic Margins. *Journal of the Geological Society*, **152**: 935–941.

Clift, P.D. and Vannucchi, P. (2004) Controls on tectonic accretion versus erosion in subduction zones: implications for the origin and recycling of the continental crust. *Reviews in Geophysics*, **42**(RG2001).

Clift, P.D., Carter, A. and Hurford, A.J. (1998) The erosional and uplift history of NE Atlantic passive margins: constraints on a passing plume. *Journal of the Geological Society of London*, **155**: 787–800.

Cloos, M. (1993) Lithospheric buoyancy and collisional orogenesis: Subduction of oceanic plateaus, continental margins, island arcs, spreading ridges, and seamounts. *Geological Society of America Bulletin*, **105**: 715–737.

Clouard, V. and Bonneville, A. (2001) How many Pacific hotspots are fed by deep-mantle plumes? *Geology*, **29**: 695–698.

Clouard, V. and Bonneville, A. (2005) Ages of seamounts, islands and plateaus on the Pacific Plate, in *Plates, Plumes, and Paradigms* (eds G.R. Foulger, J.H. Natland, D.C. Presnall and D.L. Anderson), Geological Society of America, Boulder, CO, pp. 71–90.

Coffin, M.F. and Eldholm, O. (1992) Volcanism and continental break-up: a global compilation of large igneous provinces, in *Magmatism and the Causes of Continental Break-up* (eds B.C. Storey, T. Alabaster and R.J. Pankhurst), Geological Society of London Special Publication, London, pp. 17–30.

Coffin, M.F. and Eldholm, O. (1993) Scratching the surface: estimating dimensions of large igneous provinces. *Geology*, **21**: 515–518.

Coffin, M.F. and Eldholm, O. (1994) Large igneous provinces: crustal structure, dimensions and external consequences. *Reviews in Geophysics*, **32**, 1–36.

Coltice, N., Phillips, B.R., Bertrand, H. et al. (2007) Global warming of the mantle at the origin of flood basalts over supercontinents. *Geology*, **35**: 391–394.

Comin-Chiaramonti, P., Cundari, A., Piccirillo, E.M. et al. (1997) Potassic and sodic igneous rocks from Eastern Paraguay: their origin from the lithospheric mantle and genetic relationships with the associated Paraná flood tholeiites. *Journal of Petrology*, **38**: 495–528.

Comin-Chiaramonti, P., Cundari, A., DeGraff, J.M. et al. (1999) Early Cretaceous-Tertiary magmatism in Eastern Paraguay (western Paraná basin): geological, geophysical and geochemical relationships. *Journal of Geodynamics*, **28**: 375–391.

Comin-Chiaramonti, P., Gomes, C.B., Castorina, F. et al. (2002) Anitápolis and Lages alkaline-carbonatite complexes, Santa Catarina State, Brazil: geochemistry and geodynamic implications. *Revista Brasileira de Geociências*, **32**: 639–653.

Comin-Chiaramonti, P., Marzoli, A., de Barros Gomes, C. et al. (2007) The origin of post-Paleozoic magmatism in eastern Paraguay, in *Plates, Plumes, and Planetary Processes* (eds G.R. Foulger and D.M. Jurdy), Geological Society of America, Boulder, CO, pp. 603–634.

Condomines, M., Gronvold, K., Hooker, P.J. et al. (1983) Helium, oxygen, strontium and neodymium isotopic relationships in Icelandic volcanics. *Earth and Planetary Science Letters*, **66**: 125–136.

Cordery, M.J., Davies, G.F. and Campbell, I.H. (1997) Genesis of flood basalts from eclogite-bearing mantle plumes. *Journal of Geophysical Research*, **102**: 20,179–20,197.

Courtillot, V., Davaillie, A., Besse, J. et al. (2003) Three distinct types of hotspots in the Earth's mantle. *Earth and Planetary Science Letters*, **205**: 295–308.

Courtney, R.C. and White, R.S. (1986) Anomalous heat flow and geoid across the Cape Verde Rise: evidence for dynamic support from a thermal plume in the mantle. *Geophysics Journal of the Royal Astronomical Society*, **87**: 815–867 and microfiche GJ 887/811.

Cox, K.G. (1989) The role of mantle plumes in the development of continental drainage patterns. *Nature*, **342**: 873–877.

Craig, H. and Lupton, J.E. (1976) Primordial neon, helium, and hydrogen in oceanic basalts. *Earth and Planetary Science Letters*, **31**: 369–385.

Craig, H. and Lupton, J.E. (1981) Helium-3 and mantle volatiles in the ocean and the oceanic crust, in *The Sea* (ed. Emiliani, C.), Wiley, New York, pp. 391–428.

Crawford, A.J., Falloon, T.J. and G.D.H. (1989) Classification, petrogenesis and tectonic setting of boninites, in *Boninites* (ed. A.J. Crawford), Unwin Hyman, London, pp. 1–49.

Crisp, J.A. (1984) Rates of magma emplacement and volcanic output. *Journal of Volcanology and Geothermal Research*, **20**: 177–211.

Crough, S.T. (1983) Hotspot swells. *Annual Reviews in Earth and Planetary Sciences*, **11**: 165–193.

Czamanske, G.K., Gurevitch, A.B., Fedorenko, V. et al. (1998) Demise of the Siberian plume: Paleogeographic and paleotectonic reconstruction from the prevolcanic and volcanic record, North-Central Siberia. *International Geology Reviews*, **40**(2): 95–115.

D'Orazio, M., Innocenti, F., Tonarini, S. et al. (2007) Carbonatites in a subduction system: the Pleistocene alvikites from Mt. Vulture (southern Italy). *Lithos*, **98**: 313–334.

Dahlen, F.A., Hung, S.-H. and Nolet, G. (2000) Frechet kernels for finite-frequency travel times – I. Theory. *Geophysical Journal International*, **141**: 157–174.

Dale, C.W., Pearson, D.G., Starkey, N.A. et al. (2009) Osmium isotopes in Baffin Island and West Greenland picrites: Implications for the 187Os/188Os composition of the convecting mantle and the nature of high ^{3}He/^{4}He mantle. *Earth and Planetary Science Letters*, **278**: 267–277.

Daly, R.A. (1914) *Igneous Rocks and their Origin*, McGraw Hill, New York, p. 563.

Daly, R.A. (1933) *Igneous Rocks and the Depths of the Earth*, McGraw Hill, New York, p. 598.

Dam, G., Larsen, M. and Sonderholm, M. (1998) Sedimentary response to mantle plumes: Implications from Paleocene onshore successions, West and East Greenland. *Geology*, **26**: 207–210.

Dana, J.D. (1849) Geology, in *United States Exploring Expedition* (ed. Wilkes), New York, Putnam, Philadelphia, C. Sherman, p. 756.

Darbyshire, F.A., Bjarnason, I.T., White, R.S. et al. (1998). Crustal structure above the Iceland mantle plume imaged by the ICEMELT refraction profile. *Geophysics Journal International*, **135**: 1131–1149.

Darbyshire, F.A., White, R.S. and Priestley, K.F. (2000) Structure of the crust and uppermost mantle of Iceland from a combined seismic and gravity study. *Earth and Planetary Science Letters*, **181**: 409–428.

Davies, G.F. (1988) Ocean bathymetry and mantle convection 1. Large-scale flow and hotspots. *Journal of Geophysical Research*, **93**: 10,467–10,480.

Davies, G.F. (1999) *Dynamic Earth: Plates, Plumes and Mantle Convection*, Cambridge University Press, Cambridge, pp. 458 + xi.

Davies, G.F. and Richards, M.A. (1992) Mantle convection. *Journal of Geology*, **100**: 151–206.

Davis, A.S., Gray, L.B., Clague, D.A. et al. (2002) The Line Islands revisited: New $^{40}Ar/^{39}Ar$ geochronologic evidence for episodes of volcanism due to lithospheric extension. *Geochimical, Geophysical and Geosystems*, **3**(2001GC000190), 1–28.

Dawson, P.B., Chouet, B.A., Okubo, P.G. et al. (1999) Three-dimensional velocity structure of the Kilauea caldera, Hawaii. *Geophysical Research Letters*, **26**: 2805–2808.

de Hoop, M. and van der Hilst, R. (2005) On sensitivity kernels for "wave-equation" transmission tomography. *Geophysics Journal International*, **160**: 621–633.

De Oliveira, S.G., Hackspacher, P.C., Neto, J.C.H. et al. (2000) Constraints on the evolution and thermal history of the continental platform of southeast Brazil, Sao Paulo State, using Apatite fission track analysis (AFTA). *Revista Brasileira de Geociencias*, **30**: 107–109.

DeLaughter, J., Stein, C.A. and Stein, S. (2005) Hotspots: A view from the swells. *Plates, Plumes, and Paradigms* (eds G.R. Foulger, J.H. Natland, D.C. Presnall and D.L. Anderson), Geological Society of America, Boulder, CO, pp. 257–278.

DeMets, C., Gordon, R.G., Argus, D.F. et al. (1994) Effect of recent revisions to the geomagnetic reversal time scale on estimate of current plate motions. *Geophysical Research Letters*, **21**: 2191–2194.

DePaolo, D.J. and Manga, M. (2003) Deep origin of hotspots – the mantle plume model. *Science*, **300**: 920–921.

DePaolo, D.J., Bruce, J.G., Dodson, A. et al. (2001) Isotopic evolution of Mauna Loa and the chemical structure of the Hawaiian plume. *Geochemistry, Geophysics, Geosystems*, **2**: 2000GC000139.

Deuss, A. (2007) Seismic observations of transition-zone discontinuities beneath hotspot loations, in *Plates, Plumes, and Planetary Processes* (eds G.R. Foulger and D.M. Jurdy), Geological Society of America, Boulder, CO, pp. 121–136.

Dewey, J.F. and Windley, B.F. (1988) Palaeocene-Oligocene tectonics of NW Europe, in *Early Tertiary Volcanism and the Opening of the NE Atlantic* (eds A.C. Morton and L.M. Parson), Geological Society of London Special Publication, London, pp. 25–31.

Dezes, P., Schmid, S.M. and Ziegler, P.A. (2004) Evolution of the European Cenozoic Rift System: interaction of the Alpine and Pyrenean orogens with their foreland lithosphere. *Tectonophysics*, **389**, 1–33.

Di Renzo, V., Di Vito, M.A., Arienzo, I. et al. (2007) Magmatic history of Somma–Vesuvius on the basis of new geochemical and isotopic data from a deep borehole (Camaldoli della Torre). *Journal of Petrology*, **48**: 753–784.

Dick, H.J.B., Lin, J. and Schouten, H. (2003) An ultraslow-spreading class of ocean ridge. *Nature*, **426**: 405–412.

Dickin, A.P. (1995) *Radiogenic Isotope Geology*, Cambridge University Press, Cambridge, p. 452.

Doglioni, C., Harabaglia, P., Merlini, S. et al. (1999) Orogens and slabs vs. their direction of subduction. *Earth-Science Reviews*, **45**: 167–208.

Doglioni, C., Green, D.H. and Mongelli, F. (2005) On the shallow origin of hotspots and the westward drift of the lithosphere, in *Plates, Plumes, and Paradigms* (eds G.R. Foulger, J.H. Natland, D.C. Presnall and D.L. Anderson), Geological Society of America, Boulder, CO, pp. 735–749.

Donelick, R.A., O'Sullivan, P.B. and Ketcham, R.A. (2005) Apatite Fission-Track Analysis. *Reviews in Mineralogy and Geochemistry*, **58**: 49–94.

Drury, S.A., Kelley, S.P. and Berhe, S.M. (1994) Structures related to Red Sea evolution in northern Eritrea. *Tectonics*, **13**: 1371–1380.

Du, Z.J. and Foulger, G.R. (2001) Variation in the crustal structure across central Iceland. *Geophysics Journal International*, **145**: 246–264.

Du, Z.J., Vinnik, L.P. and Foulger, G.R. (2006) Evidence from P-to-S mantle converted waves for a flat "660-km" discontinuity beneath Iceland. *Earth and Planetary Science Letters*, **241**: 271–280.

Dueker, K.G., and Sheehan, A.F. (1997) Mantle discontinuity structure from midpoint stacks of converted P to S waves across the Yellowstone hotspot track. *Journal of Geophysical Research*, **102**: 8313–8327.

Duncan, R.A. (1984) Age progressive volcanism in the New England Seamounts and the opening of the central Atlantic Ocean. *Journal of Geophysical Research*, **89**: 9980–9990.

Duncan, R.A. and Clague, D.A. (1985) Pacific plate motion recorded by linear volcanic chains, in *The Ocean Basins and Margins* (eds F.G. Stehli and S. Uyeda), Plenum, New York, pp. 89–121.

Duncan, R.A. and Hargraves, R.B. (1990) $^{40}Ar/^{39}Ar$ geochronology of basement rocks from the Mascarene Plateau, the Chagos Bank, and the Maldives Ridge, in *Proceedings ODP Scientific Results, 115* (eds R.A. Duncan, J. Backman, L.C. Peterson, et al.), Ocean Drilling Program, College Station, TX, pp. 43–51.

Dupré, B. and Allegre, C.J. (1983) Pb–Sr isotope variation in Indian Ocean basalts and mixing phenomena. *Nature*, **303**: 142–146.

Dziewonski, A.M. (2005) The robust aspects of global seismic tomography, in *Plates, Plumes, and Paradigms* (eds G.R. Foulger, J.H. Natland, D.C. Presnall and D.L. Anderson), Geological Society of America, Boulder, CO, pp. 147–154.

Dziewonski, A.M. and Anderson, D.L. (1981) Preliminary reference Earth model. *Physical Earth Planetary International*, **25**: 297–356.

Echeverría, L.M. (1980) Tertiary or Mesozoic komatiites from Gorgona island, Colombia: field relations and geochemistry. *Continental Mineral Petrology*, **73**: 253–266.

Egorkin, A.V. (2001) Upper mantle structure below the Daldyn-Alakitsk kimberlite field by nuclear explosion seismograms. *Geology of Ore Deposits*, **43**: 19–32.

Eldholm, O. and Grue, K. (1994) North Atlantic volcanic margins: Dimensions and production rates. *Journal of Geophysical Research*, **99**: 2955–2968.

Elkins-Tanton, L.T. (2005) Continental magmatism caused by lithospheric delamination, in *Plates, Plumes, and Paradigms* (eds G.R. Foulger, J.H. Natland, D.C. Presnall and D.L. Anderson), Geological Society of America, Boulder, CO, pp. 449–462.

Elkins-Tanton, L.T. (2007) Continental magmatism, volatile recycling, and a heterogeneous mantle caused by lithospheric gravitational instabilities. *Journal of Geophysical Research*, **112**: B03405, doi:10.1029/2005JB004072.

Elkins-Tanton, L.T. and Hager, B.H. (2000) Melt intrusion as a trigger for lithospheric foundering and the eruption of the Siberian flood basalt. *Geophysical Research Letters*, **27**: 3937–3940.

Ellis, M. and King, G.C.P. (1991) Structural control of flank volcanism in continental rifts. *Science*, **254**: 839–842.

England, P.C. and Molnar, P. (1990) Surface uplift, uplift of rocks, and exhumation of rocks. *Geology*, **18**: 1173–1177.

Ernst, R.E. and Buchan, K.L. (1997) Giant radiating dyke swarms: their use in identifying pre-Mesozoic large igneous provinces and mantle plumes, in *Large Igneous Provinces: Continental, Oceanic, and Planetary Volcanism* (eds J. Mahoney and M. Coffin), American Geophysical Union, Washington DC, pp. 297–333.

Ernst, R.E. and Buchan, K.L. (2001) Large mafic magmatic events through time and links to mantle plume-heads, in *Mantle Plumes: Their Identification Through Time* (eds R.E. Ernst and K.L. Buchan), Geological Society of America, Boulder, CO, pp. 483–575.

Ernst, R.E. and Buchan, K.L. (2003) Recognizing mantle plumes in the geological record. *Annual Reviews in Earth and Planetary Sciences*, **31**: 469–523.

Ernst, R.E., Head, J.W., Parfitt, E. et al. (1995) Giant radiating dyke swarms on Earth and Venus. *Earth-Science Reviews*, **39**: 1–58.

Ershov, A.V. and Nikishi, A.M. (2004) Recent geodynamics of the Caucasus-Arabia-East Africa region. *Geotectonics*, **38**: 123–136.

Evans, J.R. and Achauer, U. (1993) Teleseismic velocity tomography using the ACH method: Theory and application to continental-scale studies, in *Seismic Tomography: Theory and Applications* (eds H.M. Iyer and K. Hirahara), Chapman & Hall, London, pp. 319–360.

Fairhead, M.J. and Wilson, M. (2005) Plate tectonic processes in the south Atlantic ocean: Do we need deep mantle plumes? in *Plates, Plumes, and Paradigms* (eds G.R. Foulger, J.H. Natland, D.C. Presnall and D.L. Anderson), Geological Society of America, Boulder, CO, pp. 537–554.

Falloon, T.J., Green, D.H. and Danyushevsky, L.V. (2007a) Crystallization temperatures of tholeiite parental liquids: Implications for the existence of thermally driven mantle plumes, in *Plates, Plumes, and Planetary Processes* (eds G.R. Foulger and D.M. Jurdy), Geological Society of America, Boulder, CO, pp. 235–260.

Falloon, T.J., Danyushevsky, L.V., Ariskin, A. et al. (2007b) The application of olivine geothermometry to infer crystallization temperatures of parental liquids: implications for the temperature of MORB magmas. *Chemical Geology*, **241**: 207–233.

Farley, K.A., Natland, J. and Craig, H. (1992) Binary mixing of enriched and undegassed (primitive?) mantle components (He, Sr, Nd, Pb) in Samoan lavas. *Earth and Planetary Science Letters*, **111**: 183–199.

Farnetani, C.G. and Richards, M.A. (1994) Numerical investigations of the mantle plume initiation model for flood basalt events. *Journal of Geophysical Research*, **99**: 13,813–813,833.

Faul, U.H., Toomey, D.R. and Waff, H.S. (1994) Intergranular basaltic melt is distributed in thin, elongated inclusions. *Geophysical Research Letters*, **21**(1): 29–32.

Favela, J. and Anderson, D.L. (1999) Extensional tectonics and global volcanism. Paper presented at Editrice Compositori, Bologna, Italy.

Fee, D. and Dueker, K.G. (2004) Mantle transition zone topography and structure beneath the Yellowstone hotspot. *Geophysical Research Letters*, **31**: L18603, doi:10.1029/2004GL020636.

Ferrari, L. (2004) Slab detachment control on mafic volcanic pulse and mantle heterogeneity in central Mexico. *Geology*, **32**: 77–80.

Ferrari, L. and Rosas-Elguera, J. (1999) Alkalic (OIB type) and calc-alkalic volcanism in the Mexican Volcanic Belt: a case for plume-related magmatism and propagating rifting at an active margin? Comment to the article by Marquez, A., Oyarzun, R., Doblas, M. and Verma, S.P. *Geology*, **27**: 1055–1056.

Fitton, J.G. (1987) The Cameroon line, West Africa: a comparison between oceanic and continental alkaline volcanism, in *Alkaline Igneous Rocks* (eds J.G. Fitton and B.G.J. Upton), Geological Society of London, London, pp. 273–291.

Fitton, J.G. (2007) The OIB paradox, in *Plates, Plumes, and Planetary Processes* (eds G.R. Foulger and D.M. Jurdy), Geological Society of America, Boulder, CO, pp. 387–412.

Fitton, J.G. and Godard, M. (2004) Origin and evolution of magmas on the Ontong Java Plateau, in *Origin and Evolution of the Ontong Java Plateau* (eds J.G. Fitton, J.J. Mahoney, P.J. Wallace and A.D. Saunders), Geological Society of London, Special Publication, pp. 151–178.

Fitton, J.G. and Hardarson, B. (2008) Fertility pulses in the Iceland plume. *EOS Trans. AGU*, **89**: *Fall Meeting Supplement*, Abstract V31F-07.

Fitton, J.G., Saunders, A.D., Norry, M.J. et al. (1997) Thermal and chemical structure of the Iceland plume. *Earth and Planetary Science Letters*, **153**: 197–208.

Fitton, J.G., Mahoney, J.J., Wallace, P.J. et al. (eds) (2004a) *Origin and Evolution of the Ontong Java Plateau*, Geological Society of London, Special Publication, pp. vi + 374.

Fitton, J.G., Mahoney, J.J., Wallace, P.J. et al. (2004b) Origin and evolution of the Ontong Java Plateau: Introduction, in *Origin and Evolution of the Ontong Java Plateau* (eds J.G. Fitton, J.J. Mahoney, P.J. Wallace and A.D. Saunders), Geological Society of London, Special Publication, pp. 1–8.

Fitton, J.G., Starkey, N., Stuart, F.M. et al. (2008) Heat and Helium in the early Iceland plume. *Eos Trans. AGU, Fall Meeting Supplement*, **89**(53): Abstract V23H-07.

Fleming, A., Summerfield, M.A., Stone, J.O. et al. (1999) Denudation rates for the Southern Drakensberg Escarpment, SE Africa, derived from in-situ-produced Cosmogenic Cl-36: Initial results. *Journal of the Geological Society, London*, **156**: 209–212.

Flovenz, O.G. (1980) Seismic structure of the Icelandic crust above layer three and the relation between body wave velocity and the alteration of the basaltic crust. *Journal of Geophysics*, **47**, 211–220.

Flovenz, O.G., and Saemundsson, K. (1993) Heat-flow and geothermal processes in Iceland. *Tectonophysics*, **225**(1–2), 123–138.

Ford, C.E., Russell, D.G., Craven, J.A. et al. (1983) Olivine-liquid equilibria: temperature, pressure and composition dependence of the crystal/liquid cation partition coefficients for Mg, Fe2+, Ca, and Mn. *Journal of Petrology*, **24**: 256–265.

Forsyth, D.A., Morel-à-l'Huissier, P., Asudeh, I. et al. (1986) Alpha-Ridge and Iceland: Products of the same plume? *Journal of Geodynamics*, **6**(1–4): 197–214.

Forsyth, D.W., Harmon, N., Scheirer, D.S. et al. (2006) Distribution of recent volcanism and the morphology of seamounts and ridges in the GLIMPSE study area: Implications for the lithospheric cracking hypothesis for the origin of intraplate, non-hot spot volcanic chains. *Journal of Geophysical Research*, **111**: B11407.

Foulger, G.R. (2006) Older crust underlies Iceland. *Geophysics Journal International*, **165**: 672–676.

Foulger, G.R. (2007) The "Plate" model for the genesis of melting anomalies, in *Plates, Plumes, and Planetary Processes* (eds G.R. Foulger and D.M. Jurdy), Geological Society of America, Boulder, CO, pp. 1–28.

Foulger, G.R. and Hofton, M.A. (1998) Regional vertical motion in Iceland 1987–1992, determined using GPS survey-ing, in *Coastal Tectonics* (eds I.S. Stewart and C. Vita-Finzi), Geological Society, London, Special Publications, pp. 165–178.

Foulger, G.R. and Natland, J.H. (2003) Is "hotspot" volcanism a consequence of plate tectonics? *Science*, **300**: 921–922.

Foulger, G.R., and Anderson, D.L., (2005) A cool model for the Iceland hot spot. *Journal of Volcanology and Geothermal Research*, **141**: 1–22.

Foulger, G.R., Pritchard, M.J., Julian, B.R. et al. (2000) The seismic anomaly beneath Iceland extends down to the mantle transition zone and no deeper. *Geophysics Journal International*, **142**: Fl–F5.

Foulger, G.R., Pritchard, M.J., Julian, B.R. et al. (2001) Seismic tomography shows that upwelling beneath Iceland is confined to the upper mantle. *Geophysics Journal International*, **146**: 504–530.

Foulger, G.R., Du, Z. and Julian, B.R. (2003) Icelandic-type crust. *Geophysics Journal International*, **155**: 567–590.

Foulger, G.R., Natland, J.H. and Anderson, D.L. (2005a) Genesis of the Iceland melt anomaly by plate tectonic proc-esses, in *Plates, Plumes, and Paradigms* (eds G.R. Foulger, J.H. Natland, D.C. Presnall and D.L. Anderson), Geological Society of America, Boulder, CO, pp. 595–626.

Foulger, G.R., Natland, J.H. and Anderson, D.L. (2005b) A source for Icelandic magmas in remelted Iapetus crust. *Journal of Volcanology and Geothermal Research*, **141**: 23–44.

Fukao, Y., Masayuki, O., Tomoeki, N. and D.S.P. Group (2009) Stagnant slab: A review. *Annual Reviews in Earth and Planetary Sciences*, **37**: 19–46.

Fúlfaro, V.J. (1995) Geology of eastern Paraguay, in *Alkaline magmatism in Central-Eastern Paraguay: Relationships with coeval magmatism in Brazil* (eds P. Comin-Chiaramonti and C.B. Gomes), Edusp-Fapesp, Sao Paolo, pp. 17–30.

Gallagher, K. and Hawkesworth, C.J. (1992) Dehydration melting and the generation of contiential flood basalts. *Nature*, **358**: 57–59.

Garnero, E.J., Lay, T. and McNamara, A. (2007) Implications of lower mantle structural heterogeneity for existence and nature of whole mantle plumes, in *Plates, Plumes, and Planetary Processes* (eds G.R. Foulger and D.M. Jurdy), Geological Society of America, Boulder, CO, pp. 79–101.

Gasperini, D., Blichert-Toft, J., Bosch, D. et al. (2002) Upwelling of deep mantle material through a plate window: Evidence from geochemistry of Italian basaltic volcanics. *Journal of Geophysical Research*, **107**: 2367–2386.

Geist, D.J., Naumann, T.R., Standish, J.J. et al. (1995) Wolf Volcano, Galápagos Archipelago: melting and magmatic evolution at the margins of a mantle plume. *Journal of Petrology*, **46**: 2197–2224.

Gente, P., Dyment, J., Maia, M. et al. (2003) Interaction between the Mid-Atlantic Ridge and the Azores hot spot during the last 85 Myr: Emplacement and rifting of the hot spot-derived plateaus. *Geochemistry, Geophysics, Geosystems*, **4**(10): 8514, doi:10.1029/2003GC000527.

Georgen, J.E. (2008) Mantle flow and melting beneath oceanic ridge-ridge-ridge triple junctions. *Earth and Planetary Science Letters* **270**: 231–240.

Georgen, J.E. and Lin, J. (2002) Three-dimensional passive flow and temperature structure beneath oceanic ridge-ridge-ridge triple junctions. *Earth Planetary Science Letters*, **204**: 115–132.

Gernigon, L., Olesen, O., Ebbing, J. et al. (2009) Geophysical insights and early spreading history in the vicinity of the Jan Mayen Fracture Zone, Norwegian-Greenland Sea. *Tectonophysics*, **468**, 185–205.

Glen, W. (2005) The origins and early trajectory of mantle plume quasi-paradigm, in *Plates, Plumes, and Paradigms* (eds G.R. Foulger, J.H. Natland, D.C. Presnall and D.L. Anderson), Geological Society of America, Boulder, CO, pp. 91–118.

Goes, S., Govers, R. and Vacher, P. (2000) Shallow mantle temperatures under Europe from P and S wave tomography. *Journal of Geophysical Research*, **105**: 11,153–111,169.

Gögüs, O.H., and Pysklywec, R.N. (2008) Mantle lithosphere delamination driving plateau uplift and synconvergent extension in eastern Anatolia. *Geology*, **36**: 723–726.

Gök, R., Türkelli, N., Sandvol, E. et al. (2000) Regional wave propagation in Turkey and surrounding regions. *Geophysical Research Letters*, **27**: 429–432.

Gök, R., Sandvol, E., Türkelli, N. et al. (2003) Sn attenuation in the Anatolian and Iranian plateau and surrounding regions. *Geophysical Research Letters*, **30**: 8042.

Goldreich, P., and Toomre, A. (1969) Some remarks on polar wandering. *Journal Geophysical Research*, **74**: 2555–2567.

Gomer, B.M. and Okal, E.A. (2003) Multiple-ScS probing of the Ontong-Java Plateau. *Earth and Planetary Science Letters*. **138**: 317–331.

Gordon, R.G., Andrews, D.L., Horner-Johnson, B.C. et al. (2005a) New tests of the fixed hotspot approximation. *EOS Trans. AGU*, **86**: Abstract GP22A-06.

Gordon, R.G., Royer, J., Argus, D.F. et al. (2005b) Non-closure of the geologically instantaneous global plate motion circuit: Implications for plate non-rigidity. Paper presented at AGU Fall Meeting, American Geophysical Union, San Francisco.

Granet, M., Wilson, M. and Achauer, U. (1995) Imaging a mantle plume beneath the Massif Central (France). *Earth and Planetary Science Letters*, **136**: 281–296.

Green, D.H. and Falloon, T.J. (2005) Primary magmas at mid-ocean ridges, "hot spots" and other intraplate settings; constraints on mantle potential temperature, in *Plates, Plumes and Paradigms* (eds G.R. Foulger, J.H. Natland, D.C. Presnall and D.L. Anderson), Geological Society of America, Boulder, CO, pp. 217–248.

Green, D.H., Falloon, T.J., Eggins, S.M. et al. (2001) Primary magmas and mantle temperatures. *European Journal of Mineralogy*, **13**: 437–451.

Griffiths, R.W. and Campbell, I.H. (1990) Stirring and structure in mantle plumes. *Earth and Planetary Science Letters*, **99**: 66–78.

Griffiths, R.W. and Campbell, I.H. (1991a) Interaction of mantle plume heads with the Earths surface and onset of small-scale convection. *Journal of Geophysical Research*, **96**(B11): 18,295–18,310.

Griffiths, R.W. and Campbell, I.H. (1991b) On the dynamics of long-lived plume conduits in the convecting mantle. *Earth and Planetary Science Letters*, **103**: 214–227.

Griggs, D.T. (1939) A theory of mountain building. *American Journal of Science*, **237**: 611–650.

Gu, Y.J. and Dziewonski, A.M. (2001) Variations in thickness of the upper mantle transition zone, in long-term observations in the oceans, paper presented at 2nd OHP/ION Joint Symposium (Ocean Hemisphere network Project/International Ocean Network), International Ocean Network, Hotel Mt Fuji, Japan, January 21–27, 2001.

Gu, Y.J., Dziewonski, A.M., Weijia, S. et al. (2001) Models of the mantle shear velocity and discontinuities in the pattern of lateral heterogeneities. *Journal of Geophysical Research*, **106**: 11,169–111,199.

Gu, Y.J., Dziewonski, A.M. and Ekstrom, G. (2003) Simultaneous inversion for mantle shear velocity and topography of transition zone discontinuities. *Geophysics Journal International*, **154**: 559–583.

Gu, Y.J., An, Y., Sacchi, M. et al. (2009) Mantle reflectivity structure beneath oceanic hotspots. *Geophysics Journal International*, **178**: 1456–1472.

Gudfinnsson, G. and Presnall, D.C. (2005) Continuous gradations among primary carbonatitic, kimberlitic, melilititic, basaltic, picritic, and komatiitic melts in equilibrium with garnet lherzolite at 3–8 GPa. *Journal of Petrology*, **46**: 1645–1659.

Gudmundsson, O. (2003) The dense root of the Iceland crust. *Earth and Planetary Science Letters*, **206**: 427–440.

Gunasekera, R.C., Foulger, G.R. and Julian, B.R. (2003) Four dimensional tomography shows progressive pore-fluid depletion at The Geysers geothermal area, California. *Journal of Geophysical Research*, **108**(B3): DOI: 10.1029/2001JB000638.

Hager, B.H. (1984) Subducted slabs and the geoid: constraints on mantle rheology and flow. *Journal of Geophysical Research*, **89**(B7): 6003–6015.

Hales, T.C., Abt, D.L., Humphreys, E.D. et al. (2005) Lithospheric instability origin for Columbia River flood basalts and Wallowa Mountains uplift in northeast Oregon. *Nature*, **438**: 842–845.

Hall, A. (1996) *Igneous Petrology*, Prentice Hall, Edinburgh, p. 551.

Hamilton, W.B. (1970) Bushveld complex product of impacts? in *Symposium on the Bushveld Igneous Complex and Other Layered Intrusions* (eds J.L. Vissler and G.V. Gruenewaldt), Geological Society of South Africa Special Publication, pp. 367–379.

Hamilton, W.B. (2002) The closed upper-mantle circulation of plate tectonics, in *Plate Boundary Zones* (eds S. Stein and J.T. Freymueller), American Geophysical Union, Washington, DC, pp. 359–410.

Hamilton, W.B. (2003) An alternative Earth. *GSA Today*, **13**: 4–12.

Hamilton, W.B. (2005) Plumeless Venus preserves an ancient impact-accretionary surface, in *Plates, Plumes, and Paradigms* (eds G.R. Foulger, J.H. Natland, D.C. Presnall and D.L. Anderson), Geological Society of America, Boulder, CO, pp. 781–814.

Hamilton, W.B. (2007a) An alternative Venus, in *Plates, Plumes, and Planetary Processes* (eds G.R. Foulger and D.M. Jurdy), Geological Society of America, Boulder, CO, pp. 879–912.

Hamilton, W.B. (2007b) Driving mechanism and 3-D circulation of plate tectonics, in *Whence the Mountains? Inquiries into the Evolution of Orogenic Systems: A Volume in Honor of Raymond A. Price* (eds J.W. Sears, T.A. Harms and C.A. Evenchick), Geological Society of America, Boulder, CO, pp. 1–25.

Hamilton, W.B. (2007c) Earth's first two billion years – the era of internally mobile crust. *Geological Society of America Memoir*, **200**: 233–296.

Hansen, U., Yuen, D.A., Kroening, S.E. et al. (1993) Dynamical consequences of depth-dependent thermal expansivity and viscosity on mantle circulations and thermal structure. *Physical Earth Planetary International*, **77**(3–4): 205–223.

Hardarson, B.S., Fitton, J.G., Ellam, R.M. et al. (1997) Rift relocation – a geochemical and geochronological investigation of a palaeo-rift in northwest Iceland. *Earth and Planetary Science Letters*, **153**: 181–196.

Harpp, K. and Geist, D. (2002) Wolf-Darwin lineament and plume-ridge interaction in northern Galápagos. *Geochemistry, Geophysics, Geosystens*, **3**(11): 8504, doi: 10.1029/2002GC000370.

Harris, R.N. and McNutt, M.K. (2007) Heat flow on hot spot swells: Evidence for fluid flow. *Journal of Geophysical Research*, **112**(B03407): doi: 10.1029/2006JB004299.

Harris, R.N., Von Herzen, R.P., McNutt, M.K. et al. (2000) Submarine hydrogeology of the Hawaiian archipelagic apron 1. Heat flow patterns north of Oahu and Maro Reef. *Journal of Geophysical Research*, **105**: 21, 353–21369.

Hart, S.R. (1984) A large-scale isotope anomaly in the Southern Hemisphere mantle. *Nature*, **309**: 753–757.

Hart, S.R., Hauri, E.H., Oschmann, L.A. et al. (1992) Mantle plumes and entrainment: isotopic evidence. *Science*, **256**: 517–520.

Hartmann, W.K. and Davis, D.R. (1975) Satellite-sized planetesimals and lunar origin. *Icarus*, **24**: 504–515.

He, B., Xu, Y.-G., Ching, S.-L., Xiao, L. et al. (2003) Sedimentary evidence for rapid crustal doming prior to the eruption of the Emeishan floods basalts. *Earth and Planetary Science Letters*, **213**: 391–405.

He, B. Xu, Y.-G., Wang, Y.-M. et al. (2006) Sedimentation and lithofacies paleogeography in southwestern China before and after the Emeishan flood volcanism: new insights into surface response to mantle plume activity. *Journal of Geology*, **114**: 117–132.

He, B., Xu, Y. and Campbell, I. (2009) Pre-eruptive uplift in the Emeishan? Comment on "Re-evaluating plume-induced uplift in the Emeishan large igneous province" by Ukstins Peate & Brian, 2008. *Nature Geoscience*, **2**: 530–531.

Head, M.J., Gibbard, P.L. and Salvador, A, (2008) The Quaternary: its character and definition. *Episodes*, **31**: 1–5.

Hegarty, K.A., Duddy, I.R. and Green, P.F. (1995) The thermal history in around the Paraná Basin using apatite fission track analysis–implications for hydrocarbon occurrences and basin formation, in *Alkaline magmatism in Central-Eastern Paraguay: Relationships with coeval magmatism in Brazil* (eds P. Comin-Chiaramonti and C.B. Gomes), Edusp-Fapesp, Sao Paolo, pp. 67–84.

Helmberger, D.V., Wen, L. and Ding, X. (1998) Seismic evidence that the source of the Iceland hotspot lies at the core-mantle boundary. *Nature*, **396**: 251–255.

Helz, R.T. and Thornber, C.R. (1987) Geothermometry of Kilauea Iki lava lake, Kilauea Volcano, Hawaii. *Bulletin of Volcanology*, **49**: 651–668.

Herzberg, C. (2005) Big lessons from little droplets. *Nature*, **436**: 789–790.

Herzberg, C. and O'Hara, M.J. (2002) Plume-associated ultramafic magmas of Phanerozoic age. *Journal of Petrology*, **43**: 1857–1883.

Herzberg, C. and Gazel, E. (2009) Petrological evidence for secular coolilng in mantle plumes. *Nature*, **458**: 619–622.

Herzberg, C., Asimow, P.D., Arndt, N. et al. (2007) Temperatures in ambient mantle and plumes: Constraints from basalts, picrites, and komatiites. *Geochemistry, Geophysics, Geosystems*, **8**: Q02006, doi: 10.1029/2006GC001390.

Hess, H.H. (1962), *A history of ocean basins* (eds A.E.J. Engel, H.L. James and B.F. Leonard), Geological Society of America, Boulder, Co, pp. 599–620.

Hey, R.N., Deffeyes, K.S., Johnson, G.L. et al. (1972) The Galapagos triple junction and plate motions in the East Pacific. *Nature*, **237**: 20–22.

Hey, R.N., Sinton, J.M. and Duennebier, F.K. (1989a) Propagating rifts and spreading centres, in *Decade of North American Geology: The Eastern Pacific Ocean and Hawaii* (eds E.L. Winterer, D.M. Hussong and R.W. Decker), Geological Society of America, Boulder, CO, pp. 161–176.

Hey, R.N., Kleinrock, M.C., Miller, S.P. et al. (1989b) Sea Beam/Deep-Tow investigation of an active oceanic propagating rift system, *Journal Geophysical Research*, **91**: 3369–3393.

Hey, R.N., Sinton, J.M., Kleinrock, M.C. et al. (1992) ALVIN investigation of an active propagating rift system, Galapagos 95.5°W. *Marine Geophysical Research*, **14**: 207–226.

Hey, R.N., Martinez, A., Höskuldsson, A. et al. (2007) Propagating rift explanation for the v-shaped ridges south of Iceland, in *AGU Fall Meeting*, American Geophysical Union, San Francisco.

Hey, R.N., Martinez, F., Höskuldsson, Á. et al. (2008) Propagating rift origin of the v-shaped ridges south of Iceland, in *IAVCEI 2008 General Assembly*, Reykjavik, Iceland.

Hieronymus, C.F. and Bercovici, D. (1999) Discrete alternating hotspot islands formed by interaction of magma transport and lithospheric flexure. *Nature*, **397**: 604–607.

Hieronymus, C.F. and Bercovici, D. (2000) Non-hotspot formation of volcanic chains: control of tectonic and flexural stresses on magma transport. *Earth and Planetary Science Letters*, **181**: 539–554.

Hildreth, W. (2007) *Quaternary magmatism in the Cascades–Geologic perspectives*, US Department of the Interior.

Hill, D.P. (1989) Temperatures at the base of the seismogenic crust beneath Long Valley caldera California, and the Phlegrean Fields caldera, Italy, in *Volcanic Seismology* (eds P. Gasparini, R. Scarpa and K. Aki), Springer-Verlag, Berlin.

Hillier, J.K. (2007), Pacific seamount volcanism in space and time. *Geophysical Journal International*, **168**: 877–889.

Hillis, R.R., Holford, S.P., Green, P.F. et al. (2008) Cenozoic exhumation of the southern British Isles. *Geology*, **36**: 371–374.

Hilton, D.R., Gronvold, K., Macpherson, C.G. et al. (1999) Extreme $^{3}He/^{4}He$ ratios in northwest Iceland: constraining the common component in mantle plumes. *Earth and Planetary Science Letters*, **173**: 53–60.

Hirano, N., Takahashi, E., Yamamoto, J. et al. (2006) Volcanism in response to plate flexure, *Science*, **313**: 1426–1428.

Hoernle, K. and Schmincke, H.-U. (1993) The role of partial melting in the 15-Ma geochemical evolution of Gran Canaria: A blob model for the Canary Hotspot. *Journal of Petrology*, **34**: 599–626.

Hofmann, A.W. (1988) Chemical differentiation of the Earth: the relationship between mantle, continental crust, and oceanic crust. *Earth and Planetary Science Letters*, **90**: 297–314.

Hofmann, A.W. (1997) Mantle geochemistry: the message from oceanic volcanism. *Nature*, **385**: 219–229.

Hofmann, A.W. and White, W.M. (1982) Mantle plumes from ancient oceanic crust. *Earth and Planetary Science Letters*, **57**: 421–436.

Hofmann, A.W., Nikogosian, I.K. and Sobolev, A.V. (2000) Recycled oceanic crust observed in "ghost plagioclase" within the source of Mauna Loa lavas. *Nature*, **204**: 986–990.

Hofmeister, A.M. (2007) Thermal conductivity of the Earth's deepest mantle, in *Superplumes: Beyond Plate Tectonics* (eds D.A. Yuen, S. Maruyama, S.-I. Karato and B.F. Windley) Springer, Netherlands, pp. 269–292.

Hofmeister, A.M. and Criss, R.E. (2005) Heat flow and mantle convection in the triaxial Earth, in *Plates, Plumes, and Paradigms* (eds. G.R. Foulger, J.H. Natland, D.C. Presnall and D.L. Anderson), Geological Society of America, Boulder, CO, pp. 289–302.

Holbrook, W.S., Larsen, H.C, Korenaga, J. et al. (2001) Mantle thermal structure and active upwelling during continental breakup in the north Atlantic. *Earth and Planetary Science Letters*, **190**: 251–266.

Holford, S.P., Green, P.F., Duddy, I.R. et al. (2009) Regional intraplate exhumation episodes related to plate-boundary deformation. *Geological Society America Bulletin*, **121**: 1611–1628.

Holmes, A. (1929) Radioactivity and Earth movements. *Transactions of the Geological Society of Glasgow*, **18**: 559–606.

Holmes, A. (1944) *Principles of Physical Geology*, Thomas Nelson and Sons Ltd, London, pp. 1288 + xv.

Hooft, E.E.E., Toomey, D.R. and Solomon, S.C. (2003) Anomalously thing transition zone beneath the Galápagos hotspot. *Earth and Planetary Science Letters*, **216**: 55–64.

Hooper, P.R., Camp, V., Reidel, S. et al. (2007) The origin of the Columbia River flood basalt province: Plume versus nonplume models, in *Plates, Plumes, and Planetary Processes* (eds G R. Foulger and D.M. Jurdy), Geological Society of America, Denver, CO, pp. 635–668.

Humphreys, E.., Dueker, K.G., Schutt, D.L. et al. (2000) Beneath Yellowstone; evaluating plume and nonplume models using teleseismic images of the upper mantle. *GSA Today*, **10**: 1–7.

Humphreys, E.R. and Niu, Y. (2009) On the composition of ocean island basalts (OIB): The effects of lithospheric thickness variation and mantle metasomatism. *Lithos*, **112**: 118–136.

Hung, S.-H., Dahlen, F.A. and Nolet, G. (2000) Frechet kernels for finite-frequency traveltimes – II. Examples. *Geophysics Journal International*, **141**: 175–203.

Hung, S.-H., Shen, Y. et al. (2004) Imaging seismic velocity structure beneath the Iceland hotspot – A finite-frequency approach. *Journal of Geophysical Research*, **109**: (B08305).

Ihinger, P.D. (1995) Mantle flow beneath the Pacific plate: evidence from seamount segments in the Hawaiian-Emperor chain. *American Journal of Science*, **295**: 1035–1057.

Iitaka, T., Hirose, K., Kawamura, K. et al. (2004) The elasticity of the $MgSiO_3$ post-perovskite phase in the lowermost mantle. *Nature*, **430**: 442–445.

Ingle, S. and Coffin, M.F. (2004) Impact origin for the greater Ontong Java Plateau? *Earth and Planetary Science Letters*, **218**: 123–134.

Ito, G. (2001) Reykjanes V-shaped ridges originating from a pulsing and dehydrating mantle plume. *Nature*, **411**(6838): 681–684.

Ito, G. and Clift, P.D. (1998) Subsidence and growth of Pacific Cretaceous plateaus. *Earth and Planetary Science Letters*, **161**(1–4): 85–100.

Ito, K. and Kennedy, G.C. (1974) The composition of liquids formed by partial melting of eclogites at high temperatures and pressures. *Journal of Geology*, **82**: 383–392.

Ivanov, A. (2007) Evaluation of different models for the origin of the Siberian traps, in *Plates, Plumes, and Planetary Processes* (eds G.R. Foulger and D.M. Jurdy), Geological Society of America, Boulder, CO, pp. 669–692.

Ivanov, A.V., Demonterova, E.I., Rasskazov, S.V. et al. (2008) Low-Ti melts from the southeastern Siberian Traps Large Igneous Province: Evidence for a water-rich mantle source? *Journal of Earth Systems Science*, **117**: 1–21.

Iyer, H.M. and Hirahara, K. (eds) (1993) *Seismic Tomography: Theory and Practice*, Chapman & Hall, London, P. 848.

Iyer, H.M., Evans, J.R., Zandt, G. et al. (1981) A deep low-velocity body under the Yellowstone caldera, Wyoming: Delineation using teleseismic P-wave residuals and tectonic interpretation: Summary. *Bulletin of the Geology Society of America*, **92**: 792–798.

Jackson, E.D. and Shaw, H.R. (1975) Stress fields in central portions of the Pacific plate: delineated in time by linear volcanic chains. *Journal of Geophysical Research*, **80**: 1861–1874.

Jackson, E.D., Silver, E.A. and Dalrymple, G.B. (1972) Hawaiian-Emperor chain and its relation to Cenozoic circumpacific tectonics. *Geology Society of America Bulletin*, **83**: 601–618.

Jackson, E.D., Shaw, H.R. and Barger, K.E. (1975) Calculated geochronology and stress field orientations along the Hawaiian chain. *Earth and Planetary Science Letters*, **26**: 145–155.

Japsen, P., Bonow, J.M., Green, P.A. et al. (2006) Elevated, passive continental margins: Long-term highs or Neogene uplifts? New evidence from West Greenland. *Earth and Planetary Science Letters*, **248**: 330–339.

Ji, Y. and Nataf, H.C. (1998) Detection of mantle plumes in the lower mantle by diffraction tomography: Hawaii. *Earth and Planetary Science Letters*, **159**: 99–115.

Jicha, B.R., Scholl, D.W., Singer, B.S. et al. (2006) Revised age of Aleutian Island Arc formation implies high rate of magma production. *Geology*, **34**: 661–664.

Jóhannesson, H. and Saemundsson, K. (1998) Jardfraedikort af Islandi. Bergrunnur (Geological Map of Iceland. Bedrock Geology), Náttúrfraedistofunu Islands (Icelandic Institute of Natural History), Reykjavík.

Jones, A.P., Price, G.D., Price, N.J. et al. (2002) Impact induced melting and the development of large igneous provinces. *Earth and Planetary Science Letters*, **202**: 551–561.

Jones, A.P., Price, G.D., De Carli, P.S. et al. (2003) Impact decompression melting: a possible trigger for impact induced volcanism and mantle hotspots? in *Impact markers in the Stratigraphic Record* (eds C. Koeberl and F. Martinez-Ruiz), Springer, Berlin, pp. 91–120.

Jones, A.P., Wunemann, K. and Price, D. (2005) Modeling impact volcanism as a possible origin for the Ontong Java Plateau, in *Plates, Plumes, and Paradigms* (eds G.R. Foulger, J.H. Natland, D.C. Presnall and D.L. Anderson), Geological Society of America, Boulder, CO, pp. 711–720.

Jones, S.M., White, N. and Maclennan, J. (2002) V-shaped ridges around Iceland; implications for spatial and temporal patterns of mantle convection. *Geochemistry, Geophysics, Geosystems*, **3**(10): 2002GC000361.

Jones, S.M. (2005) Uplift associated with the North Atlantic Igneous Province, in *The Great Plume Debate* (eds I. Campbell, G.R. Foulger, J.H. Natland and W.J. Morgan). American Geophysical Union, Ft. William, Scotland.

Jordan, B.T. (2007) The mantle plume debate in undergraduate geoscience education: Overview, history, and recommendations, in *Plates, Plumes, and Planetary Processes* (eds G.R. Foulger and D.M. Jurdy), Geological Society of America, Boulder, CO, pp. 933–944.

Jourdan, F., Féraud, G., Bertrand, H. et al. (2004) The Karoo triple junction questioned: evidence from $^{40}Ar/^{39}Ar$ Jurassic and Proterozoic ages and geochemistry of the Okavango dyke swarm (Botswana). *Earth and Planetary Science Letters*, **222**: 989–1006.

Jourdan, F., Féraud, G., Bertrand, H. et al. (2005) The Karoo large igneous province: brevity, origin and relation with mass extinction in question from $^{40}Ar/^{39}Ar$ age data. *Geology*, **33**: 745–748.

Jourdan, F., Bertrand, H., Sharer, U. et al. (2007) Major-trace element and Sr-Nd-Hf-Pb isotope compositions of the Karoo large igneous province in Botswana-Zimbabwe. *Journal of Petrology*, **48**: 1043–1077.

Jourdan, F., Bertrand, H., Féraud, G. et al. (2009) Lithospheric mantle evolution monitored by overlapping large igneous provinces: Case study in southern Africa. *Lithos*, **107**: 257–268.

Julian, B.R. (2005) What can seismology say about hot spots? in *Plates, Plumes, and Paradigms* (eds G.R. Foulger, J.H. Natland, D.C. Presnall and D.L. Anderson), Geological Society of America, Boulder, CO, pp. 155170.

Julian, B.R. and Anderson, D.L. (1968) Travel times, apparent velocities and amplitudes of body waves. *Bulletin of the Seismological Society of America*, **58**(1): 339–366.

Julian, B.R. and Sengupta, M.K. (1973) Seismic travel time evidence for lateral inhomogeneity in the deep mantle. *Nature*, **242**(5398), 443–447.

Julian, B.R. and Evans, J.R. (2010) On possible plume-guided seismic waves. *Bulletin of the Seismological Society of America*, **100**: 497–508.

Jurdy, D.M. and Stoddard, P.R. (2007) The coronae of Venus: Impact, plume, or other origin? in *Plates, Plumes, and Planetary Processes* (eds G.R. Foulger and D.M. Jurdy), Geological Society of America, Boulder, CO, pp. 859–878.

Kamenetsky, V.S., Maas, R., Sushchevskaya, N.M. et al. (2001) Remnants of Gondwanan continental lithosphere in oceanic upper mantle: Evidence from the South Atlantic Ridge. *Geology*, **29**: 243–246.

Kaneoka, I. and Takaoka, N. (1978) Excess ^{129}Xe and high $^{3}He/^{4}He$ ratios in olivine phenocrysts of Kapuho lava and xenolithic dunites from Hawaii. *Earth and Planetary Science Letters*, **39**: 382–386.

Karato, S. (1993) Importance of anelasticity in the interpretation of seismic tomography. *Geophysical Research Letters*, **20**(15): 1623–1626.

Karpoff, A.M. (1980) The sedimentary deposits of Suiko seamount (Leg 55, Site 433): from the reef environment to the pelagic sedimentation, US Government Printing Office, Washington, DC, pp. 491–501.

Katsura, T., Yamada, H., Nishikawa, O. et al. (2004) Olivine-wadsleyite transition in the system (Mg,Fe)2SiO4. *Journal of Geophysical Research*, **109**: (B02209).

Katzman, R., Zhao, L. and Jordan, T.H. (1988) High-resolution, two-dimensional vertical tomography of the central Pacific mantle using ScS reverberations and frequency-dependent travel times. *Journal of Geophysical Research*, **103**: 17,933–917,971.

Kaula, W.M. (1983) Minimal upper mantle temperature variations consistent with observed heat flow and plate velocities. *Journal Geophysical Research*, **88**: 10,323–310,332.

Kay, M. (1951) *North American Geosyncline*, Geological Society of America, Boulder, CO, p. 143.

Kay, R.W. and Kay, S.M. (1993) Delamination and delamination magmatism. *Tectonophysics*, **219**: 177–189.

Kear, D. and Wood, B.L. (1957) The geology and hydrology of western Samoa. *New Zealand Geological Survey Bulletin*, **63**: 1–92.

Keller, R.A., Fisk, M.R. and White, W.M. (2000) Isotopic evidence for Late Cretaceous plume-ridge interaction at the Hawaiian hotspot. *Nature*, **405**: 673–676.

Kellogg, L.H. and Wasserburg, G.J. (1990) The role of plumes in mantle helium fluxes. *Earth and Planetary Science Letters*, **99**: 276–289.

Kellogg, L.H., Hager, B.H. and Van Der Hilst, R.D. (1999) Compositional stratification in the deep mantle. *Science*, **283**(5409): 1881–1884.

Kempton, P.D., Fitton, J.G., Saunders, A.D. et al. (2000) The Iceland plume in space and time: A Sr-Nd-Pb-Hf study of the north Atlantic rifted margin. *Earth and Planetary Science Letters*, **177**: 255–271.

Kennett, B.L.N. and Widiyantoro, S. (1999) A low seismic wave-speed anomaly beneath northwestern India: a seismic signature of the Deccan plume? *Earth and Planetary Science Letters*, **165**: 145–155.

Kerr, A.C. (2005) La Isla de Gorgona, Colombia: A petrological enigma? *Lithos*, **84**: 77–101.

Keskin, M. (2003) Magma generation by slab steepening and breakoff beneath a subduction-accretion complex: An alternative model for collision-related volcanism in Eastern Anatolia, Turkey. *Geophysical Research Letters*, **30**: 8046–8049.

Keskin, M. (2007) Eastern Anatolia: a hot spot in a collision zone without a mantle plume, in *Plates, Plumes, and Planetary Processes* (eds G.R. Foulger and D.M. Jurdy), Geological Society of America, Boulder, CO, pp. 693–722.

Keskin, M., Pearce, J.A. and Mitchell, J.G. (1998) Volcano-stratigraphy and geochemistry of collision-related volcanism on the Erzurum-Kars Plateau, North Eastern Turkey. *Journal of Volcanism and Geothermal Research*, **85**: 355–404.

Keyser, M., Ritter, J.R.R. and Jordan, M. (2002) 3D shear-wave velocity structure of he Eifel plume, Germany. *Earth and Planetary Science Letters*, **203**: 59–82.

King, G.C.P. and Ellis, M. (1990) The origin of large local uplift in extensional regions. *Nature*, **348**: 20–27.

King, S.D. (2005) North Atlantic topographic and geoid anomalies: the result of a narrow ocean basin and cratonic roots? in *Plates, Plumes, and Paradigms* (eds G.R. Foulger, JH. Natland, D.C. Presnall and D.L. Anderson), Geological Society of America, Boulder, CO, pp. 653–664.

King, S.D. and Anderson, D.L. (1995) An alternative mechanism of flood basalt formation. *Earth and Planetary Science Letters*, **136**: 269–279.

King, S.D. and Anderson, D.L. (1998) Edge-driven convection. *Earth and Planetary Science Letters*, **160**: 289–296.

King, S.D. and Ritsema, J. (2000) African hot spot volcanism: Small-scale convection in the upper mantle beneath cratons. *Science*, **290**: 1137–1140.

King, S.D. and Redmond H.D. (2007) The structure of thermal plumes and geophysical observations, in *Plates, Plumes, and Planetary Processes* (eds G.R. Foulger and D.M. Jurdy), Geological Society of America, Boulder, Co, pp. 103120,.

Kircher, A. (1664–1678) *Mundus subterraneus, quo universae denique naturae divitiae*, p. 546.

Klein, E.M. and Langmuir, C.H. (1987) Global correlations of ocean ridge basalt chemistry with axial depth and crustal thickness. *Journal of Geophysical Research*, **92**: 8089–8115.

Klosko, E.R., Russo, R.M., Okal, E.A. et al. (2001) Evidence for a rheologically strong chemical mantle root beneath the Ontong-Java Plateau. *Earth and Planetary Science Letters*, **186**: 347–361.

Kogiso, T., Hirschmann, M.M. and Frost, D.J. (2003) High-pressure partial melting of garnet pyroxenite: possible mafic lithologies in the source of ocean island basalts. *Earth and Planetary Science Letters*, **216**: 603–617.

Kohlstedt, D.L. and Holtzman, B.K. (2009) Shearing melt out of the Earth: An experimentalist's perspective on the influence of deformation on melt extraction, in *Annual Reviews of Earth Planetary Science* (ed. F. Jeanloz), Annual Reviews, Palo Alto, CA, pp. 561–593.

Kono, M. (1980) Paleomagnetism of DSDP Leg 55 basalts and implications for the tectonics of the Pacific plate, US Government Printing Office, Washington, DC, pp. 737–752.

Koppers, A.A.P. and Staudigel, H. (2005) Asynchronous bends in Pacific seamount trails: A case for extensional volcanism? *Science*, **307**: 904–907.

Koppers, A.A.P., Staudigel, H., Pringle, M.S. et al. (2003) Short-lived and discontinuous intraplate volcanism in the South Pacific: Hot spots or extensional volcanism? *Geochemistry, Geophysics, Geosystems*, **4**(10): 1089, doi: 10.1029/2003GC000533.

Koppers, A.A.P., Duncan, R.A. and Steinberger, B. (2004) Implications of a nonlinear $^{40}Ar/^{39}Ar$ age progression along the Louisville seamount trail for models of fixed and moving hot spots. *Geochemistry, Geophysics, Geosystems*, **5**: Q06L02, doi: 10.1029/2003GC000671.

Koppers, A.A.P., Russell, J.A., Jackson, M.G. et al. (2008) Samoa reinstated as a primary hotspot trail. *Geology*, **36**: 435–438.

Korenaga, J. (2005) Why did not the Ontong Java Plateau form subaerially? *Earth and Planetary Science Letters*, **234**: 385–399.

Korenaga, J. and Kelemen, P.B. (2000) Major element heterogeneity in the mantle source of the north Atlantic igneous province. *Earth and Planetary Science Letters*, **184**: 251–268.

Korenaga, J., Kelemen, P.B. and Holbrook, W.S. (2002) Methods for resolving the origin of large igneous provinces from crustal seismology. *Journal of Geophysical Research*, **107**(B9): 2178, doi:2110.1029/2001JB001030.

Krishna, K.S., Gopala Rao, D., Ramana, M.V. et al. (1995) Tectonic model for the evolution of oceanic crust in the northeastern Indian Ocean from the Late Cretaceous to the early Tertiary. *Journal of Geophysical Research*, **100**: 20,011–020,024.

Kroenke, L.W., Wessel, P. and Sterling, A. (2004) Motion of the Ontong Java plateau in the hot-spot frame of reference: 122 Ma-present, in *Origin and Evolution of the Ontong Java Plateau* (eds J.G. Fitton, J J. Mahoney, P.J. Wallace and A.D. Saunders), Geological Society of London, Special Publications, pp. 9–20.

Kruse, S.E., Liu, J.Z., Naar, D.F. et al. (1997) Effective elastic thickness of the lithosphere along the Easter seamount chain. *Journal of Geophysical Research*, **102**: 27,305–27,317.

Kuhn, T. (1962) *The Structure of Scientific Revolution*, University of Chicago Press, Chicago.

Kusznir, N.J. and Karner, G.D. (2007) Continental lithospheric thinning and breakup in response to upwelling divergent mantle flow: application to the Woodlark, Newfoundland and Iberia margins, in *Imaging, Mapping and Modelling Continental Lithosphere Extension and Breakup* (eds G.D. Karner, G. Manatschal and L.M. Pinheiro), Geological Society of London, Special Publications, pp. 389–419.

Langmuir, C.H., Klein, E.M. and Plank, T. (1992) Petrological constraints on melt formation and migration beneath mid-ocean ridges, in *Mantle Flow and Melt Generation at Mid-Ocean Ridges* (eds J. Phipps Morgan, D. Blackman and J.L. Sinton), American Geophysical Union, Washington, DC, pp. 183–280.

Langston, C.A. (1979) Structure under Mount Rainier, Washington, inferred from teleseismic body waves. *Journal of Geophysical Research*, **84**: 4749–4762.

Larsen, L.M. and Pedersen, A.K. (2000) Processes in high-Mg, high-T magmas: Evidence from olivine, chromite and glass in palaeogene picrites from West Greenland. *Journal of Petrology*, **41**: 1071–1098.

Larson, R.L. (1991) Latest pulse of Earth: Evidence for a mid-Cretaceous superplume. *Geology*, **19**: 547–550.

Laske, G., Phipps Morgan, J. and Orcutt, J.A. (2007) The Hawaiian SWELL pilot experiment – evidence for lithosphere rejuvenation from ocean bottom surface wave data, in *Plates, Plumes, and Planetary Processes* (eds G.R. Foulger and D.M. Jurdy), Geological Society of America, Boulder, CO, pp. 209–233.

Lawver, L.A. and Muller, R.D. (1994) Iceland hotspot track. *Geology*, **22**: 311–314.

Lay, T. (2005) The deep mantle thermo-chemical boundary layer: the putative mantle plume source, in *Plates, Plumes, and Paradigms* (eds G.R. Foulger, J.H. Natland, D.C. Presnall and D.L. Anderson), Geological Society of America, Boulder, CO, pp. 193–206.

Le Bas, M.J. (2000) IUGS reclassification of the high-Mg and picritic volcanic rocks. *Journal of Petrology*, **41**: 1467–1470.

Leitch, A.M., Cordery, M.J., Davies, F.G. et al. (1997) Flood basalts from eclogite-bearing mantle plumes. *South African Journal of Geology*, **100**(4): 311–318.

Lenardic, A., and Kaula, W.M. (1994) Tectonic plates, D″ thermal structure, and the nature of mantle plumes. *Journal of Geophysical Research*, **99**: 15,697–615,708.

Lenardic, A., Moresi, L-N., Jellinek, A.M. et al. (2005) Continental insulation, mantle cooling, and the surface area of oceans and continents. *Earth and Planetary Science Letters*, **234**: 317–333.

Levitt, D.A. and Sandwell, D.T. (1996) Modal depth anomalies from multibeam bathymetry: Is there a South Pacific superswell? *Earth and Planetary Science Letters*, **139**: 1–16.

Lewis, C. (2000) *The Dating Game*, 1st edn, Cambridge University Press, New York, pp. ix + 253.

Li, A.B. and Detrick, R.S. (2003) Azimuthal anisotropy and phase velocity beneath Iceland: implication for plume-ridge interaction. *Earth and Planetary Science Letters*, **214**: 153–165.

Li, X., Sobolev, S.V., Kind, R. et al. (2000a) A detailed receiver function image of the upper mantle discontinuities in the Japan subduction zone. *Earth and Planetary Science Letters*, **183**: 527–541.

Li, X., Kind, R., Priestley, K. et al. (2000b) Mapping the Hawaiian plume conduit with converted seismic waves. *Nature*, **405**(6789): 938–941.

Liégeois, J.-P., Benhallou, A., Azzouni-Sekkal, A. et al. (2005) The Hoggar swell and volcanism: Reactivation of the Precambrian Tuareg shield during Alpine convergence and West African Cenozoic volcanism, in *Plates, Plumes, and Paradigms* (eds G.R. Foulger, J.H. Natland, D.C. Presnall and D.L. Anderson), Geological Society of America, Boulder, CO, pp. 379–400.

Ligi, M., Bonatti, E., Bortoluzzi, E. et al. (1999) Bouvet triple junction in the south Atlantic: Geology and evolution. *Journal of Geophysical Research*, **104**: 29,365–29,385.

Lithgow-Bertelloni, C. and Richards, M.A. (1995) Cenozoic plate driving forces. *Geophysical Research Letters*, **22**: 1317–1320.

Lithgow-Bertelloni, C. and Guynn, J. (2004) Origin of the lithospheric stress field. *Journal of Geophysical Research*, **109**: 10.1029/2003JB002467, Art. No. B001408.

Liu, K.H. (2006) Mantle transition zone discontinuities beneath the Baikal rift and adjacent areas. *Journal of Geophysical Research*, **111**: B11301, doi: 10.1029/2005JB004099, 2006.

Luguet, A., Pearson, D.G., Nowell, G.M. et al. (2008) Enriched Pt-Re-Os isotope systematics in plume lavas explained by metasomatic sulfides. *Science*, **319**: 453–456.

Luis, J.F. and Miranda, J.M. (2008) Reevaluation of magnetic chrons in the North Atlantic between 35°N and 47°N: Implications for the formation of the Azores Triple Junction and associated plateau. *Journal of Geophysical Research*, **113**: B10105, doi: 10.1029/2007JB005573.

Lundin, E.R. and Doré, A.G. (2004) NE-Atlantic break-up: a re-examination of the Iceland mantle plume model and the Atlantic – Arctic linkage, paper presented at North-West European Petroleum Geology and Global Perspectives: 6th Conference, Geological Society, London.

Lundin, E.R. and Doré, T. (2005) The fixity of the Iceland "hotspot" on the Mid-Atlantic Ridge: observational evidence, mechanisms and implications for Atlantic volcanic margins, in *Plates, Plumes, and Paradigms* (eds G.R. Foulger, J.H. Natland, D.C. Presnall and D.L. Anderson), Geological Society of America, Boulder, CO. pp. 627–652.

Lustrino, M. (2005) How the delamination and detachment of lower crust can influence basaltic magmatism. *Earth-Science Reviews*, **72**: 21–38.

Lustrino, M. and Carminati, E. (2007) Phantom plumes in Europe and neighbouring areas, in *Plates, Plumes, and Planetary Processes* (eds G.R. Foulger and D.M. Jurdy), Geological Society of America, Boulder, CO, pp. 723–746.

Lustrino, M. and Wilson, M. (2007) The Circum-Mediterranean Anorogenic Cenozoic Igneous Province. *Earth-Science Reviews*, **81**: 1–65.

Maclennan, J. and Jones, S.M. (2006) Regional uplift, gas hydrate dissociation and the origins of the Paleocene–Eocene Thermal Maximum. *Earth and Planetary Science Letters*, **245**: 65–80.

Macpherson, C.G., Hilton, D.R., Sinton, J.M. et al. (1998) High ^{3}He/^{4}He ratios in the Manus backarc basin: implications for mantle mixing and the origin of plumes in the wesern Pacific ocean. *Geology*, **26**: 1007–1010.

Madeira, J. and Ribeiro, A. (1990) Geodynamic models for the Azores triple junction: a contribution from tectonics. *Tectonophysics*, **184**: 405–415.

Malamud, B.D. and Turcotte, D.L. (1999) How many plumes are there? *Earth and Planetary Science Letters*, **174**: 113–124.

Manea, V.C., Manea, M., Kostoglodov, V. et al. (2005) Thermal models, magma transport, and velocity anomaly estimation beneath southern Kamchatka, in *Plates, Plumes, and Paradigms* (eds G.R. Foulger, J.H. Natland, D.C. Presnall and D.L. Anderson), Geological Society of America, Boulder, CO, pp. 517–536.

Marsh, J.S., Bowen, M.P., Rogers, N.W. et al. (1992) Petrogenesis of late Archaean flood-type basic lavas from the Klipriviersberg Group, Ventersdorp Supergroup, South Africa. *Journal of Petrology*, **33**: 817–847.

Masaitis, V.L. (1983) Permian and Triassic volcanism of Siberia: problems of dynamic reconstructions. *Zapiski Vserossiiskogo Mineralogicheskogo Obshestva*, **4**: 412–425.

Mashima, H. (2005) Partial melting controls on the northwest Kyushu basalts from Saga-Futagoyama. *Island Arc*, **14**: 165–177.

Matias, A. and Jurdy, D.M. (2005) Impact craters as indicators of tectonic and volcanic activity in the Beta-Atla-Themis region, Venus, in *Plates, Plumes, and Paradigms* (eds G.R. Foulger, J.H. Natland, D.C. Presnall and D.L. Anderson), Geological Society of America, Boulder, CO, pp. 825–840.

Mazumder, R. and Sarkar, S. (2004) Sedimentation history of the Palaeoproterozoic Dhanjori Formation, Singhbhum, eastern India. *Precambrian Reserve*, **130**: 267–287.

McCoy, R.M. (2006) *Ending in Ice: The Revolutionary Idea and Tragic Expedition of Alfred Wegener*, Oxford University Press, Oxford, p. 194.

McDougall, I. and Honda, M. (1998) Primordial solar noble-gas component in the Earth: Consequences for the origin and evolution of the Earth and its atmosphere, in *The Earth's Mantle: Composition, Structure and Evolution* (ed. I. Jackson), Cambridge University Press, Cambridge, pp. 159–187.

McHone, J.G. (1996) Constraints on the mantle plume model for Mesozoic alkaline intrusions in northeastern North America. *Canadian Mineralogist*, **34**, 325–334.

McHone, J.G. (2000) Non-plume magmatism and tectonics during the opening of the central Atlantic Ocean. *Tectonophysics*, **316**: 287–296.

McHone, J.G., Anderson, D.L., Beutel, E.K. et al. (2005) Giant dikes, rifts, flood basalts, and plate tecotnics: A contention for mantle models, in *Plates, Plumes, and Paradigms* (eds G.R. Foulger, J.H. Natland, D.C. Presnall and D.L. Anderson), Geological Society of America, Boulder, CO, pp. 401–420.

McKenzie, D.P. (1989) Some remarks on the movement of small melt fractions in the mantle. *Earth and Planetary Scence Letters*, **95**: 53–72.

McKenzie, D.P. and Parker, R.L. (1967) The north Pacific: An example of tectonics on a sphere. *Nature*, **216**: 1276–1280.

McKenzie, D.P. and Weiss, N. (1975) Speculations on the thermal and tectonic history of the Earth. *Geophysics Journal of the Royal Astronomical Society*, **42**: 131–174.

McKenzie, D.P. and Bickle, J. (1988) The volume and composition of melt generated by extension of the lithosphere. *Journal of Petrology*, **29**: 625–679.

McKenzie, D.P. and O'nions, R.K. (1991) Partial melt distributions from inversion of rare Earth element concentrations. *Journal of Petrology*, **32**(5): 1021–1091.

McKenzie, D.P., Stracke, A., Blichert-Toft, J. et al. (2004) Source enrichment processes responsible for isotopic anomalies in oceanic island basalts. *Geochimica et Cosmochimica Acta*, **68**: 2699–2724.

McMillan, M.E., Heller, P.L. and Wing, S.L. (2006) History and causes of post-Laramide relief in the Rocky Mountain orogenic plateau. *Geological Society of America Bulletin*, **118**: 393–405.

McNutt, M.K. and Judge, A.V. (1990) The superswell and mantle dynamics beneath the south Pacific. *Science*, **248**: 969–975.

McNutt, M.K., Caress, D.W., Reynolds, J. et al. (1997) Failure of plume theory to explain midplate volcanism in the southern Austral Islands. *Nature*, **389**: 479–482.

Megnin, C. and Romanowicz, B. (2000) The 3D velocity structure of the mantle from the inversion of body, surface, and higher mode waveforms. *Geophysics Journal International*, **143**: 709–728.

Meibom, A. (2008) The rise and fall of a great idea. *Science*, **319**: 418–419.

Meibom, A. and Anderson, D.L. (2004) The statistical upper mantle assemblage. *Earth and Planetary Science Letters*, **217**: 123–139.

Meibom, A., Sleep, N.H. Chamberlain, C.P. et al. (2002) Re-Os isotopic evidence for long-lived heterogeneity and equilibrium processes in Earth's upper mantle. *Nature*, **419**: 705–708.

Meibom, A., Anderson, D.L., Sleep, N.H. et al. (2003) Are high $^3He/^4He$ ratios in oceanic basalts an indicator of deep-mantle plume components? *Earth and Planetary Science Letters*, **208**: 197–204.

Meibom, A., Sleep, N.H., Zahnle, K. et al. (2005) Models for noble gases in mantle geochemistry: Some observations and alternatives, in *Plates, Plumes, and Paradigms* (eds G.R. Foulger, J.H. Natland, D.C. Presnall and D.L. Anderson), Geological Society of America, Boulder, CO, pp. 347–364.

Menard, H.W. (1964) *Marine Geology of the Pacific*, McGraw-Hill, New York.

Menard, H.W. (1984) Darwin reprise. *Journal of Geophysical Research*, **89**: 9960–9968.

Menke, W. (1999) Crustal isostasy indicates anomalous densities beneath Iceland. *Geophysical Research Letters*, **26**: 1215–1218.

Menke, W. and Levin, V. (1994) Cold crust in a hot spot. *Geophysical Research Letters*, **21**(18): 1967–1970.

Menke, W., Levin, V. and Sethi, R. (1995) Seismic attenuation in the crust at the mid-Atlantic plate boundary in south-west Iceland. *Geophysics Journal International*, **122**: 175–182.

Menzies, M.A., Baker, J., Bosence, D. et al. (1992) The timing of magmatism, uplift and crustal extension: preliminary observations from Yemen. *Geological Society of London, Special Publications*, **68**: 293–304.

Menzies, M., Gallagher, K., Yelland, A. et al. (1997) Volcanic and nonvolcanic rifted margins of the Red Sea and Gulf of Aden: Crustal cooling and margin evolution in Yemen. *Geochimica et Cosmochimoca Acta*, **61**, 2511–2527.

Menzies, M., Baker, J. and Chazot, G. (2001) Cenozoic plume evolution and flood basalts in Yemen: A key to understanding older examples, Geological Society of America, Boulder, CO, pp. 23–36.

Meschede, M. (1986) A method of discriminating between different types of mid-ocean tidge basalts and continental tholeiites with the Nb-Zr-Y diagram. *Chemical Geology*, **56**: 207–218.

Mohorovičić, A. (1909) Das Beben. *Jb. met. Obs. Zagreb (Agram.)*, **9**, 1–63.

Molnar, P. and Atwater, T. (1973) Relative motion of hot spots in the mantle. *Nature*, **246**: 288–291.

Montagner, J.P. and Guillot, L. (2000) Seismic anisotropy in the Earth's mantle, in *Problems in Geophysics for the New Millennium* (eds E. Boschi, G. Ekström and A. Morelli), Editrice Compositori, Bologna, Italy, 217–254.

Montelli, R., Nolet, G., Masters, G. et al. (2004a) Global P and PP traveltime tomography: rays versus waves. *Geophysics Journal International*, **158**: 637–654.

Montelli, R., Nolet, G., Dahlen, F.A. et al. (2004b) Finite frequency tomography reveals a variety of plumes in the mantle. *Science*, **303**: 338–343.

Montelli, R., Nolet, G., Dahlen, F. et al. (2006) A catalogue of deep mantle plumes: new results from finite-frequency tomography. *Geochemistry, Geophysics, Geosystems*, **7**: Q11007, doi:11010.11029/12006GC001248.

Moore, A., Blenkinsop, T. and Cotterill, F. (2008) Controls on post-Gondwana alkaline volcanism in Southern Africa. *Earth and Planetary Science Letters*, **268**: 151–164.

Moreira, M. and Sarda, P. (2000) Noble gas constraints on degassing processes. *Earth and Planetary Science Letters*, **176**: 375–386.

Morgan, W.J. (1971) Convection plumes in the lower mantle. *Nature*, **230**, 42–43.

Morgan, W.J. (1981) Hotspot tracks and the opening of the Atlantic and Indian oceans, in *The Sea* (ed. C. Emiliani), Wiley, New York, pp. 443–487.

Morgan, W.J. (1983) Hotspot tracks and the early rifting *of the Atlantic. Tectonophysics*, **94**: 123–139.

Morgan, W.J. and Phipps Morgan, W. (2007) Plate velocities in the hotspot reference frame, in *Plates, Plumes, and Planetary Processes* (eds G.R. Foulger and D M. Jurdy), Geological Society of America, Boulder, CO, pp. 65–78.

Murakami, M., Hirose, K., Sata, N. et al. (2004) Phase transition of $MgSiO_3$ perovskite in the deep lower mantle. *Science*, **304**: 855–858.

Mutter, C.Z. and Mutter, J.C. (1993) Variations in thickness of layer 3 dominate oceanic crustal structure. *Earth and Planetary Science Letters*, **117**: 295–317.

Nakamura, E., Campbell, I.H. and Sun, S. (1985) The influence of subduction processes on the geochemistry of Japanese alkaline basalts. *Nature*, **316**(6023): 55–58.

Nakanishi, I. and Anderson, D.L. (1984) Aspherical heterogeneity of the mantle from phase velocities of mantle waves. *Nature*, **307**(5947): 117–121.

Natland, J.H. (1980) Progression of volcanism in the Samoan linear volcanic chain. *American Journal of Science*, **280-A**: 709–735.

Natland, J.H. (1989) Partial melting of a lithologically heterogeneous mantle: Inferences from crystallization histories of magnesian abyssal tholeiites from the Siqueiros Fracture Zone, in *Magmatism in the Ocean Basins* (eds A.D. Saunders and M. Norry), Geological Society, of London, Special Publication, pp. 41–77.

Natland, J.H. (2003) Capture of mantle helium by growing olivine phenocrysts in picritic basalts from the Juan Fernandez Islands, SE Pacific. *Journal of Petrology*, **44**: 421–456.

Natland, J.H. (2006) Reginald Aldworth Daly (1871–1957): Eclectic theoretician of the Earth. GSA *Today*, **16**: 24–26.

Natland, J.H. (2007) Delta-Nb and the role of magma mixing at the East Pacific Rise and Iceland, in *Plates, Plumes, and Planetary Processes* (eds G.R. Foulger and D.M. Jurdy), Geological Society of America, Boulder, CO, pp. 413–450.

Natland, J.H., and Dick, H.J.B. (2001) Formation of the lower ocean crust and the crystallization of gabbroic cumulates at a very slowly spreading ridge. *Journal of Volcanology and Geothermal Research*, **110**: 191–233.

Natland, J.H. and Winterer, E.L. (2005) Fissure control on volcanic action in the Pacific, in *Plates, Plumes, and Paradigms* (eds G.R. Foulger, J.H. Natland, D.C. Presnall and D.L. Anderson), Geological Society of America, Boulder, CO, pp. 687–710.

Nelson, D.R., Trendall, A.F., de Laeter, J.R. et al. (1992) A comparative study of the geochemical and isotopic systematics of late Archean flood basalts from the Pilbara and Kaapvaal Cratons. *Precambrian Research*, **54**: 231–256.

Nielsen, S.B., Stephenson, R. and Thomsen, E. (2007) Dynamics of Mid-Palaeocene North Atlantic rifting linked with European intra-plate deformations. *Nature*, **450**: 1071–1074.

Nilsen, T.H. (1978) Lower Tertiary laterite on the Iceland-Faeroe ridge and the Thulean land bridge. *Nature*, **274**: 786–788.

Niu, Y. and O'Hara, M.J. (2003) Origin of ocean island basalts: A new perspective from petrology, geochemistry, and mineral physics considerations. *Journal of Geophysical Research*, **108**: 2209, doi:2210.1029/2002JB002048.

Niu, Y. and O'Hara, M.J. (2008) Global correlations of ocean ridge basalt chemistry with axial depth: A new perspective. *Journal of Petrology*, **49**: 633–664.

Niu, Y., Regelous, M., Wendt, I.J. et al. (2002) Geochemistry of near-EPR seamounts: importance of source vs. process and the origin of enriched mantle component. *Earth and Planetary Science Letters*, **199**: 327–345.

Nixon, P.H. (1987) *Mantle Xenoliths*, John Wiley & Sons Ltd., Chichester UK, p. 844.

Norton, I.O. (2007) Speculations on tectonic origin of the Hawaii hotspot, in *Plates, Plumes, and Planetary Processes* (eds G.R. Foulger and D.M. Jurdy), Geological Society of America, Boulder, CO, pp. 451–470.

Nunns, A.G. (1983) Plate tectonic evolution of the Greenland-Scotland ridge and surrounding regions, in *Structure and Development of the Greenland-Scotland Ridge* (eds M.H.P. Bott, S. Saxov, M. Talwani and J. Thiede), Plenum Press, New York and London, pp. 1–30.

O'Connor, J.M., Stoffers, P., van den Bogaard, P. et al. (1999) First seamount age evidence for significantly slower African plate motion since 19 to 30 Ma. *Earth and Planetary Science Letters*, **171**: 575–589.

O'Connor, J.M., Stoffers, P., Wijbrans, J.R. et al. (2000) Evidence from episodic seamount volcanism for pulsing of the Iceland plume in the past 70 Myr. *Nature*, **408**(6815): 954–958.

O'Connor, J.M., Stoffers, P., Wijbrans, J.R. et al. (2007) Migration of widespread long-lived volcanism across the Galápagos Volcanic Province: Evidence for a broad hotspot melting anomaly? *Earth and Planetary Science Letters*, **263**: 339–354.

O'Hara, M.J. (1975) Is there an Icelandic mantle plume? *Nature*, **253**: 708–710.

Oganov, A.R. and Ono, S. (2004) Theoretical and experimental evidence for a post-perovskite phase of $MgSiO_3$ in Earth's 'D' layer. *Nature*, **430**, 445–448.

Omar, G.I. and Steckler, M.S. (1995) Fission track evidence on the initial rifting of the Red Sea: Two pulses, no propagation. *Science*, **270**: 1341–1344.

Oreskes, N. (1999) *The Rejection of Continental Drift: Theory and Method in American Earth Science*, Oxford University Press, pp. ix + 420.

Oskarsson, N., Sigvaldason, G.E. and Steinthorsson, S. (1982) A dynamic model of rift zone petrogenesis and the regional petrology of Iceland. *Journal of Petrology*, **23**: 28–74.

Ozima, M. and Igarashi, G. (2000) The primordial noble gases in the Earth: A key constraint on Earth evolution models. *Earth and Planetary Science Letters*, **176**: 219–232.

Palmason, G. (1971) *Crustal Structure Of Iceland From Explosion Seismology*, Society of Science. Isl., Reykjavik. p. 187.

Palmason, G. (1980) A continuum model of crustal generation in Iceland; Kinematic aspects. *Journal of Geophysics*, **47**: 7–18.

Panza, G.F., Peccerillo, A., Aoudia et al. (2007) Geophysical and petrological modelling of the structure and composition of the crust and upper mantle in complex geodynamic settings: the Tyrrhenian Sea and surroundings. *Earth-Science Reviews*, **80**: 1–46.

Parman, S.W. and Grove, T.L. (2005) Komatiites in the plume debate, in *Plates, Plumes, and Paradigms* (eds G.R. Foulger, J.H. Natland, D.C. Presnall and D.L. Anderson), Geological Society of America, Boulder, CO, pp. 249–256.

Parman, S.W., Grove, T.L. and Dann, J.C. (2001) The production of Barberton komatiites in an Archean subduction zone. *Geophysical Research Letters*, **28**: 2513–2516.

Parman, S.W., Kurz, M.D., Hart, S.R. et al. (2005) Helium solubility in olivine and implications for high ^{3}He/^{4}He in ocean island basalts. *Nature*, **437**: 1140–1143.

Parsons, B. and Sclater, J.G. (1977) An analysis of the variation of ocean floor bathymetry and heat flow with age. *Journal of Geophysical Research*, **82**: 803–827.

Pavlenkova, G.A. and Pavlenkova, N.I. (2006) Upper mantle structure of the Northern Eurasia from peaceful nuclear explosion data. *Tectonophysics*, **416**: 33–52.

Payne, S.J., M.R., and King, R.W. (2008) Strain rates and contemporary deformation in the Snake River Plain and surrouonding Basin and Range from GPS and seismicity. *Geology*, **36**: 647, 249–256, 650.

Peate, D.W. (1997) The Paraná-Etendeka province, in *Large Igneous Provinces: Continental, Oceanic, and Planetary Flood Volcanism*, American Geophysical Union, Washington, DC, pp. 217–245.

Peate, D.W., Hawkesworth, C.J. and Mantovani, M.S.M. (1992) Chemical stratigraphy of the Paraná lavas (South America): classification of magma types and their spatial distribution. *Bulletin of Volcanics*, **55**: 119–139.

Peccerillo, A. (1999) Multiple mantle metasomatismin central-southern Italy: geochemical effects, timing and geodynamic implications. *Geology*, **27**: 315–318.

Peccerillo, A. (2005) *Plio-Quaternary Volcanism in Italy. Petrology, Geochemistry, Geodynamics*, Springer, Heidelberg.

Peccerillo, A. and Lustrino, M. (2005) Compositional variations of Plio-Quaternary magmatism in the circum-Tyrrhenian area: Deep- versus shallow mantle processes, in *Plates, Plumes, and Paradigms* (eds G.R. Foulger, J.H. Natland, D.C. Presnall and D.L. Anderson), Geological Society of America, Boulder, CO, pp. 421–434.

Pederson, J.L., Karlstrom, K., Sharp, W. et al. (2002a) Differential incision of the Grand Canyon related to Quaternary faulting–Constraints from U-series and Ar/Ar dating. *Geologial Society of America Bulletin*, **30**: 739–742.

Pederson, J.L., Mackley, R.D. and Eddleman, J.L. (2002b) Colorado plateau uplift and erosion evaluated using GIS. *GSA Today*, **12**: 4–10.

Perlt, J., Heinert, M. and Niemeier, W. (2008), The continental margin in Iceland – A snapshot derived from combined GPS networks. *Tectonophysics*, **447**: 155–166.

Petit, C., Déverchère, J., Calais, E. et al. (2002) Deep structure and mechanical behavior of the lithosphere in the Hangai-Hovsgol region, Mongolia: new constraints from gravity modeling. *Earth and Planetary Science Letters*, **197**: 133–149.

Phillips, B.R. and Bunge, H.-P. (2005) Heterogeneity and time dependence in 3D spherical mantle convection models with continental drift. *Earth and Planetary Science Letters*, **233**: 121–135.

Phipps-Morgan, J., Morgan, W.J. and Price, E. (1995) Hotspot melting generates both hotspot volcanism and a hotspot swell? *Journal Geophysical Research*, **100**: 8045–8062.

Piccirillo, E.M. and Melfi, A.J. (1988) *The Mesozoic Flood Volcanism from the Paraná Basin (Brazil). Petrogenetic and Geophysical Aspects*, Iag-Usp, São Paulo, Brazil, p. 600.

Piccirillo, E.M., Civetta, L., Petrini, R. et al. (1989) Regional variations within the Paraná flood basalts (Southern Brazil): evidence for subcontinental mantle heterogeneity and crustal contamination. *Chemical Geology*, **75**: 103–122.

Pilet, S., Hernandez, J., Bussy, F. et al. (2004) Short-term metasomatic control of Nb/Th ratios in the mantle sources of intra-plate basalts. *Geology*, **32**: 113–116.

Pilet, S., Baker, M.B. and Stolper, E.M. (2008) Metasomatized lithosphere and the origin of alkaline lavas. *Science*, **320**: 916–919.

Pilidou, S., Priestley, K.F., Debayle, E. et al. (2005) Rayleigh wave tomography in the North Atlantic: high resolution images of the Iceland, Azores and Eifel mantle plumes. *Lithos*, **79**: 453–474.

Plomerová, J., Achauer, U., Babuška, V.L. and the BOHEMA working group (2007) Upper mantle beneath the Eger Rift (Central Europe): plume or asthenosphere upwelling? *Geophysics Journal International*, **169**: 675–682.

Pollack, H.N., Hurter, S.J. and Johnston, J.R. (1993) Heat loss from the Earth's interior: analysis of the global data set. *Reviews in Geophysics*, **31**: 267–280.

Popper, K. (1959) *The Logic of Scientific Discovery*, Hutchinson, London.

Poreda, R.J. and Farley, KA. (1992) Rare gases in Samoan xenoliths. *Earth and Planetary Science Letters*, **113**(1–2): 129–144.

Poustovetov, A. and Roeder, P.L. (2000) The distribution of Cr between basaltic melt and chromian spinel as an oxygen geobarometer. *Canadian Mineralogist*, **39**: 309–317.

Powell, J.L. (1998) *Night comes to the Cretaceous*, W.H. Freeman & Co, New York, pp. xvi + 250.

Praeg, D., Stoker, M.S., Shannon, P.M. et al. (2005) Episodic Cenozoic tectonism and the development of the NW European "passive" continental margin. *Marine and Petroleum Geology*, **22**: 1007–1030.

Presnall, D.C. and Gudfinnsson, G.H. (2005) Carbonate-rich melts in the oceanic low-velocity zone and deep mantle, in *Plates, Plumes, and Paradigms* (eds G.R. Foulger, J.H. Natland, D.C. Presnall and D.L. Anderson), Geological Society of America, Special Paper 388, pp. 207–216.

Presnall, D.C. and Gudfinnsson, G.H. (2008) Origin of the oceanic lithosphere. *Journal of Petrology*, **49**: 615–632.

Presnall, D.C., Gudfinnsson, G.H. and Walter, M.J. (2002) Generation of mid-ocean ridge basalts at pressures from 1 to 7 GPa. *Geochimica et Cosmochimica Acta*, **66**: 2073–2090.

Presnall, D.C., Gudfinnsson, G.H., Weng, Y.-H. et al. (2010) Oceanic volcanism without mantle plumes. *Journal of Petrology*, in press.

Prestvik, T., Goldberg, S., Karlsson, H. et al. (2001) Anomalous strontium and lead isotope signatures in the off-rift Oraefajokull central volcano in south-east Iceland. Evidence for enriched endmember(s) of the Iceland mantle plume? *Earth and Planetary Science Letters*, **190**(3–4): 211–220.

Price, N.J. (2001) *Major Impacts and Plate Tectonics*, Routledge, London, p. 416.

Pringle, M.S., Frey, F.A. and Mervine, E.E. (2008) A simple linear age progression for the Ninetyeast Ridge, Indian Ocean: New constraints on Indian plate tectonics and hotspot dynamics. *EOS Trans. AGU*, **89**: Fall Meet. Suppl., Abstract T54B-03.

Puchtel, I.S., Walker, R.J., Anhaeusser, C.R. et al. (2009) Re–Os isotope systematics and HSE abundances of the 3.5 Ga Schapenburg komatiites, South Africa: Hydrous melting or prolonged survival of primordial heterogeneities in the mantle? *Chemical Geology*, **262**: 391–405.

Putirka, K.D. (2005) Mantle potential temperatures at Hawaii, Iceland, and the mid-ocean ridge system, as inferred from olivine phenocrysts: Evidence for thermally driven mantle plumes. *Geochemistry, Geophysics, Geosystems*, **6**: doi:10.1029/2005GC000915.

Putirka, K.D. (2008) Excess temperatures at ocean islands: Implications fo mantle layering and convection. *Geology*, **36**: 283–286.

Putirka, K.D., Perfit, M.R., Ryerson, F.J. et al. (2007) Ambient and excess mantle temperatures, olivine thermometry, and active vs. passive upwelling. *Chemical Geology*, **241**: 177–206.

Ramberg, H. (1967) *Gravity, Deformation and the Earth's Crust*, 1st edn, Academic Press, London, p. 214

Ramberg, H. (1981) *Gravity, Deformation and the Earth's Crust*, 2nd edn, Academic Press, London, p. 452.

Rappaport, Y., Naar, D.F., Barton, C.C. et al. (1997) Morphology and distribution of seamounts surrounding Easter Island. *Journal of Geophysical Research*, **102**: 24,713–724,728.

Raymond, C.A., Stock, J.M. and Cande, S.C. (2000) Fast Paleogene motion of the Pacific hotspots from revised global plate circuit constraints, in *History and Dynamics of Plate Motions* (eds M.A. Richards, R.G. Gordon and R.D. van der Hilst), AGU Geophysical Monograph, pp. 359–375.

Reese, C.C., Viatcheslav, S.S. and Orth, C.P. (2007) Interaction between local magma ocean evolution and mantle dynamics on Mars, in *Plates, Plumes, and Planetary Processes* (eds G.R. Foulger and D.M. Jurdy), Geological Society of America, Boulder, CO, pp. 913–932.

Regelous, M., Hofmann, A.W., Abouchami, W. et al. (2003) Geochemistry of lavas from the Emperor seamounts, and the geochemical evolution of Hawaiian magmatism from 85 to 42 Ma. *Journal of Petrology*, **44**: 113–140.

Regelous, M., Niu, Y., Abouchami, W. et al. (2009) Shallow origin for South Atlantic Dupal anomaly from lower continental crust: Geochemical evidence from the Mid-Atlantic Ridge at 26°S. *Lithos*, **112**: 57–72.

Ren, Z.-Y., Ingle, S., Takahashi, E. et al. (2005) The chemical structure of the Hawaiian mantle plume. *Nature*, **436**: 837–840.

Ribe, N.M. and Christensen, U.R. (1994) Three-dimensional modeling of plume-lithosphere interaction. *Journal of Geophysical Research*, **99**: 669–682.

Ribe, N.M., Christensen, U.R. and Theissing, J. (1995) The dynamics of plume-ridge interaction, 1: Ridge-centered plumes. *Earth and Planetary Science Letters*, **134**: 155–168.

Richardson, W.P., Okal, E.A. and Van der Lee, S. (2000) Rayleigh-wave tomography of the Ontong-Java Plateau. *Physical Earth Planetary International*, **118**: 29–51.

Richter, F.M. and Parsons, B. (1975) On the interaction of two modes of convection in the mantle. *Journal of Geophysical Research*, **80**: 2529–2541.

Ringwood, A.E. (1975) *Composition and Petrology of the Earth's Mantle*, McGraw-Hill, New York, p. 618.

Ritsema, J. (2005) Global seismic maps, in *Plates, Plumes and Paradigms* (eds G.R. Foulger, J.H. Natland, D.C. Presnall and D.L. Anderson), Geological Society of America, Boulder, CO, pp. 11–18.

Ritsema, J., and Allen, R.M. (2003) The elusive mantle plume. *Earth and Planetary Science Letters*, **207**: 1–12.

Ritsema, J., van Heijst, H.J. and Woodhouse, J.H. (1999) Complex shear wave velocity structure imaged beneath Africa and Iceland. *Science*, **286**: 1925–1928.

Ritter, J.R.R. (2007) The seismic signature of the Eifel plume, in *Mantle Plumes: A Multidisciplinary Approach* (eds J.R. R. Ritter and U.R. Christensen), Springer Berlin Heidelberg, pp. 379–404.

Ritter, J.R.R., Jordan, M., Christensen, U.R. et al. (2001) A mantle plume below the Eifel volcanic fields, Germany. *Earth and Planetary Science Letters*, **186**: 7–14.

Ritter, J.R R., Jordan, M., Achauer, U., Christensen, U.R. and The Eifel. Plume Team (2002) Seismic structure and physical state of the Eifel plume, Germany. *Geophysical Research Abstracts*, **4**: 01661.

Roberge, J., Wallace, P.J., White, R.V. et al. (2005) Anomalous uplift and subsidence of the Ontong Java Plateau inferred from CO_2 contents of submarine basaltic glasses. *Geology*, **33**: 501–504.

Robinson, E.M. (1988) The topographic and gravitational expression of density anomalies due to melt extraction in the uppermost oceanic mantle. *Earth and Planetary Science Letters*, **90**: 221–228.

Robinson, J.E. and Eakins, B.W. (2006) Calculated volumes of individual shield volcanoes at the young end of the Hawaiian Ridge. *Journal of Volcanology and Geothermal Research*, **151**: 309–317.

Rocchi, S., Armienti, P., D'Orazio, M. et al. (2002) Cenozoic magmatism in the western Ross Embayment: role of mantle plume vs. plate dynamics in the development of the West Antarctic Rift System. *Journal of Geophysical Research*, **107**(B9): 2195, doi: 10.1029/2001JB000515.

Rocchi, S., Storti, F., Di Vincenzo, G. et al. (2003) Intraplate strike-slip tectonics as alternative to mantle plume activity for the Cenozoic rift magmatism in the Ross Sea region, Antarctica, in *Intraplate Strike-Slip Deformation Belts* (eds F. Storti, R.E. Holdsworth and F. Salvini), Geological Society Special Publication, Geological Society of London, London, pp. 158–171.

Rocchi, S., Armienti, P. and Di Vincenzo, G. (2005) No plume, no rift magmatism in the West Antarctic Rift, in *Plates, Plumes, and Paradigms* (eds G.R. Foulger, J.H. Natland, D.C. Presnall and D.L. Anderson), Geological Society of America, Boulder, CO, pp. 435–448.

Rona, P A. and Richardson, E.S. (1978) Early Cenozoic global plate reorganization. *Earth and Planetary Science Letters*, **40**: 1–11.

Ross, P.-S. and White, J.D.L. (2005) Mafic, large-volume, pyroclastic density current deposits from phreatomagmatic eruptions in the Ferrar Large Igneous Province, Antarctica. *Journal of Geology*, **113**: 627–649.

Ross, P.-S., Ukstins Peate, I., McClintock, M.K. et al. (2005) Mafic volcaniclastic deposits in flood basalt provinces: A review. *Journal of Volcanism and Geothermal Research*, **145**: 281–314.

Roth, J.B., Fouch, M.J., James, D.E. et al. (2008) Three-dimensional seismic velocity structure of the northwestern United States. *Geophysics Research Letters*, **35**(L15304).

Rubin, A.M. (1995) Propagation of magma-filled cracks. *Annual Reviews in Earth and Planetary Sciences*, **23**: 287–336.

Russell, S.A., Lay, T. and Garnero, E.J. (1998) Seismic evidence for small-scale dynamics in the lowermost mantle at the root of the Hawaiian hotspot. *Nature*, **396**: 255–258.

Saemundsson, K. (1979) Outline of the geology of Iceland. *Jokull*, **29**: 7–28.

Sager, W.W. (2005) What built Shatsky Rise, a mantle plume or ridge tectonics? in *Plates, Plumes, and Paradigms* (eds G.R. Foulger, J.H. Natland, D.C. Presnall and D.L. Anderson), Geological Society of America, Boulder, CO, pp. 721–734.

Sager, W.W. (2007) Divergence between paleomagnetic and hotspot model predicted polar wander for the Pacific plate with implications for hotspot fixity, in *Plates, Plumes, and Planetary Processes* (eds G.R. Foulger and D.M. Jurdy), Geological Society of America, Boulder, CO, pp. 335–358.

Sager, W.W., Jinho, K., Klaus, A. et al. (1999) Bathymetry of Shatsky Rise, northwest Pacific Ocean; implications for ocean plateau development at a triple junction. *Journal of Geophysical Research*, **104**: 7557–7576.

Sahagian, D., Proussevitch, A. and Carlson, W. (2002) Timing of Colorado Plateau uplift: Initial constraints from vesicular basalt-derived paleoelevations. *Geology*, **30**: 807–810.

Sallares, V. and Charvis, P. (2003) Crustal thickness constraints on the geodynamic evolution of hte Galápagos volcanic province. *Earth and Planetary Science Letters*, **214**: 545–559.

Sallares, V. and Calahorrano, A. (2007) Geophysical characterization of mantle melting anomaliles: A crustal view, in *Plates, Plumes, and Planetary Processes* (eds G.R. Foulger and D.M. Jurdy), Geological Society of America, Boulder, CO, pp. 507–524.

Sallares, V., Charvis, P., Flueh, E.R., Bialas, J. and The SALIERI Scientific Party (2005) Seismic structure of the Carnegie ridge and the nature of the Galapagos hotspot. *Geophysics Journal International*, **161**: 763–788.

Sandwell, D.T. and Smith, W.H.F. (1997) Marine gravity anomaly from Geosat and ERS 1 satellite altimetry. *Journal Geophysics Research*, **102**: 10,039–10,054.

Sandwell, D.T. and Fialko, Y. (2004) Warping and cracking of the Pacific plate by thermal contraction. *Journal Geophysics Research*, **109**: doi:10.1029/2004JB003091.

Sandwell, D.T., Winterer, E.L., Mammerickx, J. et al. (1995) Evidence for diffuse extension of the Pacific plate from Pukapuka ridges and cross-grain gravity lineations. *Journal Geophysics Research*, **100**: 15,087–15,089.

Sandwell, D.T., Anderson, D.L. and Wessel, P. (2005) Global tectonic maps, in *Plates, Plumes, and Paradigms* (eds G.R. Foulger, J.H. Natland, D.C. Presnall and D.L. Anderson), Geological Society of America, Boulder, CO, pp. 1–10.

Saunders, A.D., Fitton, J.G., Kerr, A.C. et al. (1997) The North Atlantic Igneous Province, in *Large Igneous Provinces* (eds J. Mahoney and M.F. Coffin), American Geophysical Union, pp. 45–93.

Saunders, A.D., England, R.W., Reichow, M.K. et al. (2005) A mantle plume origin for the Siberian traps: uplift and extension in the West Siberian Basin, Russia. *Lithos*, **79**: 407–424.

Scherstén, A., Elliott, T., Hawkesworth, C. et al. (2004) Tungsten isotope evidence that mantle plumes contain no contribution from the Earth's core. *Nature*, **427**: 234–237.

Schilling, J.-G. (1973) Iceland mantle plume: Geochemical study of Reykjanes ridge. *Nature*, **242**: 565–571.

Schilling, J.-G. (1986) Geochemical and isotopic variation along the Mid-Atlantic Ridge axis from 79°N to 0°N, in *The Western North Atlantic Region: Boulder, Colorado, Geological Society of America* (eds P.R. Vogt and B.E. Tucholke), The Geology of North America, M: 137–156.

Schilling, J.-G., Zajac, M., Evans, R. et al. (1983) Petrologic and geochemical variations along the mid-Atlantic ridge from 29°N to 73°N. *American Journal of Science*, **283**: 510–586.

Schilling, J.-G., Thompson, G., Kingsley, R. et al. (1985a) Hotspot-migrating ridge interaction in the south Atlantic. *Nature*, **313**: 187–191.

Schilling, J.-G., Sigurdsson, H., Davis, A.N. et al. (1985b) Easter microplate evolution. *Nature*, **317**: 325–331.

Schmerr, N. and Garnero, E.J. (2007) Upper mantle discontinuity topography from thermal and chemical heterogeneity. *Science*, **318**: 623–626.

Schutt, D.L. and Lesher, C.E. (2006) Effects of melt depletion on the density and seismic velocity of garnet and spinel lherzolite. *Journal of Geophysical Research*, **111**: B05401.

Searle, R. (1980) Tectonic pattern of the Azores spreading centre and triple junction. *Earth and Planetary Science Letters*, **51**: 415–434.

Searle, R.C., Francheteau, J. and Cornaglia, B. (1995) New observations on mid-plate volcanism and the tectonic history of the Pacific plate, Tahiti to Easter microplate. *Earth and Planetary Science Letters*, **131**(3–4): 395–156.

Sears, J.W. (2007) Lithospheric control of Gondwana breakup: Implications of a trans-Gondwana icosahedral fracture system, in *Plates, Plumes, and Planetary Processes* (eds G.R. Foulger and D.M. Jurdy), Geological Society of America, Boulder, CO, pp. 593–601.

Self, S., Jay, A.E., Widdowson, M. et al. (2008) Correlation of the Deccan and Rajahmundry Trap lavas: Are these the longest and largest lava flows on Earth? *Journal of Volcanology and Geothermal Research*, **172**(3–19): 3–9.

Sengor, A.M.C. (2001) Elevation as indicator of mantle-plume activity, in *Mantle Plumes: Their Identification Through Time* (eds R.E. Ernst and K.L. Buchan), Geological Society of America, Boulder, CO, pp. 183–225.

Sengor, A.M.C., Ozeren, S., Zor, E. et al. (2003) East Anatolian high plateau as a mantle-supported, N-S shortened domal structure. *Geophysical Research Letters*, **30**: 8045.

Shapiro, M.N. (2005) Kinematics of the Campanian–Maastrichtian island arcs in northeastern Asia in light of drilling results on the Emperor Seamounts. *Geotectonics*, **39**: 408–415.

Sharma, K.K. (2007) K-T magmatism and basin tectonism in western Rajasthan, India, results from extensional tectonics and not from Réunion plume activity, in *Plates, Plumes, and Planetary Processes* (eds G.R. Foulger and D.M. Jurdy), Geological Society of America, Boulder, CO, pp. 775–784.

Sharp, W.D. and Clague, D.A. (2006) 50-Ma initiation of Hawaiian-Emperor bend records major change in Pacific plate motion. *Science*, **313**: 1281–1284.

Shaw, H.R. (1973) Mantle convection and volcanic periodicity in the Pacific; evidence from Hawaii. *Geological Society of America Bulletin*, **84**: 1505–1526.

Shen, Y., Solomon, S.C., Bjarnason, I.T. et al. (1998) Seismic evidence for a lower-mantle origin of the Iceland plume. *Nature*, **395**: 62–65.

Shen, Y., Solomon, S.C., Bjarnason, I.T. et al. (2002) Seismic evidence for a tilted mantle plume and north-south mantle flow beneath Iceland. *Earth and Planetary Science Letters*, **197**: 261–272.

Sheth, H.C. (1999a) Flood basalts and large igneous provinces from deep mantle plumes: fact, fiction, and fallacy. *Tectonophysics*, **311**: 1–29.

Sheth, H.C. (1999b) A historical approach to continental flood basalt volcanism: insights into pre-volcanic rifting, sedimentation, and early alkaline magmatism. *Earth and Planetary Science Letters*, **168**: 19–26.

Sheth, H.C. (2005a) Were the Deccan flood basalts derived in part from ancient oceanic crust within the Indian continental lithosphere? *Gondwana Research*, **8**: 109–127.

Sheth, H.C. (2005b) From Deccan to Réunion: No trace of a mantle plume, in *Plates, Plumes, and Paradigms* (eds G.R. Foulger, J.H. Natland, D.C. Presnall and D.L. Anderson), Geological Society of America, Boulder, CO, pp. 477–502.

Sheth, H.C. (2007a) Plume-related regional pre-volcanic uplift in the Deccan Traps: Absence of evidence, evidence of absence: Discussion, in *Plates, Plumes, and Planetary Processes* (eds G.R. Foulger and D.M. Jurdy), Geological Society of America, Boulder, CO, pp. 803–813.

Sheth, H.C. (2007b) "Large Igneous Provinces (LIPs)": Definition, recommended terminology, and a hierarchical classification. *Earth-Science Reviews*, **85**: 117–124.

Silveira, G. and Stutzmann, E. (2002) Anisotropic tomography of the Atlantic ocean. *Physical Earth Planetary International*, **132**: 237–248.

Silveira, G., Stutzmann, E., Griot, D.-A. et al. (1998) Anisotropic tomography of the Atlantic Ocean from Rayleigh surface waves. *Physical Earth Planetary International*, **106**: 257–273.

Silver, P.G., Behn, M.D., Kelley, K. et al. (2006) Understanding cratonic flood basalts. *Earth and Planetary Science Letters*, **245**: 190–201.

Sipkin, S.A. and Jordan, T.H. (1980) Regional variation of Q_{ScS}. *Bulletin of the Seismological Society of America*, **70**: 1071–1102.

Sleep, N.H. (1987) Lithospheric heating by mantle plumes. *Geophysics Journal of the Royal Astronomical Society*, **91**: 1–12.

Sleep, N.H. (1990) Hotspots and mantle plumes: Some phenomenology. *Journal of Geophysical Research*, **95**: 6715–6736.

Sleep, N.H. (1996) Lateral flow of hot plume material ponded at sublithospheric depths. *Journal of Geophysical Research*, **101**: 28,065–28,083.

Sleep, N.H. (2002) Ridge-crossing mantle plumes and gaps in tracks. *Geochemica, Geophysocs, Geosystem*, **3**: art. no. 8505.

Sleep, N.H. (2004) Thermal haloes around plume tails. *Geophysics Journal International*, **156**: 359–362.

Smith, A.D. (2003) Critical evaluation of Re-Os and Pt-Os isotopic evidence on the origin of intraplate volcanism. *Journal of Geodynamics*, **36**: 469–484.

Smith, A.D. (2005) The streaky mantle alternative to mantle plumes and its bearing on bulk-Earth geochemical evolution, in *Plates, Plumes, and Paradigms* (eds G.R. Foulger, J.H. Natland, D.C. Presnall and D.L. Anderson), Geological Society of America, Boulder, CO, pp. 303–326.

Smith, A.D. (2007) A plate model for Jurassic to Recent intraplate volcanism in the Pacific Ocean basin, in *Plates, Plumes, and Planetary Processes* (eds G.R. Foulger and D.M. Jurdy), Geological Society of America, Boulder, CO, pp. 471–496.

Smith, A.D. (2009) The fate of subducted oceanic crust and the origin of intraplate volcanism, in *The Lithosphere: Geochemistry, Geology and Geophysics* (eds J.E. Anderson and R.W. Coates), Nova Science Publishers, Inc., pp. 123–140.

Smith, A.D. and Lewis, C. (1999) The planet beyond the plume hypothesis. *Earth-Science Reviews*, **48**: 135–182.

Smith, W.H.F. and Sandwell, D.T. (1997) Global sea floor topography from satellite altimetry and ship depth soundings. *Science*, **277**: 1957–1962.

Smyth, J.R. and Frost, D.J. (2002) The effect of water on the 410-km discontinuity: An experimental study. *Geophysical Research Letters*, **29**(10): 1485, doi: 10.1029/2001GL014418.

Sobolev, A.V., Hofmann, A.W., Sobolev, S.V. et al. (2005) An olivine-free mantle source of Hawaiian shield basalts. *Nature*, **434**: 591–597.

Sobolev, A.V., Hofmann, A.W., Kuzmin, D.V. et al. (2007) The amount of recycled crust in sources of mantle-derived melts. *Science*, **316**: 412–417.

Stein, C.A. and Abbott, D. (1991) Heat-flow constraints on the south-Pacific superswell. *Journal of Geophysical Research*, **96**: 16,083–16,100.

Stein, C.A. and Stein, S. (1992) A model for the global variation in oceanic depth and heat-flow with lithospheric age. *Nature*, **359**: 123–129.

Stein, C.A. and Stein, S. (1993) Constraints on Pacific midplate swells from global depth-age and heat flow-age models, in *The Mesozoic Pacific: Geology, Tectonics, and Volcanism* (ed. M.S. Pringle), American Geophysical Union, Washington, DC, pp. 53–76.

Stein, C.A. and Stein, S. (2003) Sea floor heat flow near Iceland and implications for a mantle plume. *Astronomy & Geophysics*, **44**: 1.8–1.10.

Stein, C.A. and Von Herzen, R.P. (2007) Potential effects of hydrothermal circulation and magmatism on heatflow at hotspot swells, in *Plates, Plumes, and Planetary Processes* (eds G.R. Foulger and D.M. Jurdy), Geological Society of America, Boulder, CO, pp. 261–274.

Steinberger, B. (2000) Plumes in a convecting mantle: Models and observations for individual hotspots. *Journal of Geophysical Research-Solid Earth*, **105**(B5): 11,127–11,152.

Steinberger, B., Sutherland, R. and O'Connell, R.J. (2004). Prediction of Emperor-Hawaii seamount locations from a revised model of global plate motion and mantle flow. *Nature*, **430**: 167–173.

Stofan, E.R. and Smrekar, S.E. (2005) Large topographic rises, coronae, large flow fields, and large volcanoes on Venus, in *Plates, Plumes, and Paradigms* (eds G.R. Foulger, J.H. Natland, D.C. Presnall and D.L. Anderson), Geological Society of America, Boulder, CO, pp. 841–861.

Stoker, M.S., Praeg, D., Shannon, P.M. et al. (2005) Neogene evolution of the Atlantic continental margin of NW Europe (Lofoten Islands to SW Ireland): anything but passive, in *Petroleum Geology: North-West Europe and Global Perspectives – Proceedings of the 6th Petroleum Geology Conference* (eds A.G. Doré and B.A. Vining), Geological Society, London, pp. 1057–1076.

Stracke, A., Zindler, A, Salters, V.J.M. et al. (2003) Theistareykir revisited. *Geochemistry, Geophysics Geosystems*, **4**: 10.1029/2001GC000201.

Stracke, A., Hofmann, A.W. and Hart, S.R. (2005) FOZO, HIMU, and the rest of the mantle zoo. *Geochemistry, Geophysics, Geosystems*, **6**(Q05007).

Stuart, F.M., Lass-Evans, S., Fitton, J.G. et al. (2003) High ^{3}He/^{4}He in picritic basalts from Baffin Island: the role of a mixed reservoir in mantle plumes. *Nature*, **424**: 57–59.

Stuart, W.D., Foulger, G.R. and Barall, M. (2007) Propagation of Hawaiian-Emperor volcano chain by Pacific plate cooling stress, in *Plates, Plumes, and Planetary Processes* (eds G.R. Foulger and D.M. Jurdy), Geological Society of America, Boulder, CO, pp. 497–506.

Sun, S.S. (1980) Lead isotopic study of young volcanic rocks from mid-ocean ridges, ocean islands and island arcs, *Philosophical Transactions of the Royal Society of London, Series A*, **297**: 409–445.

Tackley, P.J. (1998) Three-dimensional simulations of mantle convection with a thermo-chemical basal boundary layer. *The Core-Mantle Boundary Region*, **28**: 231–253.

Takahashi, E., Nakajima, K. and Wright, T.L. (1998) Origin of the Columbia River basalts: Melting model of a heterogeneous plume head. *Earth and Planetary Science Letters*, **162**: 63–80.

Tarduno, J.A. and Cottrell, R.D. (1997) Paleomagnetic evidence for motion of the Hawaiian hotspot during formation of the Emperor seamounts. *Earth and Planetary Science Letters*, **153**: 171–180.

Tarduno, J.A., Duncan, R.A., Scholl, D.W. et al. (2007) The Emperor seamounts: Southward motion of the Hawaiian hotspot plume in the Earth's mantle. *Science*, **301**: 1064–1069.

Tarduno, J.A., Bunge, H.-P., Leep, N. et al. (2009) The bent Hawaiian-Emperor hotspot track: Inheriting the mantle wind. *Science*, **324**: 50–53.

Tauzin, B., Debayle, E. and Wittlinger, G. (2008) The mantle transition zone as seen by global Pds phases: No clear evidence for a thin transition zone beneath hotspots. *Journal of Geophysical Research*, **113**: B08309, doi: 10.1029/2007JB005364.

Taylor, B. (2006) The single largest oceanic plateau: Ontong Java–Manihiki–Hikurangi. *Earth and Planetary Science Letters*, **241**: 372–380.

Taylor, H.P., Jr. (1968) The oxygen isotope geochemistry of igneous rocks. *Continental Mineral Petrology*, **19**: 1–71.

Tejada, M.L.G., Mahoney, J.J., Castillo, P.R. et al. (2004) Pin-pricking the elephant: Evidence on the origin of the Ontong Java Plateau from Pb-Sr-Hf-Nd isotopic characteristics of ODP Leg 192 basalts, in *Origin and Evolution of the Ontong Java Plateau* (eds G. Fitton, J. Mahoney, P. Wallace and A. Saunders), Geological Society of London, London.

Thompson, R.N., Dickin, A.P., Gibson, I.L. et al. (1982) Elemental fingerprints of isotopic contamination of Hebridean Palaeocene mantle-derived magmas by Archaean sial. *Continental Mineral Petrology*, **79**(2): 159–168.

Toksöz, M.N., Minear, J.W. and Julian, B.R. (1971) Temperature field and geophysical effects of a downgoing slab. *Journal of Geophysical Research*, **76**(5): 1113–1138.

Torsvik, T.H., van der Voo, R. and Redfield, T.F. (2002) Relative hotspot motions versus True Polar Wander. *Earth and Planetary Science Letters*, **202**: 185–200.

Tozer, D.C. (1973) Thermal plumes in the Earth's mantle. *Nature*, **244**: 398–400.

Trampert, J. and Spetzler, J. (2006) Surface wave tomography: Finite frequency effects lost in the null space. *Geophysics Journal International*, **164**: 394–400.

Trampert, J., Deschamps, F., Resovsky, J. et al. (2004) Probabilistic tomography maps chemical heterogeneities throughout the lower mantle. *Science*, **306**: 853–856.

Turcotte, D.L. (1974) Membrane tectonics. *Geophysical Journal of the Royal Astronomical Society*, **36**: 33–42.

Turner, S.P., Hawkesworth, C., Gallagher, K. et al. (1996) Mantle plumes, flood basalts, and thermal models for melt generation beneath continents: Assessment of a conductive heating model and application to the Paraná. *Journal of Geophysical Research*, **101**: 11,503–11,518.

Ukstins Peate, I. and Bryan, S.E. (2008) Re-evaluating plume-induced uplift in the Emeishan large igneous province. *Nature Geoscience*, **1**: 625–629.

Ukstins Peate, I. and Bryan, S.E. (2009) Pre-eruptive uplift in the Emeishan? *Nature Geoscience*, **2**: 531–532.

Ukstins Peate, I., Larsen, M. and Lesher, C.E. (2003) The transition from sedimentation to flood volcanism in the Kangerlussuaq Basin, East Greenland: Basaltic pyroclastic volcanism during initial Palaeogene continental break-up. *Journal of the Geological Society of London*, **160**: 759–772.

Ukstins Peate, I., Baker, J.A., Al-Kadasi, M. et al. (2005) Volcanic stratigraphy of large-volume silicic pyroclastic eruptions during Oligocene Afro-Arabian flood volcanism in Yemen. *Bulletin of Volcanology*, **68**: 135–156.

Ulmer, P. (2001) Partial melting in the mantle wedge – the role of H_2O in the genesis of mantle-derived "arc-related" magmas. *Physical Earth Planetary International*, **127**: 215–232.

Vacher, P., Mocquet, A. and Sotin, C. (1998) Computation of seismic profiles from mineral physics: the importance of the non-olivine components for explaining the 660 km depth discontinuity. *Physical Earth Planetary International*, **106**: 275–298.

Van Ark, E. and Lin, J. (2004) Time variation in igneous volume flux of the Hawaii-Emperor hot spot seamount chain. *Journal of Geophysical Research*, **109**(B11401).

van Balen, R.T., van der Beek, P.A. and Cloetingh, S.A.P.L. (1995) The effect of rift shoulder erosion on stratal patterns at passive margins: Implications for sequence stratigraphy. *Earth and Planetary Science Letters*, **134**: 527–544.

van der Hilst, R.D. and de Hoop, M.V. (2005) Banana-doughnut kernels and mantle tomography. *Geophysics Journal International*, **163**: 956–961.

van Wijk, J.W., Huismans, R.S., Ter Voorde, M. et al. (2001) Melt generation at volcanic continental margins: no need for a mantle plume? *Geophysical Research Letters*, **28**: 3995–3998.

van Wijk, J.W., van der Meer, R. and Cloetingh, S.A.P.L. (2004) Crustal thickening in an extensional regime: application to the mid-Norwegian Vøring margin. *Tectonophysics*, **387**: 217–228.

VanDecar, J.C., James, D.E. and Assumpcao, M. (1995) Seismic evidence for a fossil mantle plume beneath south America and implications for plate driving forces. *Nature*, **378**: 25–31.

Vink, G.E. (1984) A hotspot model for Iceland and the Voring Plateau. *Journal of Geophysical Research*, **89**: 9949–9959.

Vinnik, L.P., Du, Z. and Foulger, G.R. (2005) Seismic boundaries in the mantle beneath Iceland: a new constraint on temperature. *Geophysics Journal International*, **160**: 533–538.

Vita-Finzi, C., Howarth, R.J., Tapper, S.W. et al. (2005) Venusian craters, size distribution , and the origin of coronae, in *Plates, Plumes, and Paradigms* (eds G.R. Foulger, J.H. Natland, D.C. Presnall and D.L. Anderson), Geological Society of America, Boulder, CO, pp. 815–823.

Vogt, P.R. (1971) Asthenosphere motion recorded by the ocean floor south of Iceland *Earth and Planetary Science Letters*, **13**: 153–160.

Vogt, P.R. and Jung, W.-Y. (2005) Paired (conjugate) basement ridges: Spreading axis migration across mantle heterogeneities? in *Plates, Plumes, and Paradigms* (eds G.R. Foulger, J.H. Natland, D.C. Presnall and D.L. Anderson), Geological Society of America, Boulder, CO, pp. 555–580.

Vogt, P.R. and Jung, W.-Y. (2007) Origin of the Bermuda volcanoes and Bermuda Rise: History, observations, models, and puzzles, in *Plates, Plumes, and Planetary Processes* (eds G.R. Foulger and D.M. Jurdy), Geological Society of America, Boulder, CO, pp. 553–592.

Vollmer, R. (1989) On the origin of the Italian potassic magmas. A discussion contribution. *Chemical Geology*, **74**: 229–239.

von Herzen, R.P., Cordery, M.J., Detrick, R.S. et al. (1989) Heat-flow and the thermal origin of hot spot swells – the Hawaiian swell revisited. *Journal of Geophysical Research*, **94**: 13,783–13,799.

von Herzen, R.P., Detrick, R.S., Crough, S.T. et al. (1982) Thermal origin of the Hawaiian swell – heat-flow evidence and thermal models. *Journal of Geophysical Research*, **87**: 6711–6723.

Walker, G.P.L. (1963) The Breiddalur central volcano eastern Iceland. *Quarterly Journal of the Geological Society of London*, **119**: 29–63.

Walker, G.P.L. (1990) Geology and volcanology of the Hawaiian Islands. *Pacific Science*, **44**: 315–347.

Watts, A.B., Weissel, J.K., Duncan, R.A. et al. (1988) Origin of the Louisville ridge and its relationship to the Eltanin fracture zone system. *Journal of Geophysical Research*, **93**: 3051–3077.

Wawerzinek, B., Ritter, J.R.R., Jordan, M. et al. (2008) An upper-mantle upwelling underneath Ireland revealed from non-linear tomography. *Geophysics Journal International*, **175**: 253–268.

Weaver, B.L. (1991) The origin of ocean island basalt end-member compositions: Trace element and isotopic constraints. *Earth and Planetary Science Letters*. **104**: 381–397.

Weber, M., Bock, G. and Budweg, M. (2007) Upper mantle structure beneath the Eifel from receiver functions, in *Mantle Plumes: A Multidisciplinary Approach* (eds J.R.R. Ritter and U.R. Christensen), Springer Berlin Heidelberg, pp. 405–415.

Weeraratne, D.S., Forsyth, D.W., Yang, Y. et al. (2007) Rayleigh wave tomography beneath intraplate volcanic ridges in the South Pacific. *Journal of Geophysical Research*, **112**(B06303).

Wegener, A.L. (1915) *Die Entstehung der Kontinente und Ozeane (The origin of continents and oceans)*, Friedrich Vieweg und Sohn, Braunschweig.

Wegener, A.L. (1924) *Die Entstehung der Kontinente und Ozeane (The origin of continents and oceans)*, 3rd edn, Friedrich Vieweg und Sohn, Braunschweig, pp. xx + 212.

Wernicke, B. (1981) Low-angle normal faults in the Basin and Range Province: nappe tectonics in an extending orogen. *Nature*, **291**: 645–648.

Wessel, P. (1997) Sizes and ages of seamounts using remote sensing: Implications for intraplate volcanism. *Science*, **277**: 802–805.

Wessel, P. and Kroenke, L.W. (2009) Observations of geometry and ages constrain relative motion of Hawaii and Louisville plumes. *Earth and Planetary Science Letters*, **284**: 467–472.

Wessel, P., Harada, Y. and Kroenke, L.W. (2006) Toward a self-consistent, high-resolution absolute plate motion model for the Pacific. *Geochemistry, Geophysics, Geosystems*, **7**(Q03L12).

White, J.D.L. and McClintock, M.K. (2001) Imense vent complex marks flood-basalt eruption in a wet, failed rift: Coombs Hills, Antarctica. *Geology*, **29**: 935–938.

White, R.S. and McKenzie, D.P. (1989) Magmatism at rift zones: The generation of volcanic continental margins and flood basalts. *Journal of Geophysical Research*. **94**: 7685–7729.

White, R.S. and McKenzie, D.P. (1995) Mantle plumes and flood basalts. *Journal of Geophysical Research*, **100**: 17,543–17,585.

White, R.S., Bown, J.W. and Smallwood, J.R. (1995) The temperature of the Iceland plume and origin of outward-propagating V-shaped ridges. *Journal of the Geological Society of London*, **152**: 1039–1045.

White, W.M. (2005) *Geochemistry*, an on-line textbook.

White, W.M. and McBirney, A.R. (1993) Petrology and geochemistry of the Galápagos Islands: portrait of a pathological mantle plume. *Journal of Geophysical Research*, **98**: 19,533–19,563.

Wilson, D.S. and Hey, R.N. (1995) History of rift propagation and magnetization intensity for the Cocos-Nazca spreading center. *Journal of Geophysical Research*, **100**: 10,041–10,056.

Wilson, J.T. (1963) A possible origin of the Hawaiian Islands. *Canadian Journal of Physics*, **41**: 863–870.

Wilson, M. and Downes, H. (1991) Tertiary–Quaternary extension related alkaline magmatism in western and central Europe. *Journal of Petrology*, **32**: 811–849.

Winterer, E.L. and Sandwell, D.T. (1987) Evidence from en-echelon cross-grain ridges for tensional cracks in the Pacific plate. *Nature*, **329**: 534–537.

Wirth, R. and Rocholl, A. (2003) Nanocrystalline diamond from the Earth's mantle underneath Hawaii. *Earth Planetary and Science Letters*, **211**: 357–369.

Wolfe, C.J., Bjarnason, I.T., VanDecar, J.C. et al. (1997) Seismic structure of the Iceland mantle plume. *Nature*, **385**: 245–247.

Wolfe, C.J., Solomon, S.C., Silver, P.G. et al. (2002) Inversion of body-wave delay times for mantle structure beneath the Hawaiian islands: results from the PELENET experiment. *Earth and Planetary Science Letters*, **198**: 129–145.

Wood, B.J. (1995) The effect of H_2O on the 410-kilometer seismic discontinuity. *Science*, **268**: 74–76.

Worthington, T.J., Hekinian, R., Stoffers, P. et al. (2006) Osbourn Trough: Structure, geochemistry and implications of a mid-Cretaceous paleospreading ridge in the South Pacific. *Earth and Planetary Science Letters*, **245**: 685–701.

Wright, T.J., Ebinger, C., Biggs, J. et al. (2006) Magma-maintained rift segmentation at continental rupture in the 2005 Afar dyking episode. *Nature*, **442**: 291–294.

Xu, Y.-G., Chung, S.-L., Jahn, B.M. et al. (2001) Petrologic and geochemical constraints on the petrogenesis of Permian-Triassic Emeishan flood basalts in southwestern China. *Lithos*, **58**: 145–168.

Yamamoto, M., Morgan, W.J. and Morgan, J.P. (2007a) Global plume-fed asthenosphere flow: (1) motivation and model development, in *Plates, Plumes, and Planetary Processes* (eds G.R. Foulger and D.M. Jurdy), Geological Society of America, Boulder, CO, pp. 165–188.

Yamamoto, M., Morgan, W.J. and Morgan, J.P. (2007b) Global plume-fed asthenosphere flow: (2) Application to the geochemical segmentation of mid-ocean ridges, in *Plates, Plumes, and Planetary Processes* (eds G.R. Foulger and D.M. Jurdy), Geological Society of America, Boulder, CO, pp. 189–208.

Yamasaki, T. and Gernigon, L. (2009), Styles of lithospheric extension controlled by underplated mafic bodies. *Tectonophysics*, **468**: 169–184.

Yang, T., Shen, Y., Van der Lee, S. et al. (2006) Upper mantle structure beneath the Azores hotspot from finite-frequency seismic tomography. *Earth and Planetary Science Letters*, **250**: 11–26.

Yasuda, A., Fujii, T. and Kurita, K. (1994) Melting phase relations of an anhydrous mid-ocean ridge basalt from 3 to 20 GPa; Implications for the behavior of subducted oceanic crust in the mantle. *Journal of Geophysical Research*, **99**: 9401–9414.

Yaxley, G.M. (2000) Experimental study of the phase and melting relations of homogeneous basalt + peridotite mixtures and implication for the petrogenesis of flood basalts. *Continental Mineral Petrology*, **139**: 326–338.

Yaxley, G.M. and Green, D.H. (1998) Reactions between eclogite and peridotite: Mantle refertilisation by subduction of oceanic crust. *Schweiz. Min. Pet. Mitteil.*, **78**: 243–255.

Yoder, H.S., Jr. and Tilley, C.E. (1962) Origin of basalt magmas: an experimental study of natural and synthetic rock systems. *Journal of Petrology*, **3**: 342–532.

Yuan, H. and Dueker, K.G. (2005) Teleseismic P-wave tomogram of the Yellowstone plume. *Geophysical Research Letters*, **32**(L07304).

Yuen, D.A., Cserepes, L. and Schroeder, B.A. (1998) Mesoscale structures in the transition zone: Dynamical consequences of boundary layer activities. *Earth Planets Space*, **50**: 1035–1045.

Zhao, D., Tian, Y., Lei, J. et al. (2009) Seismic image and origin of the Changbai intraplate volcano in East Asia: Role of big mantle wedge above the stagnant Pacific slab. *Physical Earth Planetary International*, **173**: 197–206.

Zhou, M., Malpas, J., Song, X., et al. (2002) A temporal link between the Emeishan large igneous province (SW China) and the end-Guadalupian mass extinction. *Earth and Planetary Science Letters*, **196**: 113–122.

Ziegler, P.A. (1992) European Cenozoic rift system. *Tectonophysics*, **208**: 91–111.

Ziegler, P.A. and Dèzes, P. (2007) Cenozoic uplift of Variscan Massifs in the Alpine foreland: Timing and controlling mechanisms. *Global Planetary Change*, **58**: 237–269.

Zindler, A. and Hart, S.R. (1986) Chemical Geodynamics. *Annual Reviews in Earth Planetary and Sciences*, **14**: 493–571.

Zorin, Y.A., Turutanov, E.K., Kozhevnikov, V.M. et al. (2006) The nature of Cenozoic upper mantle plumes in East Siberia (Russia) and Central Mongolia. *Russian Geology and Geophysics*, **47**: 1056–1070.

Index

Page numbers in *italics* represent figures, those in **bold** represent tables.

Plates vs. Plumes: A Geological Controversy, 1st edition. By Gillian R. Foulger.
Published 2010 by Blackwell Publishing Ltd.

Printed and bound by CPI Group (UK) Ltd, Croydon, CR0 4YY

27/10/2024

14580388-0003